The Ecology of Loch Lomond

Developments in Hydrobiology 101

Series editor
H. J. Dumont

The Ecology of Loch Lomond

Edited by

K.J. Murphy, M.C.M. Beveridge & R. Tippett

Reprinted from Hydrobiologia, vol. 290 (1994)

SPRINGER SCIENCE+BUSINESS MEDIA, B.V.

Library of Congress Cataloging-in-Publication Data

A C.I.P. Catalogue record for this book is available from the Library of Congress.

ISBN 978-94-010-4326-7 ISBN 978-94-011-0758-7 (eBook)
DOI 10.1007/978-94-011-0758-7

Printed on acid-free paper

Contents

Foreword

'You'll take the high road and I'll take the low road
And I'll be in Scotland afore ye.
Where me and my true love will never meet again
On the bonnie bonnie banks of Loch Lomond.'

These words were reputedly written by a Scottish soldier awaiting execution in an English jail after fighting for Bonnie Prince Charles and the Jacobite cause in 1745. Loch Lomond has long held a special place in the hearts of all the people of Scotland not only for its historic significance but also for the beauty of its countryside. Less widely known is the ecological importance of the area. It is the largest stretch of freshwater in Britain and the only loch to be crossed by the Highland Boundary Fault, one of the country's most important geological features.

The University of Glasgow has recognised the importance of this loch situated so near to the main campus and since 1948 has operated a field station on its shores. In the mid 1960's the Field Station was re-established at Rowardennan as an all year round facility with laboratory and living accommodation for resident research workers and visiting field courses. In 1992 a symposium was held to celebrate the 25th anniversary of the opening of the new station and this volume presents the proceeding of that symposium. The programme of invited presentations was planned to review the state of our knowledge of Loch Lomond and its catchment.

In addition to thanking all contributors to the volume for all the help they have given to the editors, we particularly acknowledge the following for refereeing the papers:

B. Bluck, Department of Geology, University of Glasgow, U.K.
C. Boylen, Rensselaer Freshwater Institute, New York, U.S.A.
J. Caffrey, Central Fisheries Board, Dublin, Ireland.
R. Doughty, Clyde River Purification Board, East Kilbride, U.K.
N. Giles, The Game Conservancy, Fordingbridge, U.K.
J. Hamilton, University of Paisley, U.K.
A. Ibbotsen, Institute of Aquaculture, University of Stirling, U.K.
D. Mackay, Norhtheast River Purification Board, Aberdeen, U.K.
G. Pender, Department of Civil Engineering, University of Glasgow, U.K.
I. Pulford, Department of Chemistry, University of Glasgow, U.K.
B. Rorslett NIVA, Oslo, Norway
P. Wade, ICOLE, University of Loughborough, U.K.

K.J. Murphy
M.C.M. Beveridge
R. Tippett

Hydrobiologia **290**: ix–x, 1994.
K. J. Murphy, M. C. M. Beveridge & R. Tippett (eds), The Ecology of Loch Lomond.

Glasgow University Field Station at Rossdhu, Loch Lomond: the first British University Freshwater Field Station

Peter S. Maitland[1] & James D. Hamilton[2]
[1] *Fish Conservation Centre, Easter Cringate, Stirling FK7 9QX, Scotland, UK;*
[2] *65 Main Road, Elderslie, Johnstone PA5 9AZ, Scotland, UK*

Introduction

In the context of the 25th Anniversary of the University Field Station at Rowardennan it was natural to recall the earlier Glasgow University Field Station at Rossdhu and the contribution which it and those who worked there made to knowledge of the limnology of Loch Lomond and its catchment.

Although some investigations of the limnology of Loch Lomond had been carried out previously, the initiation in 1938 by Dr Harry Slack of a field course in limnology was the stimulus for much subsequent research. Interrupted by the Second World War, the course was resumed in 1946, working out of Balmaha. Later in 1946, field laboratories were assembled from ex-army huts on the west side of the loch at Rossdhu. This was a novel venture and for many years it remained the only such station in a British University. Much of the baseline research on Loch Lomond was carried out from Rossdhu by various workers, many of whom pursued subsequent careers in freshwater biology.

The Rossdhu Field Station

Following the success of the early field courses, in the autumn of 1946 field laboratories were built on the west side of Loch Lomond in the policies of Rossdhu House (Yonge, 1947). The site on the shore of the loch was provided by Sir Iain Colquhoun of Luss and some funding was provided by a second benefactor, Sir Harold Bowden. The station was built, furnished and equipped at relatively small expense – advantage being taken of the abundance of war surplus materials. Other equipment was that which had been used in the Challenger expeditions and in the bathymetric survey of Scottish lochs (Murray & Pullar, 1910); it had later been deposited in the Hunterian Museum by Sir John Murray. With these simple resources Harry Slack and his many colleagues carried out much of the basic research on Loch Lomond from the Rossdhu Station, but eventually in 1964, thanks to the efforts of Sir Maurice Yonge and the generosity of the Nuffield Foundation, a new University Field Station was completed on the opposite side of the loch, 5 km south of Rowardennan.

During the existence of the Rossdhu Field Station many scientists spent substantial periods of time there and much significant research was accomplished. This included the first major studies of the sediments, zoobenthos, plankton and fish of Loch Lomond (Slack, 1957). In addition, the first major attempt to study biting midges in Scotland and the feasibility of controlling them was carried out here (Kettle, 1952). Alongside this extensive work on Loch Lomond, studies of other waters in its catchment – both lochs and rivers – were also carried out (e.g. Hamilton, 1958; Maitland, 1962). As well as these research activities, several hundred students attended field courses at Rossdhu over the period of its existence (Slack, 1958).

The staff

The main participants at the Rossdhu Field Station in work related to freshwater ecology were as follows: H. D. Slack (sediments and zoobenthos), R. M. Badcock (stream benthos), A. D. Berrie (molluscs), W. Burton (equipment design), M. A. Chapman (zooplankton), J. A. Downes (midge control), A. N. Fedoruk (zoobenthos), J. F. Flannagan (limnology), F. W. K. Gervers (fish), J. D. Hamilton (fish and zooplankton), A. E. Henderson (molluscs), D. S. Kettle (midge control), J. W. H. Lawson (midge control), W. W. Macdonald (midge control), P. S. Maitland (fish and stream benthos), A. H. Parker (midge control), W. D. Russell Hunter (molluscs), D. B. C. Scott (fish), A. Todd (limnology), A. H. Weatherley (fish), A. C. J. Weerekoon (zoobenthos) and P. K. H. Yeoh (stream benthos).

Unfortunately it was not possible to obtain details of all these workers but most were covered in the poster which this paper summarises. The poster also showed the two major research vessels used at Rossdhu – the 'Ivan' and the 'Fiona'.

Dr Harry Slack was clearly the leading light in the development and construction of both the old Field Station at Rossdhu and the new one at Rowardennan (Maitland, 1983). The prospect of extending the knowledge of the ecology of the fresh waters of Scotland on so broad and firm a foundation as that laid down by the classical work of the bathymetric survey at the beginning of the century was a cogent factor in his initiating a novel course in freshwater biology at Loch Lomond in 1938 for science students at Glasgow University. At first, all equipment had to be carried from the University to the loch and installed in boats which were hired there at Balmaha. Unfortunately, this start was interrupted by the Second World War, but in the spring of 1946, immediately after his demobilisation, Harry again organised student field work on Loch Lomond, out of Balmaha and, as noted above, this soon led to the development of the Rossdhu Field Station.

Conclusions

The poster display on which this paper is based presented memories of most of those who worked at Rossdhu and recorded their achievements then and later in their careers. The items displayed (which included a photograph of each worker), together with other material collected during the preparation of the poster will form the basis of an archive which will be held at the present University Field Station at Rowardennan.

It is believed that, in its lifetime, the Rossdhu Field Station fulfilled the objectives and hopes of those who created it. 'Valuable as teaching centres, the laboratories will be at least as important as bases for research work over a wide range of subjects, some purely scientific, others, it is hoped, of value in the improvement of fisheries or in the control of insect pests.' (Yonge, 1947).

Acknowledgments

The authors are pleased to thank all those who responded to the request for information: without their help there would have been no poster and only an inadequate archive.

References

Hamilton, J. D., 1958. On the biology of *Holopedium gibberum* Zaddach (Crustacea: Cladocera). Verh. int. Ver. Limnol. 14: 435–439.

Kettle, D. S., 1952. Midge (*Culicoides*) control in Scotland. Trans. Int. Congr. Ent. 9: 935–938.

Maitland, P. S. 1962. *Bathynella natans*, new to Scotland. Glasg. Nat. 18: 175–176.

Maitland, P. S., 1983. Harry Dawson Slack. Year Book, Roy. Soc. Edin. 1983: 180–183.

Murray, J. & L. Pullar, 1910. Bathymetrical survey of the freshwater lochs of Scotland. Edinburgh, Challenger.

Slack, H. D., 1957. Studies on Loch Lomond. I. Glasgow, Blackie.

Slack, H. D., 1958. Studying Loch Lomond. Scott. Field. 105, 668: 30–31.

Yonge, C. M., 1947. Field studies on Loch Lomond. Field. 1947: 1.

Hydrobiologia **290**: xi–xv, 1994.
K. J. Murphy, M. C. M. Beveridge & R. Tippett (eds), The Ecology of Loch Lomond.

An introduction to Loch Lomond

Roger Tippett
Department of Zoology, The University, Glasgow G12 8QQ, UK

Key words: Loch Lomond, morphometry, thermal stratification

Loch Lomond is the largest area of freshwater in Great Britain. It is situated near Glasgow, the largest centre of population in Scotland and more than half the population of Scotland, about 2.6 million people, live within one hour's journey time of the shores (Hamilton, 1988).

It has a long and interesting place in the history of the study of limnology, and in 1946 Glasgow University established a research station on the west shore at Rossdhu. In 1965 this was moved to the present site near the Ross Peninsula at Rowardennan on the east shore. The twenty fifth anniversary of the opening of the new station was celebrated by a symposium on the Limnology of Loch Lomond and its catchment, held at Stirling University in 1992. This volume comprises papers from the presentations at that symposium, and this introduction aims at giving a general description of the conditions within the loch basin and its catchment. This background information may help in understanding the papers from the symposium.

In all the major morphometric features, Loch Lomond is one of the most impressive of Britains lakes. Table 1 presents comparative data for Lochs Lomond, Ness, Awe and Morar taken from Murray & Pullar (1910) suitably metricated. From these data it can be seen to be the largest surface area, second largest volume, although only a little more than one third the volume of Loch Ness, and third longest and deepest.

Loch Lomond lies between Lat. 56 °N and 56 °19′N, and Long. 4 °30′W and 4 °43′W. A detailed bathymetric chart of the loch was published by the Admiralty in 1861 and this still forms the basis for modern navigation maps. A further bathymetric survey was carried out by Murray & Pullar (1910). The loch is divided into three main basins with a number of subsidiary ones (Fig. 1). The northernmost basin is the Ardlui basin, extending to 50 m depth and as far South as Island I Vow (3 km). The Tarbet or North basin extends from Island I Vow southwards for a distance of about 14 km. It has a maximum width of 1.5 km and maximum depth of 190 m. This basin is thus of a typically Piedmont type (Hutchinson, 1957). The Luss (mid) basin extends 5.5 km south from Rowardennan to Inch Lonaig, has a maximum breadth of approximately 3.5 km and extends to 60 m depth. The Strathcashel basin is effectively a sub-basin of the Luss basin, is about 40 m deep and lies to the South and East of Inch Lonaig.

Table 1. Comparative morphometric data from four of Scotland's largest lochs (rank in brackets).

	Lomond	Ness	Awe	Morar
Length (km)	36.25	38.75	40.75	18.7
	(3)	(2)	(1)	(10)
Maximum depth (m)	190.5	230.5	93.86	310.9
	(3)	(2)	(15)	(1)
Mean depth (m)	37.1	132.4	32.1	86.8
	(16)	(1)	(20)	(2)
Area (km^2)	70.27	55.76	38.02	26.37
	(1)	(2)	(3)	(5)
Volume ($\times 10^6$ m^2)	2652	7520	1242	2329
	(2)	(1)	(5)	(3)

The northern limit of the South or Fault basin is the chain of islands including Inch Tavannach, Moan, Cruin and Fad. From there it extends 8 km South to the outflow via the river Leven at Balloch. It has a

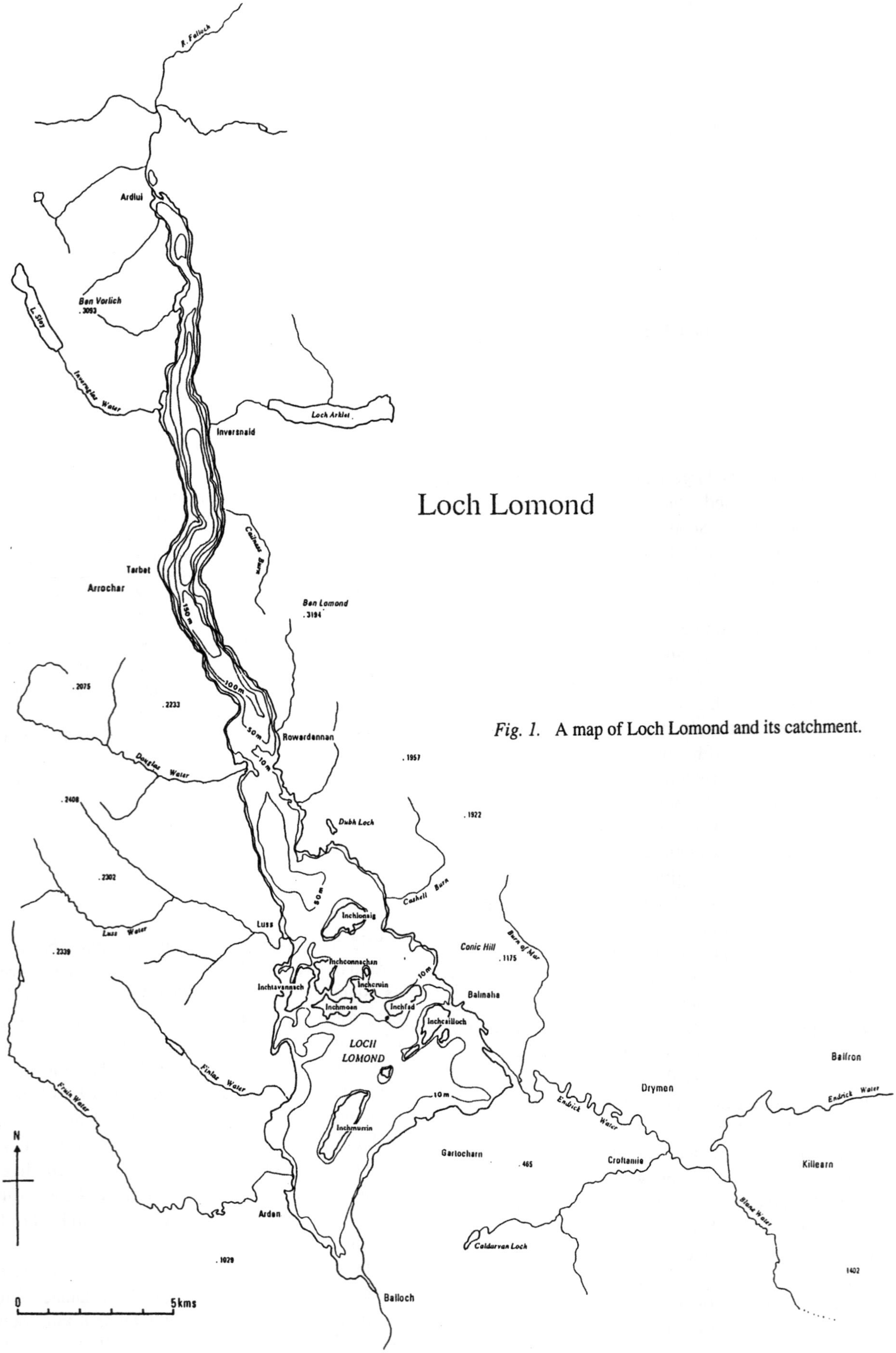

Loch Lomond

Fig. 1. A map of Loch Lomond and its catchment.

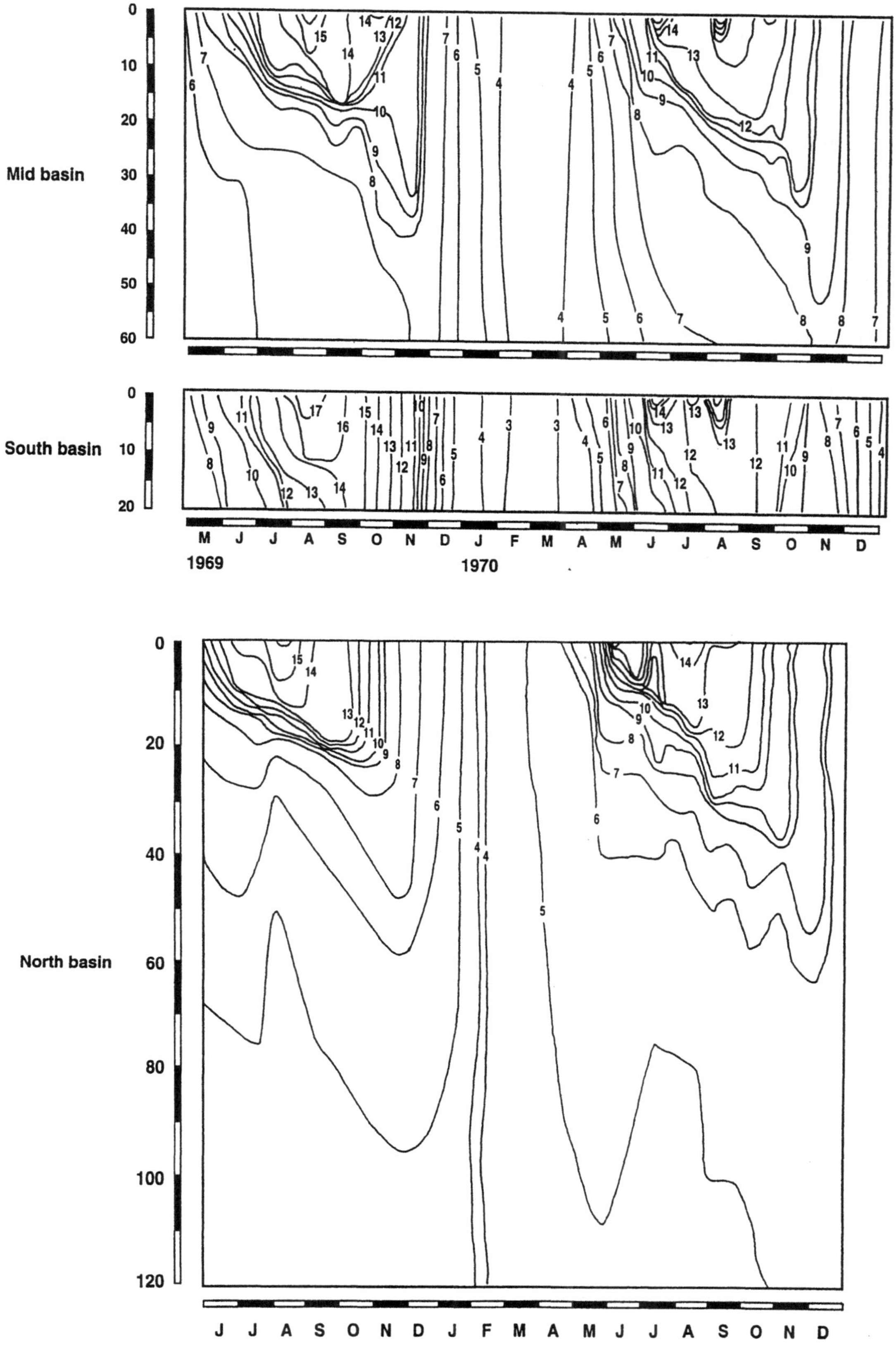

Fig. 2. Seasonal changes in temperature over the period May 1969 to December 1970 in the North, Mid and South basins of Loch Lomond.

maximum width of 7 km and nowhere is this basin deeper than 23 m.

Thus the loch is formed of a chain of discrete basins of increasing width and decreasing depth from North to South. The morphology is largely influenced by the Highland Boundary Fault (H.B.F.) which crosses the South basin. Loch Lomond is the only large lake to lie across this important geological feature of Scotland, and this gradient of environmental conditions is one reason why Loch Lomond is such an interesting subject of study.

The total drainage area of the catchment is of the order of 700 Km^2 of which about 40% lies South of the H.B.F. In this region the land is not mountainous and nowhere rises above 250 m. Annual rainfall is of the order of 1500 mm per year (30 year average), and runoff factor is about 65%.

North of the fault, however, the topography is much more mountainous with much of the land within 2 km of the loch shore being above 500 m, and a number of peaks over-shadowing the North basin rising to more than 900 m elevation (Ben Lomond, Ben Vane & Ben Vorlich). Rainfall in this region is greater with a 30 year average exceeding 2500 mm and runoff factor in excess of 80%.

The two main inflows are the Falloch Water, flowing in at the North of the Ardlui basin, and the Endrick Water flowing in at the eastern extremity of the South basin, the largest part of its catchment lying South of the H.B.F. Outflow is via the River Leven at Balloch at the southernmost extremity of the loch, and this flows into the Clyde Estuary at Dumbarton, 6 km downstream. There is also an abstraction for water by the Central Scotland Water Development Board at Ross Priory, 2 km S.W. of the mouth of the Endrick Water. Although there is little direct information on the hydrological input into Loch Lomond, Table 2 presents a generalised picture of the origins of water flowing into the loch.

Table 2. Inflow waters into Loch Lomond. (Based on data supplied by Clyde River Purification Board).

Origin	Mean inflow ($m^3 \times 10^6\ y^{-1}$)	Percentage
Rivers North of H.B.F.	654	50.6
Small streams North of H.B.F.	214	16.6
Rivers South of H.B.F.	278	21.5
Rainfall on loch surface	146	11.3
Total	1292	100

Thus it can be seen that little more than one fifth of the inflow comes from South of the H.B.F., although 40% of the catchment lies in that region. From these data it can be calculated that the mean retention time is just over 2 years.

Loch Lomond was one of the first lakes in the world on which the temperature of natural freshwaters was investigated. In 1812 Jardine used a type of reversing thermometer to measure temperatures down to a depth of 100 fathoms (Leslie, 1838). Buchanan (1886) and Murray & Pullar (1910) also studied the temperature in the loch, but although the features of thermal stratification were observed, its significance was not discussed. Slack (1957) has presented the most detailed discussion of temperature in Loch Lomond and the results presented in this introduction are in broad agreement with that work. Isotherms from the three sampling stations in the North, Mid and South basins over a period of 20 months from May 1969 to December 1970 are presented in Fig. 2.

The temperature of the water in the South basin may fall to less than 3 °C in winter, and this is the only region of the lake which may freeze. Although sheltered parts such as Balmaha Bay freeze for some days in most winters, 1963 was the last year in which ice cover was at all extensive. in 1969–79 winter it was below 4 °C for the three months between mid December and mid March. The water temperature in the mid basin was below 4 °C for only 6 weeks in that winter and the lowest temperature recorded was 3.7 °C. The temperature of the water in the North basin did not fall significantly below 4 °C at all during the winter, so that for the 4 month period from the end of November to the end of March this was the warmest region of the loch. The winter of 1970–71 was milder than in the previous year, and all regions were about 2 °C warmer in December than at the same time in the previous year.

Evidence of the beginning of the spring heating of the water in 1970 became apparent in the South basin in the latter half of January, but was observed two or three weeks later in the two northern basins. The summer temperature of the surface water rose to about 16 °C in all regions, but local heating of surface water in the South and Mid basins to 19 °C and 18 °C respectively was observed during prolonged periods of warm weather. Stable thermal stratification does not occur in the South basin and the greatest temperature difference is 2 °C throughout the water column.

Stable thermal stratification occurs during summer in both northern basins, and becomes demonstrable at the end of May or beginning of June. The temperature of the hypolimnion in the Mid basin rises to above 7 °C (8 °C in 1970) by the time the stratification is breaking down in November. In the North basin however there is much less heating in the hypolimnion, the temperature being less than 6 °C at a depth equivalent to the bottom of the Mid basin, and the bulk of the hypolimnion does not heat above 5 °C. Complete uniformity of temperature with depth is not achieved until late January. Thus the temperature of water is uniform throughout the water column of the South basin for up to 8 months of the year, in the Mid basin for up to 6 months and in the North basin for not more than 3 months.

References

Buchanan, J. Y., 1886. Distribution of temperature in Loch Lomond during the autumn of 1885. Proc. Roy. Soc. Edinb., 13: 403–428.

Hamilton, J. D., 1988. Recent influences on the ecology of Loch Lomond, Scotland. Verh. int. Ver. Limnol. 23: 403–413.

Hutchinson, G. E., 1957. A Treatise on Limnology. Volume 1: Geography, Physics and Chemistry. John Wiley & Sons Inc., New York, 1015 pp.

Leslie, J., 1838. Climate. in Encyclopaedia Britannica, Edinburgh, 7th ed., 6: 743-764.

Murray, J. & L. Pullar, 1910. A Bathymetrical Survey of the Freshwater Lochs of Scotland. Edinburgh: Challenger. Vol 1.

Slack, H. D., 1957. Studies on Loch Lomond: Vol. 1. Blackie, Glasgow.

Hydrobiologia **290**: 1–12, 1994.
K. J. Murphy, M. C. M. Beveridge & R. Tippett (eds), The Ecology of Loch Lomond.

August Thienemann and Loch Lomond – an approach to the design of a system for monitoring the state of north-temperate standing waters

Brian Moss[1], Penny Johnes[1,3] & Geoffrey Phillips[2]
[1]*Dept. of Environmental & Evolutionary Biology, P.O., Box 147, Brownlow St. University of Liverpool, Liverpool L69 3BX, UK*
[2]*National Rivers Authority, Eastern Region, Ipswich, UK*
Present address: Dept. of Geography, University of Reading, Whiteknight, Reading, UK

Abstract

Limnologists had an early preoccupation with lake classification. It gave a necessary structure to the many chemical and biological observations that were beginning to form the basis of one of the earliest truly environmental sciences. August Thienemann was the doyen of such classifiers and his concept with Einar Naumann of oligotrophic and eutrophic lakes remains central to the world-view that limnologists still have.

Classification fell into disrepute, however, as it became clear that there would always be lakes that deviated from the prescriptions that the classifiers made for them. Continua became the *de rigeur* concept and lakes were seen as varying along many chemical, biological and geographic axes. Modern limnologists are comfortable with this concept. That all lakes are different guarantees an indefinite future for limnological research.

For those who manage lakes and the landscapes in which they are set, however, it is not very useful. There may be as many as 300 000 standing water bodies in England and Wales alone and maybe as many again in Scotland. More than 80 000 are sizable (>1 ha). Some classification scheme to cope with these numbers is needed and, as human impacts on them increase, a system of assessing and monitoring change must be built into such a scheme. Although ways of classifying and monitoring running waters are well developed in the UK, the same is not true of standing waters. Sufficient understanding of what determines the nature and functioning of lakes exists to create a system which has intellectual credibility as well as practical usefulness. This paper outlines the thinking behind a system which will be workable on a north European basis and presents some early results.

Introduction

The richness of science, as in the arts and humanities. lies in the relationship between its people and their material. For that common reason the much regretted division of science and the humanities into C. P. Snow's (1959) 'Two Cultures' represents only a stage in development among people made defensive by insecurity. Gerard Manley Hopkins (1844–1889) celebrated the inflow streams of Loch Lomond, by area Britain's largest lake, in his famous poem '*Inversnaid*':

'O let them be left, wildness and wet;
Long live the weeds and the wilderness yet'

and Harry Slack (1957) celebrated the basin in '*Studies on Loch Lomond* 1' in a no less profound way. Others have followed on both the roads, adding to our knowledge and contributing towards a fusion of explanation and feeling that can be the only formula for the management of any acceptable future world.

August Thienemanm never visited Loch Lomond but he would have understood Hopkins & Slack equally. He was born in Germany on Sept 7 1882 of a family of bibliophiles and successively educated at the Universities of Greifsweld, Innsbruck, Heidelberg and Munster. From 1907 to 1917 he worked at Munster's Agricultural Experiment Station where, amid the practicalities of fish culture he founded a department of Hydrobiology before embarking on a forty year spell as Director of the Limnological Station at Plon. During most of this time he was also Professor of Hydrobiology at Kiel. When he died in 1957 he had 460 publications to his credit and the joint foundership of the International Association for Limnology.

Thienemann approached limnology through the benthic invertebrates and by 1909 had conceived two types of lakes – those characterised by *Tanytarsus* in the sub-alpine regions of Germany and those by *Chironomus* in the lowlands uncovered by the postglacial retreat of the Baltic Sea. In 1915 he discovered examples of each of the two types in two small lakes or maare on either side of a ridge in the Eifel district. He linked the two types with the oxygen regime in the hypolimnion. Thus Thienemann early became preoccupied with the creation of a structure for the understanding of lakes, a classification system.

A few years later he came into contact with a Swede, Einar Christian Leonard Naumann, nine years his junior, who was to develop parallel interests in a similar career (Rodhe, 1975). Naumann came from a medical family, joined the fisheries laboratory at Aneboda after a formal education at the University of Lund and started a limnology unit at that University. In 1927 he became Professor of Limnology there and could list 300 publications by the time of his death in 1934 al the early age of 43. He was the co-founder with Thienemann of the International Association. The nature of the founding and their relationship is described by Rodhe (1974, 1975).

Rodhe knew both men and contrasts their natures – the urbane, balanced and tolerant Thienemann, much influenced by Goethe, deeply rooted in western humanity and Naumann, excited, energetic, extremist and passionate. Naumann loved schemes and models and liked to preside over an order created from his intellect. He came independently to the same conclusions as Thienemann about two sorts of lakes, though his tools were phytoplanktonic. His field sites in upland Scandinavia and the western shores of the Baltic mirrored those of Thienemann in Germany and he recognised the links between geology and surroundings, water chemistry and plankton. In 1917 he introduced the concepts of oligotrophy and eutrophy to replace the hitherto geographical designation of Alpine and Baltic lakes. He borrowed the terms from Weber (1907) who had coined them to describe the changing base status of mires as they succeeded from supply by mineral rich ground water to mineral poor rain water. A little later Thienemann accepted Naumann's ideas and as the information from chemistry, plankton, benthos and oxygen fused together the central theme of limnology even today had emerged. Moreover, the basis for a more complex ordering of information about lakes was born.

Naumann (1931) and his successors went on to create more complex classifications, adding argillotrophic, dystrophic, acidotrophic, siderotrophic and others to the original eutrophic and oligotrophic. Thienemann was more cautious, regarding the categories as hypothetical abstractions rather than realities and developing (1931) the concepts of harmonic and disharmonic for those that did or did not quite fit the abstraction. Perhaps it was his wide culture that discouraged him from taking too mechanistic an approach or perhaps it was his findings on an expedition with Ruttner and Feuerborn in 1928/29 to Indonesian lakes (Thienemann, 1931) that infertile lakes sometimes had deoxygenated hypolimnia. Had he visited Loch Lomond at the time he would have delighted in the different basins in the same lake, one decidedly oligotrophic the other tending to eutrophic. But Naumann's successors might have created yet another category for it. They soldiered on and lake classification (*Seetypenlehre*) was the major topic at the International Congress for Limnology in Finland in 1956:

'The challenge of lake types is the great catalyst of limnological research. It has amalgamated the various branches of lake study into limnology. There is hardly any truly limnological problem that lacks connection with the Seetypenlehre' H. J. Elster (1958)

But the concept of lake types was flawed and attention was turning to processes in lakes. Some time previously Pearsall (1921), working in the Cumbrian Lake District had created the idea of continua among lakes in his arrangement of the English Lakes into a series from the least fertile rocky lakes to the more fertile silted ones. His concept embraced the idea of a continuum in time as well as in space. Lakes were supposed to evolve from one state to another. The different lakes were seen as stages in this succession like zones through a sand dune system. Later work has shown that the natural eutrophication he envisaged is very unusual (Moss, 1988) and that in undisturbed catchments a progressive oligotrophication is more likely as the soils are leached of nutrients faster than rock weathering can resupply them. Nonetheless Pearsall's scheme was an early tunnel under the foundations of the idea that, given enough information, lakes could be placed in a discrete number of classes.

From the 1960s onwards it became clear that lakes are essentially individual. They vary along many axes, each occupying a place in a multidimensional space defined by variables that depend on geography and morphometry, hydrology, rock and soil chemistry, accidents of colonization and environmental change.

To be sure many relationships between the variables have been securely described (Vollenweider, 1968; Schindler, 1978). The wisdom and insight that Thienemann and Naumann showed in linking geology, water chemistry, hypolimnion deoxygenation and benthic fauna is highlighted in the statistical relationships that can now be drawn using more data than these pioneers could ever have dreamed of despite their lengthy publication lists. But the concept of discrete types is dead though its ghost familiarly haunts too many courses in ecology and freshwater biology.

The state of lakes in the twentieth century

Lakes suffer from gravity. They are very vulnerable because they lie at the bottoms of catchment basins and their water and sediment qualities inevitably reflect the imposition of many human activities on the natural system. There are sometimes human impacts even in the most remote location without any people at all living in the catchment and without any direct human use of the land. Effects of acid precipitation are one example and eutrophication and organic pollution by fur seals in some Antarctic lakes may be another. Colony size of the seals has increased perhaps because of reduction of the whale stocks which otherwise would compete with the seals for krill.

Many lakes close to urban areas are severely eutrophicated ultimately from sewage effluent and lowland agriculture. Upland ones often suffer from acidification, large ones from recreation pressure and the smallest standing waters from the threat of being filled in to create more uniform fields for agricultural machines.

Loch Lomond itself can sport a varied list of existing and threatened impacts (Hamilton, 1988). These include recreational abuse such as bank erosion by boats, suspended solid pollution, eutrophication from an increasing summer tourist population, the impacts of afforestation and forestry activities, water abstraction and water level regulation, water transfer from other basins, potential acidification of the northern inflows, the dumping of road works spoil, a proposed pumped storage hydroelectric generation scheme, and the introduction of fish such as the ruffe (*Gymnocephalus cernuus* L.), which may threaten the coregonid powan (*Coregonus lavaretus* (L.)) for which the loch is one of a very few British sites. Equally long lists, or ones with more intensive expression of the major impacts, could be drawn up for all of the better known lakes in England and Wales and many also in Scotland and Northern Ireland.

There has long been a concern for water quality in Great Britain and indeed in mainland Europe also. Following the effects of organic pollution in the mid 19th Century, culminating in cholera epidemics among the increasing urban populations, the need for monitoring schemes and regular measurements in rivers was recognised. This legacy persists today in a preoccupation with oxygen concentrations, biochemical oxygen demand and ammonia levels coupled with a variety of basically similar systems of assessing the benthic invertebrate fauna. The British authorities have never worried too much about lakes, however, except where they have been used for water supply. Even then quite severe eutrophication symptoms have not been seen as problems because their effect on water quality could be mitigated by technological treatment (Collingwood, 1977).

Other countries, particularly those with more lakes than the UK, have been more sensitive to the state of their lakes. The effects of eutrophication were picked up in the 1940s in the USA by Hasler (1947), those of acidification in Norway at least ten years before the existence of the problem was admitted in the UK. In his later years, August Thienemann himself became very preoccupied with the threats to continental lake waters. He developed (1956) a six point programme for their safeguarding nearly forty years ago. The points are still totally appropriate: inform the public of the vulnerability of water; employ limnologists as well as engineers on water projects; create strict laws for the safeguard of water; treat the lake and river system as a unity regardless of political boundaries; educate engineers in ecology; and increase the support for fundamental research, for such is the basis for practice.

In 1989 the highlighting by the press (National Rivers Authority, 1990) of the deaths of dogs and sheep caused by toxic cyanophyte blooms at Rutland Water was one of the factors that emphasised the need to have some sort of system of assessing the state of standing waters and changes in it. It was not the only factor. The Nature Conservancy Council had been developing ways of determining suitable lake sites for conservation for some time and the general tenor of European Community legislation in the Nitrates Directive (91/676/EEC) and Urban Wastewaters Treatment Directive (91/271/EEC) has required the designation of areas sensitive to eutrophication. Controversies like that of whether to ban water skiing on lakes like Windermere and the creation of the Norfolk Broads area as a *de facto* National Park have all contributed. The Royal Commission on Environmental Pollution is currently reporting on eutrophication and acidification. But what should the nature of a monitoring scheme be? Clearly that used for rivers is inappropriate for lakes and any scheme needs a system of categories. '*Seetypen*', the German concept of discrete lake types, however, does not find favour with modern limnologists.

Approaches to a classification and monitoring scheme for lakes

Lakes vary along many axes-they are multivariate systems. With the advent of easily usable multivariate analytical methods an obvious approach to coping with continuous variation, whilst delineating the groupings that are necessary to give some sort of order to a practicable system, would be simply to collect as many data as possible from as many lakes as possible and to have TWINSPAN or some other method sort out some groupings. We believe this would not be a sensible approach. Such groupings would represent only the present state of the lakes; they would have nothing to say about how much they have deviated from some previous state. Lakes, for example, that are, but always have been, highly eutrophic would be grouped with lakes that have recently become eutrophic due to human activities; similarly the naturally and the artificially acidified. The essential need of a monitoring scheme is not to classify lakes *per se* but to assess the degree to which they have changed as a basis for decisions about what ought to be or might be done to remedy the situation. The most important consideration is thus to establish a basis of comparison between the lake at present, and at any time in the future, with some past baseline state.

The first problem then becomes what should be the baseline state. The nature of the lake immediately at its origin might be taken. For most British lakes this would be a turbid state, milky with suspended clay from glacial meltwater, and lacking much in the way of living organisms. The state the lake had reached once the glaciers had fully retreated, and the water cleared, might be a better baseline. But this was only the start of a lengthy slow natural change which in the absence of human activity would have resulted in a progressive oligotrophication (Moss, 1988). There would be no particular reason for selecting the state in 10 000 BP any more than 1000 BP. About 5000 BP, however, represents the start of human induced changes in British lakes for it was about then that the Neolithic cultivators began the process of woodland clearance that was to herald enormous changes in land use and hence nutrient loading. Such a base is an attractive one. Its demerit, however, is that determination of it for a large number of lakes would require thousands of person-years of palaeolimnological expertise whilst the need for a monitoring scheme is immediate. There are perhaps half a million bodies of water in the UK of all sizes, something like 5500 above 4 ha and 81 000 above about 1 ha (Smith & Lyle, 1979). Furthermore if the baseline state is to have the significance that it might be reattainable by suitable restoration, a prehistoric target is unlikely to be practically or politically acceptable. Something more recent is desirable.

Owing to organic pollution, the river systems in urban Britain changed greatly from the mid nineteenth century onwards or even earlier (Eaton, 1989) but anecdotal evidence suggests that the lakes were not seriously affected by human activities until much later. For lakes the pace of change accelerated markedly after the Second World War (1939–1945). The latter catalysed the enormous intensification of agriculture that has brought many changes to the countryside overall (Green, 1981); it led to a marked movement of the human population to the towns as mechanisation took over agriculture and in turn this led to much more of the sewage being centrally treated and its effluents delivered directly to the rivers and lakes instead of to the subsoil. Post-war prosperity has meant increased energy consumption and consequent acidification of rain and snow. Furthermore the novel advent of paid holidays proved a great boost to the tourist industry. A fall in the relative cost of cars led also to a greater use of the countryside. Because the impact of these changes has been recognised and widely deprecated, the immediate pre-war period emerges as a sensible baseline for determining change in lakes.

How then might the baseline states of a large number of lakes be easily and cheaply determined and how might change from the baseline be determined?

A three-axis continuum model for north temperate lakes

There is a very large body of information about lakes and what determines the nature of their ecosystems. Many variables are involved-physical, chemical and biological with a multiplicity of their interactions (Table 1). There is even an element of randomness for accident is a significant determinant of community composition. In freshwaters this is particularly so because of their often island-like nature and the brief time that has elapsed for colonisation since the last retreat of the continental glaciation. The latter considerations suggest that a structure for lake monitoring should not rest on too fine a set of taxonomic considerations. It was the complexity of community composition that largely led the former *Seetypenlehre* into disrepute (Brinkhurst, 1974).

The nature of the medium, water, overwhelmingly sets the nature and constraints of the aquatic habitat. A scheme that is ultimately physico-chemically based, with a cement formed by overall biological features, is thus likely to be practicable and robust. Other than accident, the variables in Table 1 can be grouped into those that are morphometric, hydrological, major chemical and nutrient. It is possible thus to conceive a three dimensional space with axes related to morphometry, major ions, and key nutrients productivity in which each lake or other standing water would occupy a unique point. Such a concept satisfies the intellectual requirement to cope with the continuous variation that lake systems express but an infinity of possibilities is not practicable in a scheme intended for use by statutory regulatory authorities. Continua can be divided into segments, however, and a division of each major axis into, for example, five segments or classes would give a three dimensional matrix with 125 boxes or classification units. It might seem that this is a retreat to the complexity of the early classification schemes of Naumann and his followers – or worse!. However such a scheme would not be a typology with named units defined by preconception. It would be a practical compromise with the accepted case of continuous variation simultaneously along several axes. And it would use a set of major determinants whose importance few, if any, in the limnological community, would doubt.

The baseline state of a lake within the continuum could then be defined by the coordinates of the box in which it fell and any subsequent changes in its state described by its movement from one box to another. The degree of change might then be quantified by the degree to which additional variables, both chemical and biological, differ between the baseline and present states.

The problem then becomes one of: (i) determining the most suitable variable and unit for each of the three major axes; (ii) deciding the most useful segmentation along the major axes; (iii) determining the variables to be used to characterise the nature of the lakes contained within each box in the baseline state; (iv) determining the baseline states of a very large number of lakes in the immediately pre war period, when limnological studies had barely begun in Great Britain: and (v) determining the practicable scope of using the scheme.

Determination and segmentation of the major axes

Morphometry embraces considerations of basin area, maximum depth, distribution of depth, shoreline development, altitude and catchment area to lake area ratio among others. Linked with it are the hydrological variables of discharge and retention time and the seasonal distribution of these. The latter are determined by latitude and altitude, and precipitation to evaporation ratios. There is no simple formula which will express all of these in a linear way along an axis. However many of the structural variables are linked. Lakes of large area tend to be deep, thermally stratifying and with poorly developed littoral zones. Small lakes are likely to be shallow, unstratified and littorally dominated. Maximum depth or volume are thus reasonable variables to define the morphometric axis and are highly correlated with one another. Neither is closely linked with retention time, however. Large lakes in wet areas may have as short a retention time as small lakes in dry areas. To cope with this it is necessary to create several versions of the morphometry axis corresponding with different flushing regimes To maintain practicability these should be as few as possible and we propose three, corresponding with retention times of one week to one month, one month to one year and greater than one year. Water bodies with retention times less than one week are considered to be rivers. And to avoid the problem of inclusion of large river systems with inevitably long overall retention times because of their length, a proviso is made that a water body must have a ratio of its longest axis to its shortest of less than 20:1 to be included.

The major ion axis is perhaps the easiest to define because of the generally close relationship among the major cations and anions in a wide range of freshwaters (Rodhe, 1948). There is some variation in this relationship so no individual ionic measure such as pH, Ca, or alkalinity would be suitable. Conductivity, however is a widely measured and understood variable which integrates the total ionic content and seems the most suitable basis for this axis.

More difficult is determination of a suitable productivity axis. There are many uncertainties in determining

Table 1. Some variables characterising freshwater bodies

(a) GEOGRAPHICAL
Altitude, latitude, Proximity to the sea
(b) MORPHOMETRIC
Area, Depth, Volume, Shoreline development, Maximum to mean depth ratio, Catchment to basin area ratio. Degree and nature of thermal stratification.
(c) HYDROLOGICAL
Precipitation, Evaporation, Effective precipitation, Flushing rate (Retention time)
(d) WATER CHEMISTRY
Major ion concentration and balance, pH, Key nutrient loadings and concentrations, Trace element availability, Chemical stratification, Degree of aeration, Nature and concentrations of labile and refractory organic substances, Sediment type
(e) BIOLOGICAL
Productivity and community composition of pelagic, littoral and profundal communities. Diversity. Interactions among and between communities. Predation and competition, symbiosis and parasitism.
(f) RANDOM AND ACCIDENTAL
Presence or absence of particular species, Degree of isolation, Past floods and droughts, Fluctuating climate, Degree of attainment of equilibrium after glaciation. Disturbance. Introduction of exotic species.

productivity directly, but there is a wide acknowledgement of its ultimate link with nutrient loading, particularly of phosphorus and in some cases of nitrogen. The effectiveness of a particular loading depends on the hydrology. For a given loading (in mass delivered per unit area of lake per unit time), the effective concentration of the nutrient depends on the retention time of the water mass in the lake. Vollenweider (1968) effectively expressed this in his equation:

$$[m] = L/z(s+r),$$

where m is the mean concentration of nutrient in the lake, L is the areal loading, z is the mean depth, r is the flushing rate (the reciprocal of retention time) and s is the rate of loss of nutrient to the sediment, itself a function of depth and retention time. These hydrological considerations are taken into account by choice of the appropriate retention time option along the morphometric axis. This then allows total loading rate to be used as the unit for the productivity axis. In general phosphorus loading will determine productivity in lakes that have not suffered extreme eutrophication. There are, nonetheless, some lakes in which unusually high supplies of naturally occurring phosphorus in the catchment make nitrogen the key limiting nutrient to the crop size of the phytoplankton and ultimately to productivity in general. Some of the Cheshire/Shropshire Meres in England are good examples (Table 2). This eventuality necessitates two alternative axes of nitrogen and phosphorus loading for the productivity axis. Selection of which one to use in a particular instance depends on the available N:P ratio in the loadings. Ratios greater than 9:1 by weight would indicate use of the phosphorus axis and of less than 9:0, of the nitrogen axis.

Segmentation of the axes (Fig. 1) cannot be based on any fundamental considerations because of the nature of the continua which the axes express. The segments (classes) are determined for practical convenience. However they should have a generally exponential pattern because small changes at the lower end of the continua are likely to have much greater effects than similar changes at the upper ends.

The nature of the box variables

The coordinates of a particular lake along the main axes define the units or boxes of the classification scheme. With the three alternative morphometric axes and the two alternative productivity axes all segmented into five classes each there are 125 potential boxes in the scheme, each with two nutrient and three retention time options. Movement from one box in the baseline state to another will be the indication that a lake has changed and the degree of movement will give a measure of the degree of change. This will often be a fairly coarse measure and will not in itself indicate whether

Table 2. Limnological data from the Shropshire/Cheshire Meres-Winter 1991/92

	Cond. (mS cm^{-1})	Sol react P ($\mu g\ l^{-1}$)	Total P ($\mu g\ l^{-1}$)	Nitrate N ($mg\ l^{-1}$)	Amm N ($mg\ l^{-1}$)	Cphyll *a* ($mg\ l^{-1}$)
Whitemere	316	1238	1704	0.25	0.68	5.7
Crosemere	474	241	304	0.23	0.49	8.4
Colemere	243	473	546	0.28	1.02	1.91

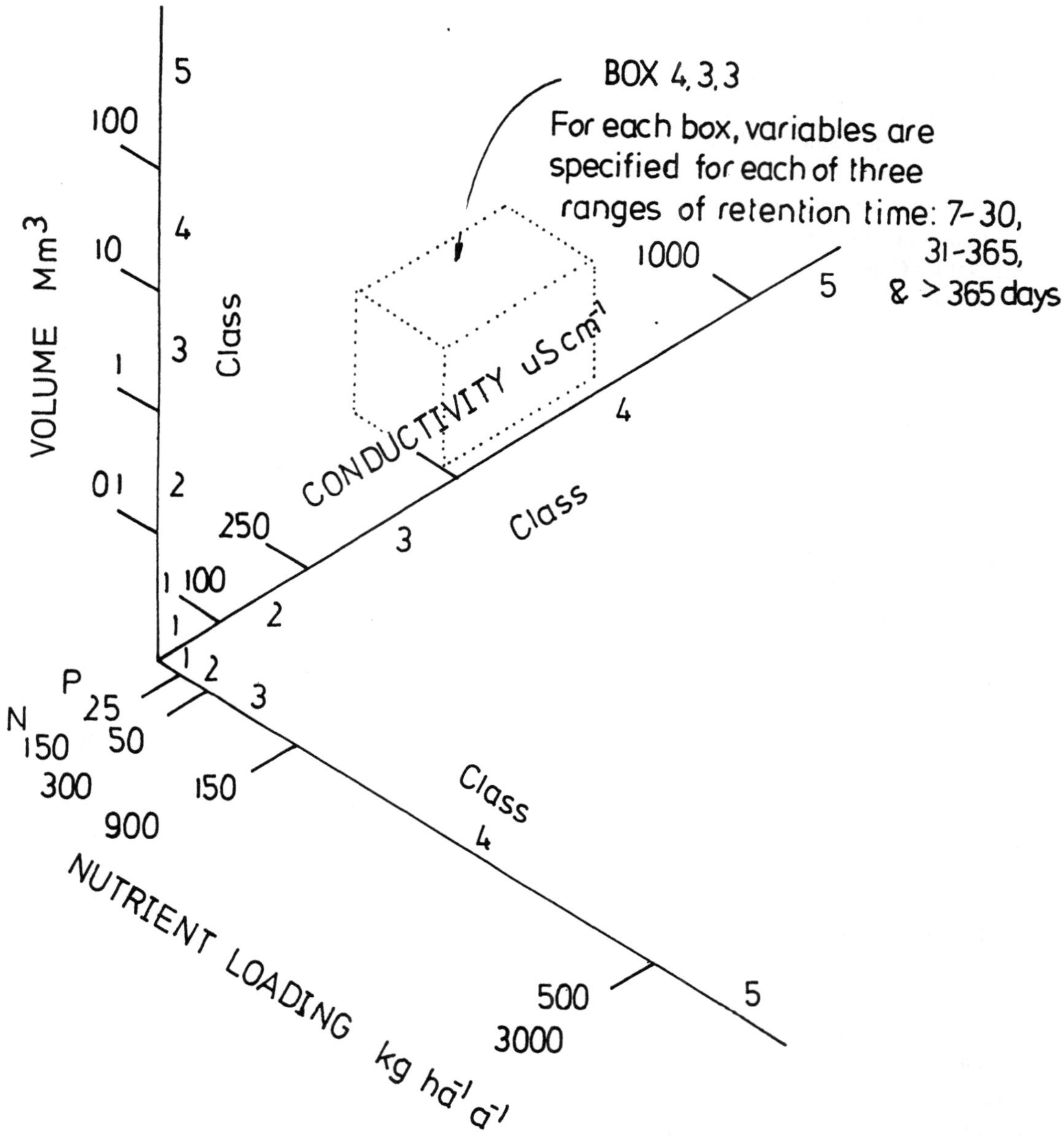

Fig. 1. Proposed axes and segments along the axes for a three dimensional matrix containing north-temperate lakes.

the change is an important one in terms of water quality or conservation value. Sensitivity to these will come in a comparison between the variables contained within the boxes. These variables are intended to describe the characteristic features of the group of lakes corresponding to each box position as parsimoniously and reliably as possible. They might include, for example, chemical variables like the range of potential maximum concentrations of total P and total N, the range of potential maximum winter concentrations of soluble reactive P and of nitrate N, the range of sulphate (particularly because of changes linked with acidification) and perhaps other major ion concentrations, pH and alkalinity. They will include measures of the maximum likely phytoplankton crop as chlorophyll *a* and the existence of visible algal surface blooms. Overall measures of ecological nature will include presence or absence of coregonid or native salmonid fish species, presence or absence of substantial beds of submerged plants in appropriate parts of the littoral zone and use of the Nature Conservancy Council's trophic rating score based on submerged plants (Palmer, 1989; Palmer *et al.*, 1992). This is the only part of the scheme requiring identification expertise and submerged plants have been chosen because of their ease of identification as well as sensitivity to lake state. Other box variables are likely to include maximum depth or water storage capacity because some reservoirs are filling in rapidly whilst the levels of others have been artificially increased. The variables quoted for the boxes will often be ranges and will be determined from existing lakes (in a small number of cases where lakes still exist in their baseline state), from well established models, for example relating phosphorus to chlorophyll *a*, from historic records and, where necessary, by interpolation in the first instance. Degree of change when a lake has moved from one box to another will be quantified by an index measuring the number of box variables that have changed and the percentage change.

Table 3. Changes in the mean summer conductivities (μS/cm) of some of the Cheshire, Staffordshire and Shropshire meres between 1976 and 1991. Data for 1976 are largely taken from Beales (1976) and quoted by Reynolds (1979). Data for 1979 and 1987 are taken from Wigginton & Palmer (1989) and 1991 data are original.

Mere	1976	1979	1987	1991
Berrington Pool	292	290	415	403
Betton Pool	256	243	315	361
Bomere	94	100	145	144
Colemere	289	303	310	248
Crosemere	373	340	390	471
Fenemere	570	507	690	736
Hatchmere	406	380	460	503
Ossmere		310	460	476
Quoisley Mere	522	450	600	616
Tatton Mere	496	400	515	514
Whitemere	239	240	315	320
Petty Pool	337	–	–	482
Cop Mere	404	–	–	488
Combermere	40	–	–	477
Chapel Mere	559	–	–	705
Betley Mere	609	–	–	713

Determining the baseline states

Determining the baseline states of lakes is fundamental to both the philosophy and the practicality of the scheme. The coordinates of each lake along each of the three major axes – volume (and hydrology), conductivity, and phosphorus or nitrogen loading in the immediate pre-war period must be determined. In almost all cases directly-determined data do not exist and any attempt to reconstruct them through palaeolimnological work would be far too slow and expensive. Dating techniques are also only approximate for this period. There are indirect methods however. They give approximations which are likely to be reasonably accurate and which give anchor points for the scheme. By analogy they are the reference points surveyors might use in making a map. The exact location of them is less important than that they exist and stay fixed.

It is unlikely that the volume of most natural lakes has changed more than trivially since 1930 and a current knowledge of area and mean depth will sufficiently locate the lake in the scheme. Reservoirs may have changed in volume particularly in upland catchments where peat erosion has been a significant problem (Labadz, Burt & Potter, 1991; Phillips, Yalden & Talling, 1981) and for these the constructed volume can be used. Average hydrology also changes little over periods of a few decades. Retention times can be determined from the catchment area, the effective precipitation and the lake volume. Meteorological data are readily available.

Conductivity is more likely to have changed either through the effects of acidification in the uplands or of land drainage and marine seepage in coastal areas.

Table 4. Hindcasted (1931) and present (1991) conductivities (mS/cm) for five U.K. lakes. Grafham Water is an artificial reservoir created in the 1960s. Values hindcasted for 1931 are thus hypothetical for the catchment that then existed

Lake	Hindcasted	Observed	Change
Slapton Ley	135–350	300	No
Grafham Water*	550–850	944	Yes
Crosemere	350–600	471	No
Windermere	20–100	72	No
Loch Lomond	50–150	132	No

There is evidence also of an effect of the dry summers of the 1980s in increasing conductivities in some inland areas (Table 3). We propose therefore to hindcast the conductivity from data independently of the present conductivity. A considerable body of data relates rock type to the chemistry of water running off it. From the geological maps a broad division among major rock types can be made and consequent conductivities apportioned among them. For lakes within 50 km of the coast, sea spray, redistributed by wind, may significantly increase the conductivity of rain and an allowance can readily be made for this. For immediately coastal lakes in the lowlands the effects of natural seepage and historic land drainage cannot easily be calculated independently and in these few cases resort to past records will be necessary to determine the baseline conductivity. The hindcasted determinations will be ranges rather than single values but preliminary calculations (Table 4) give results that accord with expectation.

Determination of the baseline nutrient loading (and of that on subsequent dates) might seem to be the most difficult problem. However a great deal of data have now accumulated which correlate delivery rates of N and P with land usage and stock holdings. Data also exist in relation to human populations also. Thus if the land usage of a catchment, the headage of stock of various kinds and the human population are known, an estimate of nutrient loading can be made. The land use of the British Isles has been documented annually in the Agricultural Census Returns and these data are held in the Public Records Offices at Edinburgh and Kew. The records are on a parish basis but the situation for a catchment is unlikely to be greatly different from the sum of the parishes which contain it. Corrections can be made to the export coefficients for N and P dependent on changing fertilization practice from the annual recommendations given by the Ministry of Agriculture. Fisheries and Food and the Scottish Agriculture Department. The human population has been censused on a ten-year basis and these data are also available on a parish basis. An example of calculation of loading rates is given in Table 5 and Johnes (1990) shows that the method gives a very close correspondence to directly measured rates of loading for a large catchment in southern England. Table 6 gives some examples of hindcasted values for 1931 and for 1987 for a variety of lakes. Nineteen-thirty-one has been chosen as the baseline pre-war date for the scheme because it just predates the extreme intensification of agriculture and because a human population census was taken that year. Nineteen-eighty-seven is the year of the most recent agricultural census for which parish based data are available to the public. Choice of the N or P alternatives along the productivity axis is determined by the N to P ratio of the loading rates.

Changes in lake quality between 1931 and 1987 are shown in Table 7 for a variety of British lakes. The baseline and 1987 states are defined by the class coordinates of the boxes they occupied in the two years (Fig. 1). Occasionally there has been a change in conductivity but most have changed along the productivity axis as might be expected from the present pervasiveness of eutrophication. Switches between N and P limitation are also apparent.

The practicable scope of the scheme

Any scheme for monitoring changes in lake state must be usable for a large number of water bodies in a short period. Even when small ponds, important to conservation and to the landscape and highly vulnerable as they are, are discounted, there are about 50 000 standing waters greater than 1 ha in area in England and Wales and a further 30 000 in Scotland. These are underestimates of the number of water bodies that are significant. Many smaller bodies are of concern to the public (the New Forest ponds for example) or are parts of statutorily designated nature reserves of various kinds. Such smaller pools might add another 10 000 making a total of 90 000 significant sites. Let us say that statutory authorities were able to allocate 100 officers on a UK basis to the maintenance of a lake monitoring scheme and let us require the state of the water bodies to be monitored every five years. There would thus be about 22 500 person-days available per year to cover the 90 000 water bodies. This amounts to only just over 1 person-day available per water body

Table 5. Example of the calculation of nutrient loading for current (1987) N and P loads on Slapton Ley in Devon. Values are derived from compilations of export coefficients compiled by Johnes (1991) for different land use types and livestock, from the Agricultural Census Returns for 1987, and for people from the most recent Official Human Population Census. Values for domestic stock and for humans are corrected to give the actual loads delivered to the lake by allowance for retention in the body, on the land, loss through denitrification, or in sewage sludge disposed of elsewhere.

Source	N input kg ha^{-1}.a	P input kg ha^{-1}.a	N export (%)	P export kg ha^{-1}.a	N export kg/a	P export kg/a
Perm grass	160	16	5	0.1	16700	209
Temp grass	244	24	5	0.3	12600	310
Cereals	135	20.7	12	0.8	14200	689
Root crops	93	7.9	20	0.8	2180	94
Field veg.	132	38	20	0.65	2430	60
Oilseed rape	190	27	30	0.8	5230	2.7
Small fruit	112	35	20	0.65	0	0.58
Bare fallow	0	0	100 kg/ ha.a	0.2	11400	4.4
Rough grazing	10	0	13 kg/ha.a	0.02	751	1.2
Woodland	20	0	13 kg/ha.a	0.02	2280	3.6
	kg/hd.a	kg/hd.a	(%)	(%)		
Cattle	70	7.7	16.2	2.9	68200	7430
Pigs	19	5.6	14.5	2.6	3360	1010
Sheep	8.9	1.5	17	3	23100	3900
Poultry	0.3	0.2	15.3	2.7	78	52
Horses	77	11	16.2	2.9	Negl	Negl
People			2.14 kg/hd.a	0.09 kg/hd.a	4300	189
Total load on lake					167t	14t

Table 6. Hindcasting of 1931 and 1987 loads of phosphorus and nitrogen for some UK lakes. 'Stock' includes the human population.

	1931			1987			
	Stock	Land	Total	Stock	Land	Total	Change (%)
P load t/a							
Slapton Ley	6.4	1.35	7.75	12.6	1.38	14	82
Barton Broad	12.1	5.5	17.6	43.1	9.1	52.2	197
Ennerdale	3.2	0.25	3.5	4.0	0.26	4.23	22
Windermere	22.6	1.77	24.3	90.6	1.78	92.4	280
N load t/a							
Slapton Ley	44.8	12.6	57.4	99	67.8	167	191
Barton Broad	73.9	28.6	103	147	238	385	274
Ennerdale	19.1	4.48	23.6	23.6	1.69	25.2	7
Windermere	166	40	206	645	119	763	270

over the 5 year cycle. Such numbers are notional but they make the point that a comprehensive scheme must be as economical as possible in the time it allocates per water body.

Table 7. Initial classification results for some UK lakes. Numbers are the class coordinates along the three major axes (see Fig. 1). Axis 1 is that of Volume, 2 that of conductivity and 3, nutrient loading. N and P in the Axis 3 columns indicate the hindcasted key limiting nutrient. Values for reservoirs for 1931 are the hypothetical ones for the potential catchment. All reservoirs indicated with an asterisk were constructed before 1931.

	1931			1987		
Lake	Axis 1	Axis 2	Axis 3	Axis 1	Axis 2	Axis 3
Slapton Ley	3	2	3N	3	3	4N
Loe Pool	3	3	4N	3	3	4P
*Chew Valley	4	3	1P	4	3	3N
*Blagdon	3	3	2P	3	3	3N
*Bough Beech	3	4	5P	3	4	5N
*Grafham	4	3	1P	4	4	3P
*Farmoor	3	4	2P	3	4	4P
*Cow Green	4	1	1P	4	1	1P
Crosemere	2	3	3N	2	3	3N
Rostherne M	3	3	4N	3	3	5N
Esthwaite	3	2	4N	3	2	5N
Windermere	5	1	1P	5	1	3P

How much time would the scheme suggested above take? It has the merit that the all-important baseline state can be determined largely as a desk exercise. Lake area can be determined from large scale maps and bathymetric data are already available for many lakes. For others, a bathymetric survey would need to be carried out once only. Retention time can be calculated with the further help of meteorological information already available in data banks. Baseline conductivity is also calculated from available geological and other maps and data for the calculation of the 1931 baseline nutrient loadings for 50–100 sites can be extracted in a day from the Public Record Offices. A whole suite of adjacent water bodies could be handled together and calculations can be automated with suitable computer software so it is difficult to estimate the time needed for setting up the scheme. We estimate perhaps 100 person-years. Thereafter the scheme is designed to require two day-visits per water body, one in winter and one in summer per five year cycle for collection of the box-variable data. Again several lakes might be dealt with in a day. Central analytical services are assumed for chemical analysis of the samples.

Limnologists will argue that the approach is minimalist and that seasonal variation might make single winter and summer samples unrepresentative. This danger is acknowledged but will be dealt with in the way that box variables in the baseline state will be quoted as ranges or upper limits. Afficionados of particular lakes will also protest that the particular features of 'their' lake are not detected in the general scheme. For example, a particular rare species of flatworm or race of fish is not recorded or the effects of a particular specific impact monitored. However, if changes are taking place in particular species these will generally result from a larger scale change which the scheme will detect. And unusual impacts will also either be manifestly obvious or, if not, are likely to be reflected in one or other of the box variables. The great beauty of limnology has been its steadfastly holistic approach which has revealed and explained a network of interconnections. From the map of these, much can be deduced. And once some indication of change has been picked up by this scheme there can be more detailed investigation if necessary.

From the accumulating results of such a long term monitoring exercise many advantages, both intellectual and practical, must arise. We will be able to see an overall pattern of change and its magnitude and will thus have a rational and detailed case for promoting restoration of damaged lakes and unwisely managed catchments. For example, only the remotest of the lakes in Table 7 are unlikely to have undergone significant eutrophication. Limnologists have known or suspected this for many years but lack of a systematic synoptic picture has meant that eutrophication has entered the official and public consciousness only very recently. The case Loch Lomond, ostensibly a lake tucked safely away from the urban and agricultural depredations of midland and southern England, when multiplied by a thousand or ten thousand or even more, must add up to an incontrovertible argument for a much wiser management and use of our land surfaces and their dustbins, the water bodies.

Thus the ghost of August Thienemann visits Loch Lomond. He was wise in being among the first to see the connections within and between lakes and in seeing the threats to them. He was wise enough not to go too far along the path of a lake classification by rigid pigeonholing, but recognised continua in the form of the disharmonic lakes that did not fit the original type categories. And it is from him that the inspiration for the scheme we have proposed is derived – but perhaps with some of the passion of Einar Naumann also!

Acknowledgments

The work of developing the Lake Classification and Monitoring Scheme from our original ideas has been financed by the National Rivers Authority. Presently however the work is at the stage of a research project and should not be taken to represent the Authority's official view in any way. Data for the Shropshire and Cheshire Meres were obtained through a research contract placed by English Nature, the Nature Conservancy Council for England.

References

Beales, P. W., 1976. Palaeolimnological studies on a Shropshire mere. Ph.D. Thesis, University of Cambridge.

Brinkhurst, R. O., 1974. The Benthos of Lakes. Macmillan, London, 190 pp.

Collingwood, R. W., 1977. A survey of eutrophication in Britain and its effects on water supplies. Technical Report TR40. Water Research Centre, Medmenham.

Eaton, J. W., 1989. Ecological aspects of water management in Britain. J. appl. Ecol. 26: 835–849.

Elster, H. J., 1958. Das limnologische Seetypensystem, Ruckblick und Ausblich. Verh. int. theor. angew. Limnol. 13: 101–120.

Green, B., 1981. Countryside Conservation. Unwin Hyman, London.

Hamilton, J. D., 1988. Recent influences on the ecology of Loch Lomond, Scotland. Verh. int. Ver. theor. angew. Limnol. 23: 403–413.

Hasler, A. D., 1947. Eutrophication of lakes by domestic drainage. Ecology 28: 383–395.

Johnes, P. J., 1990. An investigation of the effects of land use upon water quality in the Windrush catchment. D. Phil Thesis, Univ. of Oxford.

Labadz, J. C., T. P. Burt & A. W. R. Potter, 1991. Sediment yield and delivery in the blanket peat moorland of the South Pennines. Earth Surf. Proc. Land 16: 255–271.

Moss, B., 1988. Ecology of Freshwaters, 2nd edn., Man and Medium. Blackwell Scientific, Oxford, 417 pp.

National Rivers Authority, 1990. Toxic Blue-Green algae. Water Quality Series 2: 1–125.

Naumann, E., 1917. Undersokningar over fytoplankton och under den pelagiska regionen forsiggaende gyttje- och dybildningar inom vissa syd- och mellansvenska urbergsvatten. Kungl. Svenska Vetenskapsakademiens Handlingar 56: 1–165.

Naumann, E., 1931. Limnologische Terminologie. Urban und Schwarzenberg, Berlin, 776 pp.

Palmer, M., 1989. A botanical classification of standing waters in Great Britain. Nature Conservancy Council Research and Survey Report 19.

Palmer, M., S. L. Bell & I. Butterfield, 1992. A botanical classification of standing waters in Britain. Applications for conservation and monitoring. Aquat. Cons. 2: 125–144.

Pearsall, W. H., 1921. The development of vegetation in the English Lakes, considered in relation to the general evolution of glacial lakes and rock basins. Proc. R. Soc. 92: 259–284.

Phillips, J., D. Yalden & J. Tallis, 1981. Peak District Moorland Erosion Study. Phase I Report. Peak Park Joint Planning Board, Bakewell.

Reynolds, C. S., 1979. The limnology of the eutrophic meres of the Shropshire-Cheshire plain: a review. Field Studies 5: 93–173.

Rodhe, W., 1948. Environmental requirements of freshwater plankton algae. Experimental studies in the ecology of phytoplankton. Symb. bot upsal. 10: 1–149.

Rodhe, W., 1974. The International Association of Limnology: Creation and function. Mitt. int. Ver. theor. angew. Limnol. 20: 44–70.

Rodhe, W., 1975. The SIL founders and our fundament. Verh. int. Ver. theor. angew. Limnol. 19: 16–25.

Schindler, D. W., 1978. Factors regulating phytoplankton production and standing crop in the world's freshwaters. Limnol. Oceanogr. 23: 478–486.

Slack, H. D., 1957. Studies on Loch Lomond I. University of Glasgow and Blackie and Son, Glasgow, 133 pp.

Smith, I & A. Lyle, 1979. Distribution of Freshwaters in Great Britain. Institute of Terrestrial Ecology, Cambridge, 30 pp.

Snow, C. P., 1959. The Two Cultures. Cambridge University Press. Cambridge, 27 pp.

Thienemann, A., 1909. Vorlaufige Mitteilung uber Probleme und Ziele der biologischen Erforschung der neun Westfalischen Talsperen. Ber. Versamml. Bot. Zool. Ver. Rheinl. – Westf. Jg 1909: 101–108.

Thienemann, A., 1915. Physikalische und chemische Untersuchungen in den Maaren der Eifel II. Verh. Naturhist. Ver. preuB. Rheinlande u. Westfalens 71: 273–389.

Thienemann, A., 1931. Tropische Seen und Seetypen. Arch. hydrobiol. Suppl. 9: 205–231.

Thienemann, A., 1956. Das Wasser als Sorge Europas. Universitas 11: 849–858.

Vollenweider, R. A., 1968. Scientific fundamentals of the eutrophication of lakes and flowing waters with particular reference to nitrogen and phosphorus as factors in eutrophication. OECD, Paris, 159 pp.

Weber, C. A., 1907. Aufbau und Vegetation der Moore Norddeutschlands. Beiblatt zu den Botanischen Jahrbuchem 90: 19–34.

Wigginton, M. & M. Palmer, 1989. Survey of Shropshire and Cheshire meres 1987. Nature Conservancy Council England Field Studies Unit Report 59.

Hydrobiologia **290**: 13–20, 1994.
K. J. Murphy, M. C. M. Beveridge & R. Tippett (eds), The Ecology of Loch Lomond.

Geology of the Loch Lomond catchment

James G. MacDonald
University of Glasgow, Dept of Adult & Continuing Education, 59 Oakfield Avenue, Glasgow G12 8LW, UK

Key words: Loch Lomond, catchment, bedrock

Abstract

Loch Lomond is unusual in its geographical position and relationship to underlying geological structures. To the north-west of the Highland Boundary Fault the bedrock is composed mainly of the Dalradian Supergroup, a sequence of regionally metamorphosed Late Precambrian marine sediments. Immediately to the south-east of the Highland Boundary Fault the Highland Bonder Complex includes fragments and slices of ophiolitic origin and sediments with a range of Ordovician ages. The Dalradians may have originated in a remote location on the northern margin of Gondwanaland. To the south-east of the Highland Border the bedrock lies within the Midland Valley of Scotland and consists entirely of Upper Palaeozoic rocks of Devonian and Carboniferous age. The nature of the bedrock has an influence on the environment of Loch Lomond with respect to slope stabilities and water quality.

Introduction

Loch Lomond is unusual amongst Scottish Lochs in that it combines the position and characteristics of a Highland and Lowland loch. It is also somewhat unusual in its near northsouth orientation. Many highland lochs are obviously related in their position either to the major structural grain of the country, like Loch Awe which is orientated along a major fold axis in the Caledonian mountain belt, or Loch Ness on the line of tbe Great Glen Fault; others, such as Loch Earn and Loch Katrine, are related to the pattern of Tertiary drainage initiated by eastwards tilting of the Scottish crust. Loch Venachar, although straddling the Highland Boundary Fault, also conforms with the pattern of Tertiary drainage. Loch Lomond, by contrast, cuts across the main structural trends and the pattern of Tertiary drainage.

The contrast between the northern, narrow entrenched highland part of Loch Lomond north of Luss and the much wider shallower southern portion is closely related to the bedrock geology (Fig. 1). The fundamental geological divide is along the line of the Highland Boundary Fault which crosses the loch from Balmaha southwest to Arden. To the north of this line most of the catchment lies within the area of the Dalradian Supergroup. The rocks of this part of the Scottish crust were initially formed more than 600 million years ago in late Proterozoic times as greywackes, a suite of marine turbidites which were deformed in Late Precambrian times and again during the Ordovician Period at the time of the formation of the Caledonian Mountain Belt. At the same time they were subjected to regional metamorphism (mainly low to high greenschist facies) to form the belts of slate, schistose grit and schist exposed at the present day (Rogers *et al.*, 1989). They have the north-east to south-west structural trend characteristic of the Caledonides in Scotland.

To the south-east of the Highland Boundary Fault the catchment is within the area of the Midland Valley of Scotland. Except for the area immediately adjacent to the Highland Boundary Fault the bedrock consists entirely of rocks of Devonian and Lower Carboniferous age including Lower and Upper Old Red Sandstone, basal Carboniferous sediments of the Inverclyde Group and the overlying volcanic rocks of the Clyde Plateau. In the Loch Lomond area, the Midland Valley sediments from the Upper Old Red Sandstone upwards, overlap the Highland Boundary Fault so that in places they rest directly on the Dalradians.

The Highland Boundary Fault is one of the fundamental fracture lines of the crust in the British Isles. Its locus coincides with a narrow belt of rocks of Ordovician or older age collectively known as the Highland

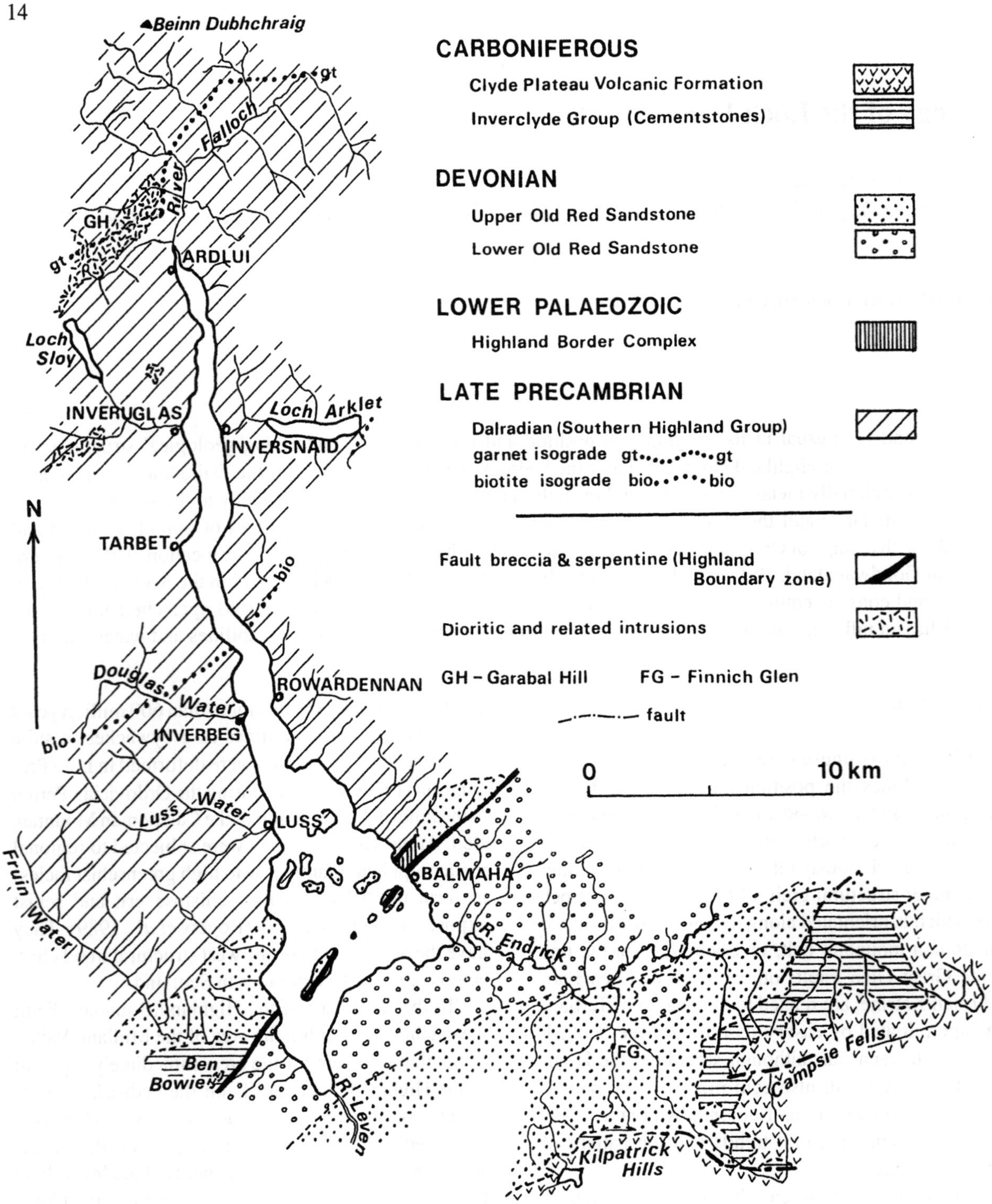

Fig. 1. Bedrock geology of the Loch Lomond catchment.

Border Complex. These rocks have in places been highly sheared and altered as the result of the intense stresses associated with compressional and shear tectonics following the emplacement of mobile segments of continental crust (terranes) from the south.

In addition to the main groups of rocks described above, igneous activity resulted in the emplacement of a suite of sheets and dykes of lamprophyre and porphyrite with occasional larger masses, especially the Garabal Hill Complex, in Late Silurian times. These

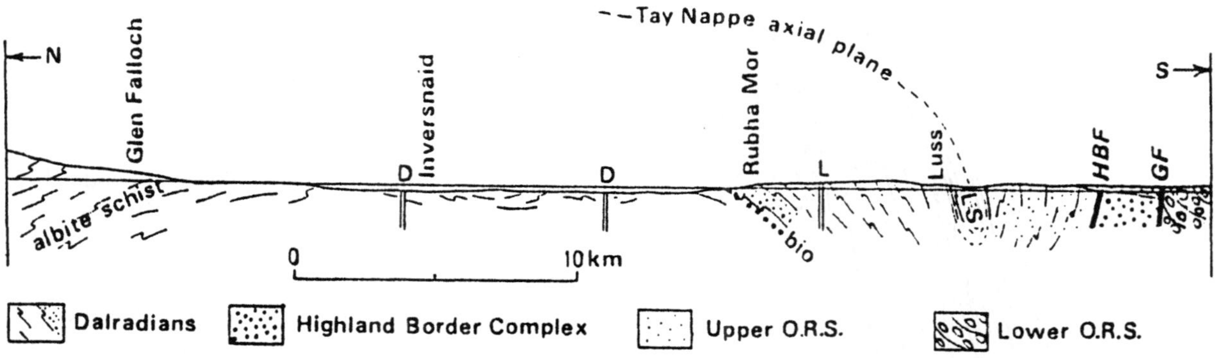

Fig. 2. Diagrammatic north-south geological cross section of the Loch Lomond catchment. D – dolerite dyke; L – lamprophyre (Inverbeg-Rowardennan dyke); bio – biotite isograde, the Dalradians to the south of this are in the chlorite zone, to the north the biotite zones extends as far as Glen Falloch north and west of which garnet makes its first appearance (see Fig. 1); LS – Luss Slates in the downturned nose of the Tay Nappe; HBF – Highland Boundary Fault; GF – SW extension of the Gualann Fault. North of the Gualann Fault the Upper Old Red Sandstone overlaps the Highland Boundary fault and rests directly on the Dalradians. The precise position of the Highland Boundary Fault is uncertain here.

intrusions are minor manifestations of the same post tectonic igneous episode which is expressed in other parts of the Dalradian area in major volcanic centres such as those in the Etive area and Glen Coe. The Dalradian schists have also been intruded by late Carboniferous quartz dolente dykes.

The Dalradians

The Dalradian rocks in the Loch Lomond area vary in composition, in the degree of metamorphism and in the complexity of the folding. Most of the area is occupied by the Ben Ledi Grits which vary in metamorphic texture from schist to schistose grit and phyllite. At Luss, however, a broad belt of slates form the finest grained part of the succession. They lie in the downturned nose of the Tay Nappe, a major recumbant fold (Fig. 2). Between the Luss slates and the Highland Boundary Fault lies the Leny Grit. Its outcrops include some coarse grained pebbly lithologies with clasts in excess of 18 mm. Less coarse grained schistose grits outcrop to the north of Luss and probably represent the extension of this lithology round the nose of the downturned fold. Graded bedding in the schistose grits indicates that the succession to the north of Luss has been inverted (Shackleton, 1958) thus the upper limb of the Tay nappe has been eroded away and, north of Luss, only the inverted lower limb is preserved.

Structural complexity increases northward from Luss. Between Luss and Glen Douglas a strongly developed strain slip cleavage disrupts the traces of earlier folding (Fig. 3). This is most clearly seen in lithologies of intermediate grain size, whereas in the phyllites, with their finer grain size and high content of chlorite, the rocks are more strongly sheared. By contrast the coarser schistose grits sometimes occur as discrete, relatively undeformed units, which retain recognisable graded bedding inherited from original sedimentation. Northwards from Glen Douglas the rocks are increasingly deformed by further fold phases although evidence of the earlier deformation is often still clearly visible (Fig. 4). These latter structures are well exposed in cuttings beside the old road on the west bank of Loch Lomond north of Inverbeg between Rudha Mor and Rudha Dubh (Bowes, in Lawson & Weedon, 1992).

The northwards increase in structural complexity is accompanied by an increase in metamorphic grade reflecting rising temperatures and pressures during metamorphism at progressively deeper levels within the fold belt. This is indicated by the replacement of chlorite by biotite. Close to the Highland Boundary Fault the rocks lie within the chlorite zone (Tilley, 1925). Near Rowardennan biotite first makes its appearance (Mather, 1970) and most of the rest of the Dalradians within the Loch Lomond catchment lie within the biotite zone except in the extreme north where the garnet zone is encountered. In Glen Falloch and the adjacent area there has been retrogressive metamorphism of biotite zone rocks to produce albite porphyroblast schists of lower greenschist facies equivalent in grade to the chlorite zone (Watkins, 1983).

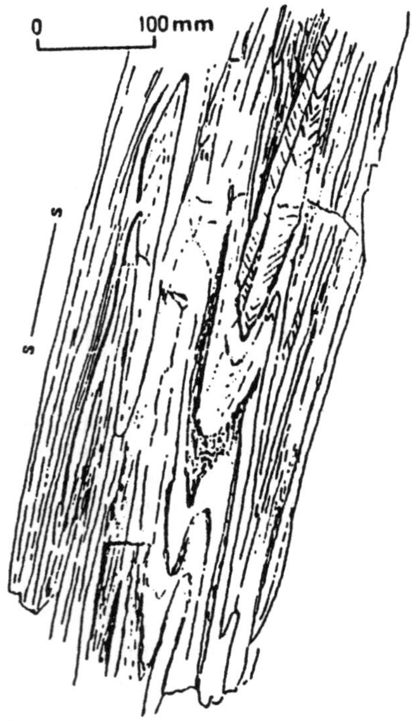

Fig. 3. Strain slip cleavage (orientation indicated by s-s) and folds in Ben Ledi Grits exposed on a near vertical east-facing road cutting 3.4 km north of Luss (from a photograph taken in 1979). Folded bedding remnants are visible on the right hand side of the drawing. The herring-bone pattern at the top right is a vestige of an earlier foliation indicating that the rocks have been folded at least twice. This locality has been destroyed by recent road widening operations.

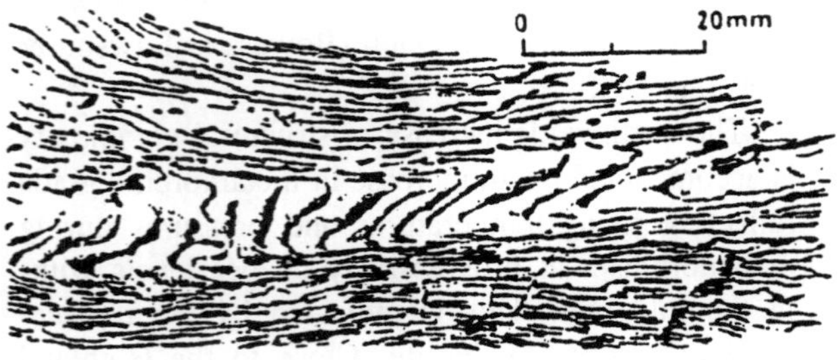

Fig. 4. Early foliation and folds, of the type illustrated in Fig. 3, preserved in a coarse grained lithology on a flat lying outcrop amongst complexly folded schists near Rubha Mor, north of Inverbeg.

Post tectonic igneous intrusions

The most extensive outcrops of igneous rocks within the Dalradian area occur in the Garabal Hill area to the west of Inverarnan (Nockalds, 1940). The Garabal Hill – Glen Fyne Complex includes lithologies ranging from ultrabasic rocks to gabbro, and appinitic diorite to granite. The main granite outcrops lie outside the Loch Lomond catchment in the Glen Fyne area. The typically calc-alkaline affinities of these rocks indicate that they are likely to be late Silurian (about 400 Ma) in age and contemporaneous with the major volcanic centres, such as Etive and Glen Coe, which are situated along the line of the Dalradian outcrop from the southwest Highlands to the northeast of Scotland.

Smaller dioritic intrusions of appinitic and related types occur to the northeast of Garabal Hill on both sides of Glen Falloch (Anderson, 1935) and in the Inveruglas area. There are also many minor sheets and dykes of porphyrite and lamprophyre cutting the Dalradians. The dykes tend to follow the typical Caledonian NW-SE trend. These intrusions frequently contain xenoliths of deeper seated material a notable example being the Inverbeg Rowardennan dyke (Anderson & Tyrrell 1936; Dempster & Bluck, 1991) which has been quarried on both sides of the loch.

The Dalradians, the Highland Boundary Fault and the Highland Border Complex

The prominent positive linear topographic feature produced by Conic Hill above Balmaha, the islands of Inchcailloch and Inchmurrin and its prolongation through the gap between Ben Bowie and Killoeter ridge about three kilometers east of Helensburgh has generally been regarded as marking the line of the Highland Boundary Fault. Recent reappraisal of the detailed geology of the Balmaha area (Bluck *in* Lawson & Weedon, 1992) indicates that the boundary fault, in the strict sense, is situated more than a kilometer to the NW and outcrops at Arrochymore Point.

The ridge referred to above is composed of hard resistant conglomerates of Lower Old Red Sandstone age which have been tilted to a near vertical position against a fracture referred to by Bluck (*ibid*) as the Gualann Fault. The conglomerates belong to the Midland Valley sedimentary trough and between the Gualann Fault and the Highland Boundary Fault lies the Highland Border Complex (Fig. 5).

The Highland Border Complex (Curry *et al.*, 1984) consists of many different rock types. At the Gualann Fault a serpentine conglomerate and breccia contains clasts including serpentine, gabbro, dolerite, spilite, sandstone and quartzite. The clasts in this zone range in age from 1863 Ma (Dempster & Bluck, 1989) to about 500 Ma. At Arochymore Point serpentine and sheared serpentine of unknown age are probably faulted against Dalradians. Between the two faults the Arochymore Point serpentine and associated chert is succeeded by

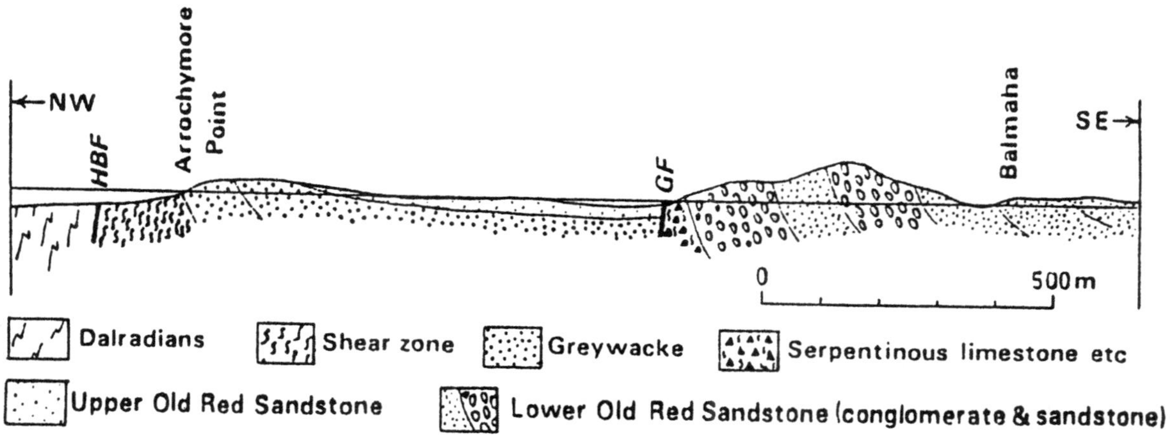

Fig. 5. Cross section of the Highland Border in the Balmaha area. HBF – Highland Boundary Fault, GF – Gualann Fault. The shear zone SE of the Highland Boundary Fault consists of highly deformed serpentine and other components of the ophiolitic complex which forms a basement to the greywackes; together they comprise the Highland Border Complex in the Balmaha area along with the varied lithologies of the Gualann Fault zone. The basal Lower Old Red Sandstone rests unconformably on the Highland Border Complex in this area.

sandstones which, although fossils have not been found in them, resemble similar rocks in the Aberfoyle area which contain chitinozoa indicating a Caradoc (upper Ordovician) age.

Although the origins of the Highland Border Complex are uncertain it is clear that the serpentine, splilite, gabbro, dolerite, black shale and chert, a classic ophiolitic association, are disrupted slices and fragments of oceanic lithosphere and their marine sedimentary cover. These in turn are succeeded unconformably by Upper Ordovician marine shelf sediments (sandstones and limestones). Older Ordovician marine sediments have been found within the Highland Border Complex at Aberfoyle (Curry *et al.*, 1992, Ingham *et al.*, 1985).

The Ordovician sediments of the Highland Border Complex were deposited in a marine basin. At the same time the Dalradians were being subjected to deformation and rapid uplift (Bluck & Dempster, 1991, Dempster, 1985) with consequent erosion pruducing an estimated 10 to 30 km^3 of sediment per km^2 of the Dalradian land surface (Bluck in Lawson & Weedon, 1992). No evidence has been found that any of this sediment found its way into the Highland Border Complex basin. This suggests that the Dalradians and the Highland Border Complex could not have been in close proximity in Ordovician times. On the other hand close comparisions can be made between the tectonic history of the Dalradians and parts of the marginal cratonic areas of north and west Africa which at the time was part of the ancient southern continental grouping, Gondwanaland. It has been concluded (Bluck & Dempster, 1991) that the Dalradians became detached from their original location between 600 and 570 Ma and were transported laterally to dock with what was to become the northern part of Scotland between 550 and 530 Ma (Lower to Middle Cambrian), and did not approach anywhere close to their final position before 435 Ma, in the Silurian Period. Even after this the area was to be subjected to a further, compressive, Upper Palaeozoic event before the present day distribution of terranes was finally established.

Upper Palaeozoic Rocks

These occur principally to the south of the Highland Boundary Fault although there is clear evidence that some encroachment onto the Dalradians took place during the deposition of the Upper Old Red Sandstone. The Lower Old Red Sandstone, however, rests unconformably only on rocks of the Highland Border Complex. In the vicinity of the Highland border the Lower Old Red Sandstone consists of very coarse conglomerates interbedded with sandstone beds. The conglomerates contain boulders composed principally of metamorphic quartzite with subordinate lava and occasional clasts derived from the Highland Border Complex (Bluck in Lawson & Weedon, 1992). There is a lack of evidence that the Dalradians were a source for any of the clasts in the Lower Old Red Sandstone sediment. The conglomerates are fans of sediment which appear to have been derived from the NE, east and SE. They are typical of the deposits of fault bounded basins.

The conglomerate fans do not persist upwards in the succession and they thin laterally to the southeast. They are close to being vertical in the Highland border area but the dip of the Lower Old Red Sandstone decreases to the SE where only sandstone is exposed in most of the area of the Endrick catchment. The steep dip of the Lower Old Red Sandstone in the vicinity of the Highland Border implies a significant compressive tectonic event following its deposition. Middle Old Red Sandstone is absent throughout the Midland Valley and the Upper Old Red Sandstone rests unconformably on all the older formations of the area. This implies that the Dalradians were not finally emplaced in their present position until mid-Devonian times.

The Upper Old Red Sandstone in the Loch Lomond area consists of light red sandstones and breccias. NW of the Gualann Fault at Balmaha breccias outcropping on the shore of the Loch contain abundant clasts derived from the Dalradians a short distance to the NW. Where it is exposed to the SE the Upper Old Red Sandstone is usually fine grained as in the thick sequence exposed in the Finnich gorge (NS 495848).

There is no clear break between the Upper Old Red Sandstone and the overlying basal Carboniferous Inverclyde Group. Near the top of the Upper Old Red Sandstone there is a transition from typical red beds into caliche bearing horizons in which all or most of the red colour disappears. This is finally succeeded by the basal Carboniferous cementstones and shales in the Ballagan Beds which occur widely through the area below the Clyde Plateau lavas.

The Clyde Plateau Volcanic Formation forms the uplands of the Campsie and Kilpatrick Hills with NW facing escarpments from which the head waters of the Endrick river system flow. Rock types include volcanic ash, lavas and intrusions belonging to a transitional basaltic association (MacDonald & Whyte, 1981; Whyte & MacDonald, 1974) with compositions ranging from olivine basalt to trachyte.

The only other solid rocks which occur within the Loch Lomond catchment are a number of quartz dolerite dykes of Upper Carboniferous age. These occur within the Midland Valley where they have an east-west orientation. Where they cut the Dalradians their trend tends to be closer to that of the characteristic Caledonian grain of the country rocks but in Glen Falloch they follow the more typical late Carboniferous east-west trend.

Environmental influences of Bedrock Geology

In the Dalradian area folding combined with the development of strong foliation, particularly in the more pelitic lithologies, and the occurrence of steep slopes on hillsides produces possible hazards from slope instability. Where foliation directions coincide with hill slopes there is a potential for landslips due to detachment of rock masses at splits along foliation planes opened up by a combination of freeze-thaw and water pore pressure. Similar effects commonly lead to the sliding off of unconsolidated Quaternary deposits.

This was recognised during the construction of the new road on the west bank of the loch between Luss and Tarbet, particularly in the area immediately to the south of Rubha Mor where large slabs of schistose grit have an orientation coinciding in places with the slope of the hillside. Account of this has been taken by the building contractors who have stabilized the rock by the insertion of large numbers of rock bolts and the diversion of the drainage around the top of the cuttings. The long term effectiveness of these measures depends on maintenance of the drainage and monitoring of the condition of the rock bolts.

Another area where there could be a potential hazard from rock slides is the NE slopes above Loch Sloy reservoir. Difficulties were encountered during construction of the dam due to the presence of many faults and complex folding in the biotite schists at the site (North of Scotland Hydro-Electric Board, 1950). The raising of the water level of Loch Sloy probably does not contribute to any direct increase in hazard but there are some very large blocks near the top of the steep slope of Ben Vorlich above the dam which are detached from the main outcrops. There may be a long term risk that they will ultimately slip off as the result of freeze thaw activity, especially if there is any deterioration in climatic conditions.

On the east side of the loch there is evidence of unstable conditions on westward facing slopes south of Inversnaid. The general orientation of the Tay Nappe in this area is horizontal but there are later folds which contribute to smaller scale undulations of the rock surface so that locally the dip of the foliation is down hill although generally at angles less than the angle of the hill slope. In places the glacial drift has become detached from the smooth surface of the underlying bed rock with resulting mass transfer down slope.

Such movements are generally on a small scale but occasionally have resulted in more spectacular effects. About 4 km south of Inversnaid a break in the forest

Table 1. Average discharge of water into Loch Lomond from the catchments of the River Falloch and River Endrick in 1990 (Clyde River Purification Board). The total area of the Loch Lomond catchment is about 525 km^2.

	Catchment area (km^2)	Average discharge (m^3/s)
River Falloch	80.3	7.7
River Endrick	219.9	8.6

runs steeply down slope. The break may be original, at least in part, due to thin soil having discouraged planting which took place during 1962–1964. At the top of the steep part of the slope the till covering the bed rock is slipping progressively and there is evidence of recent slope failure. At several points down the break trees have been toppled and in some cases splintered. Some of this damage is probably due to erosion of the sides of the gully but in other cases the trees have been hit by large boulders, up to about 16 tonnes weight, bounding down the hillside. Little rock debris occurs at the loch side as the slope continues steeply below water level and the debris is presumably dispersed at depth.

The quality of the water entering the loch is likely to be influenced by the topography and hydrology of the catchments. In the Dalradian area the rivers and streams are mostly short and flow over relatively impermeable rocks. It can be expected that water flowing in from these sources would have a low dissolved mineral content. The great bulk of the Midland Valley part of the catchment lies within the Endrick system which drains a relatively large area (Table 1). It rises on the edges of the uplands of the Clyde Plateau Volcanic Formation consisting mainly of basic lavas, briefly traverses the carbonate rich Inverclyde Group and flows more gently across the Upper and Lower Old Red Sandstone. However much of the area is blanketed by superficial Quaternary deposits which could well influence water quality more than bedrock.

Any variations in the chemistry of the water in the loch that might be attributable to contrasts in the geology of the catchment ought to be made more obvious by the position of the two most abundant sources of inflow. The River Falloch, at the northern end of the loch, draining a relatively small catchment (Table 1) which lies totally within the Dalradian area, has an average discharge of about 90% of that of the River Endrick. The only significant river which drains both Dalradian and Old Red Sandstone areas is the Fruin Water which enters the loch to the south of the line of the Highland Boundary Fault. Thus although the Highland catchment of Loch Lomond is not very much greater in size than that in the Midland Valley a much greater volume of water entering the loch is from sources in the Dalradian area and the entry to the loch of water which has drained from areas of Palaeozoic rocks is concentrated in the southern, shallow end of the loch close to its outlet, the River Leven, at Balloch.

Acknowledgments

The author is grateful to Professor Brian J. Bluck for his patience during discussions on the relationships of the Dalradians, the Highland Boundary Fault and the Highland Border Complex. Thanks are also due to David Wilkinson who drew the author's attention to the area of slope instability south of Inversnaid.

References

Anderson, J. G. C., 1935. Intrusions of the Glen Falloch Area. Geol. Mag. 72: 458–68.

Anderson, J. G. C. & G. W. Tyrrell, 1936. Xenolithic minor intrusions in the Loch Lomond district. Trans Geol. Soc. Glasg. 14: 373–84.

Bluck, B. J. & T. J. Dempster, 1991. Exotic metamorphic terranes in the Caledonides: Tectonic history of the Dalradian block, Scotland. Geology, 19: 1133–6.

Curry, G. B., B. J. Bluck, C. J. Burton, J. K. Ingham, D. J. Siveter & A. Williams, 1984. Age, evolution and tectonic history of the Highland Border Complex, Scotland. Trans r. Soc. Edinb. 75: 113–33.

Clyde River Purification Board, 1990. Report for the year ending 31st December 1990.

Dempster, T. J. & B. J. Bluck, 1989. The age and origin of boulders in the Highland Border Complex: constraints on terrane movement. J. Geol. Soc. London. 146: 377–9.

Dempster, T. J. & B. J. Bluck, 1991. Xenoliths in the lamprophyre dykes of Lomondside: constraints on the nature of the crust below the southern Dalradian. Scot. J. Geol. 27: 157–65.

Flett, J. S., 1929. The Inverbeg Lamprophyre. In Summary of Progress for 1928, part 2. Mem. Geol. Surv. 29–35.

Ingham, J. K., G. B. Curry & A. Williams, 1986. Early Ordovician Dounans Limestone Fauna, Highland Border Complex, Scotland. Trans r. Soc. Edinburgh. Earth Sci. 76: 481–513.

Lawson, J. D. & D. S. Weedon, 1992. Geological excursions around Glasgow and Girvan. Geol. Soc. Glasgow.

MacDonald, J. G., 1974. Geology. In Tippett, R. (ed.), A Natural History of Loch Lomond, 7–23. University of Glasgow Press.

MacDonald, J. G., 1979. A Field Guide to some Dalradian Structures on Loch Lomondside. Trans Geol. Soc. Glasg. Session 120: 14–17.

MacDonald, J. G. & F. Whyte, 1981. Petrochemical evidence for the genesis of a Lower Carboniferous transitional basaltic suite in the Midland Valley of Scotland. Trans r. Soc. Edinb: Earth Sci. 72: 75–88.

Mather, J. D., 1970. The Biotite Isograde and the Lower Greenschist Facies in the Dalradian Rocks of Scotland. J. Petrol. 11: 253–75.

Nockolds, S. R., 1940. The Garabal Hill – Glen Fyne Igneous Complex. Quart. J. Geol. Soc. London. 94: 451–510.

North of Scotland Hydro-electric Board, 1950. The Loch Sloy Hydro-electric Scheme. The Engineer. 190: 57–60.

Rodgers, G., T. J. Dempster, B. J. Bluck & P. W. G. Tanner, 1989. A high precision U-Pb age for the Ben Vuirich granite: implications for the evolution of the Scottish Dalradian Supergroup. J. Geol. Soc. London 146: 389–92.

Shacklton, R. M., 1958. Downward-facing structures ofthe Highland Border. Quart. J. Geol. Soc. London. 113: 361–392.

Tilley, C. E., 1925. A preliminary survey of metamorphic zones in the Southern Highlands of Scotland. Quart. J. Geol. Soc. London. 81: 100–12.

Watkins, K. P., 1983. Petrogenesis of Dalradian albite porphyroblast schists. J. Geol. Soc. London. 140: 601–18.

Whyte, F. & J. G. MacDonald, 1974. Lower Carboniferous Vulcanicity in the Northern Part of the Clyde Plateau. Scott. J. Geol. 10: 187–198.

Hydrobiologia **290**: 21–28, 1994.
K. J. Murphy, M. C. M. Beveridge & R. Tippett (eds), The Ecology of Loch Lomond.

Aspects of the hydrology and hydrography of Loch Lomond

J. C. Curran & T. Poodle
Clyde River Purification Board, Rivers House, Murray Road, East Kilbride, Glasgow G75 0LA, UK

Key words: hydrology, hydrography, flow, current, sediment, trend, Loch Lomond, Endrick Water

Abstract

A review is made of the rather limited studies that have been conducted into the available hydrological data for the catchment, and of the sparse hydrographic data relating to the Loch itself.

Introduction

A computer search has been conducted into 8 bibliographic databases (Aqualine, Enviroline, ASFA, AFEE, Delft Hydro, Pollution, Energyline, Conference Papers), but not a single reference has been located from these sources throughout the past ten to twenty years relating either to the hydrology or hydrography of the Loch Lomond catchment. Considerable archived data are available relating to the climatology and hydrology of the catchment but little analysis has been conducted other than for the major undertakings of the Sloy hydro and the Ross Priory abstraction schemes.

The ecosystem of the Loch and its catchment depends critically on the physical characteristics of the environment. In addition, potential shoreside developments may be affected significantly by, for example, Loch water levels, sediment balance, wind climate, wave action, shore erosion, dispersion and currents. Clearly there is a requirement for a synthesis of all available physical data, which would go well beyond the limited bounds of this paper, and for a longer-term programme of detailed investigations.

Rainfall and river flows

Rainfall within the Loch Lomond catchment varies between 1300 mm per year near Drymen and 3600 mm per year on the upper slopes of Ben Ime above Loch Sloy (Met Office, 1977). Annual evaporation is estimated to vary between 450 mm per year in the south of the area and 400 mm in the north (Met Office) while catchments at higher elevations possibly experience only 250 mm (Reynolds, 1969).

The Clyde River Purification Board (CRPB) operates four river gauging stations and five rainfall stations within the catchment (Fig. 1). Table 1 indicates that approximately 40% of the total catchment contributing to Loch Lomond is gauged and also displays some representative data for each of the rivers.

Artificial influences

Curran & Poodle (1992) provide some details of the various artificial influences on the hydrology of the Loch. Figure 1 shows that there are five abstractions, all for potable supply, and two additions, one for power generation at Loch Sloy and one for the enhancement of a potable abstraction on the Endrick Water. The effect on the overall water balance amounts to a net loss of around 2.3% of flow. Since the standard deviation of the mean annual flow in the River Leven is 12% then the artificial component is relatively minor.

Trends in runoff

It has been shown (Curran, 1991) that west central Scotland has experienced a 30% increase in mean annual rainfall over the period 1970 to 1990. The Loch Lomond catchment also shows evidence of a strong

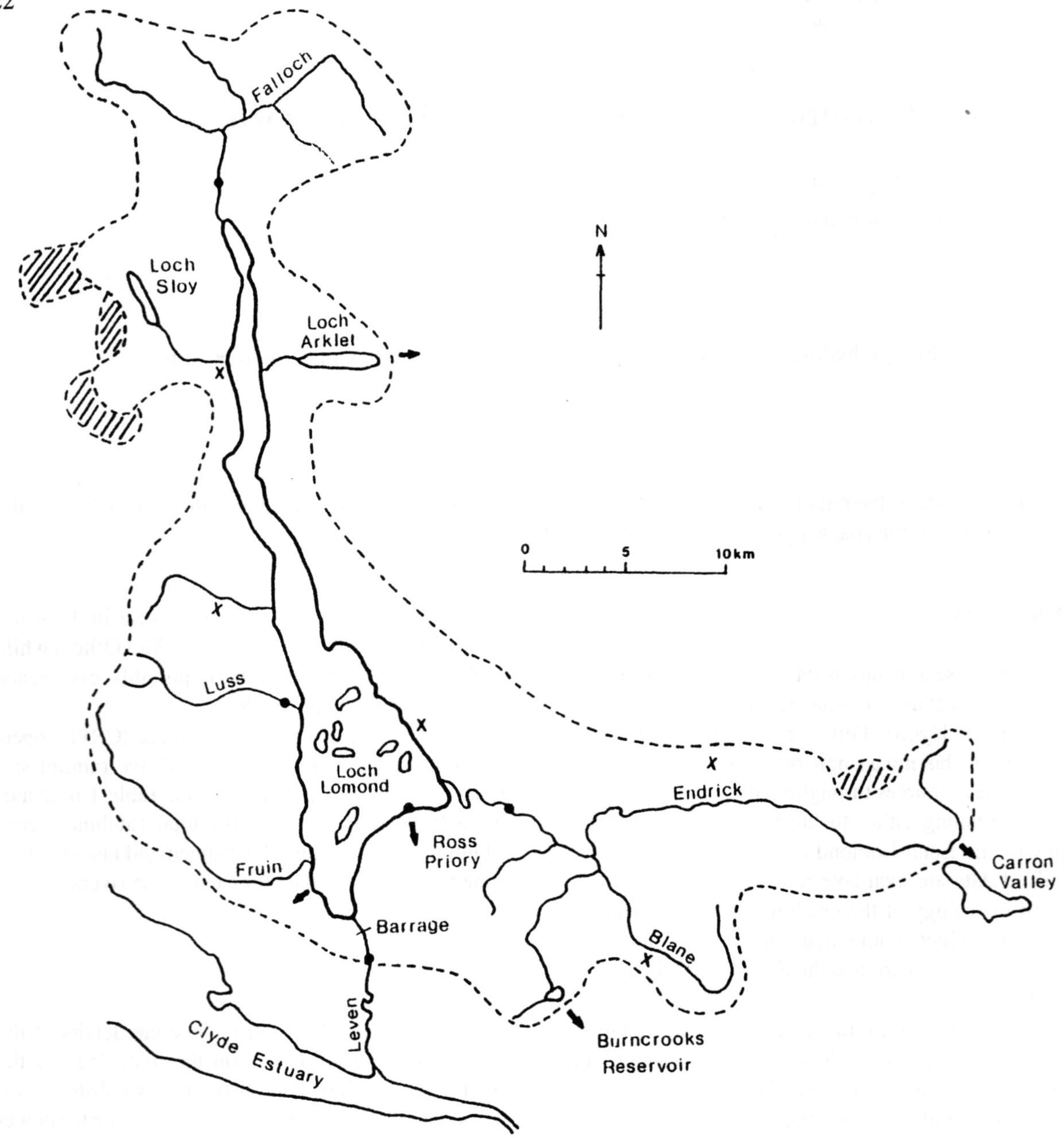

Fig. 1. An illustrative map of Loch Lomond and its sub-catchments. • CRPB river gauging stations; × CRPB rainfall gauges; ↑ major abstraction points; ▨ intercepted external catchments

trend in precipitation; linear regression of annual totals at the Arrochymore raingauge (grid ref. NS 415 918) suggests an increase of (25 ± 7)% over the period 1972 to 1990 with the linear trend verified at the 99.8% significance level. The mean annual flows in the Endrick Water have increased by (36 ± 11)% over the period 1969 to 1990, with trend verified at 99.0% significance (Fig. 2). There is evidence that the rainfall has been disproportionately augmented during the winter months when surface evaporation is very low. Virtually all precipitation during the winter appears as runoff in rivers whereas, during summer, a significant proportion is lost through evaporation. For this reason the percentage increase in mean river flow is in excess of that exhibited by the rainfall.

The increase in runoff has inevitably led to generally higher water levels in the Loch. Linear regression suggests that the mean annual level has risen by (0.2 ±

Table 1. Summary of CRPB gauging stations and records

Station	River	Grid Ref	Record start	Catchment area (km^2)	Flows ($m^3\ s^{-1}$) Av.	Mean annual Min.	Mean annual Max.	95‰
85003	Falloch	NS321197	1970	80.3	5.5	0.15	182.3	0.22
85004	Luss	NS356929	1975	35.3	2.6	0.13	67.8	0.18
85002	Endrick	NS485866	1963	219.9	7.1	0.54	115.0	0.57
85001	Leven	NS394803	1963	784.3	41.8	8.4	119.8	8.10

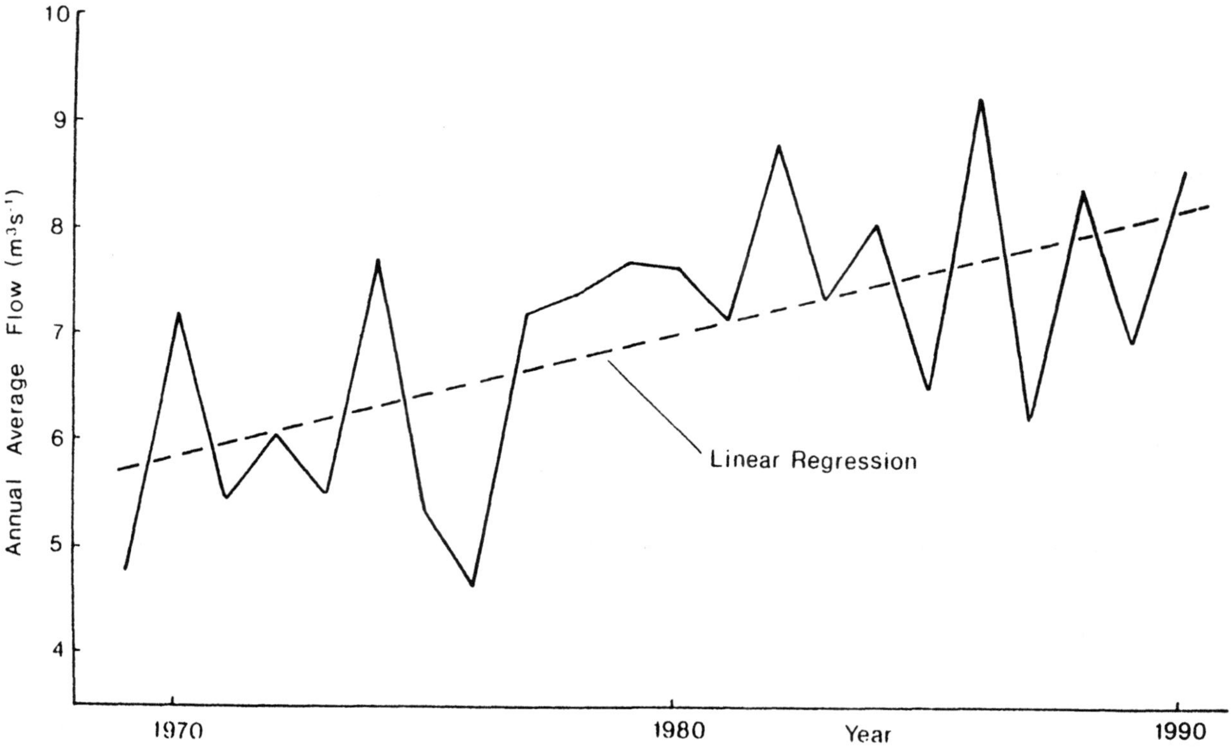

Fig. 2. Annual average flow by year of the Endrick Water, 1969–1990.

0.1)m over the period 1977 to 1990 (Curran & Poodle, 1992).

Despite the increase in rainfall, no evidence has been found of a corresponding increase in the incidence of high flows or flood events in any of the Lomond catchment streams. Over the years there have been visible changes in the Endrick catchment regarding field drainage and forestry but no detectable changes have been observed in the response time, or flashiness, of the Endrick Water (Curran & Poodle, 1992).

Hydrography

If detailed studies have previously been undertaken into the density structure, turbulence and circulation within the Loch, then virtually nothing has recently been published. A review of 8 bibliographic computer databases again failed to provide any references within the last ten to twenty years.

Temperature structure

Slack (1957) reports temperature profiles at various locations within the Loch which suggest a uniform temperature through depth of around 5 to 6 °C dur-

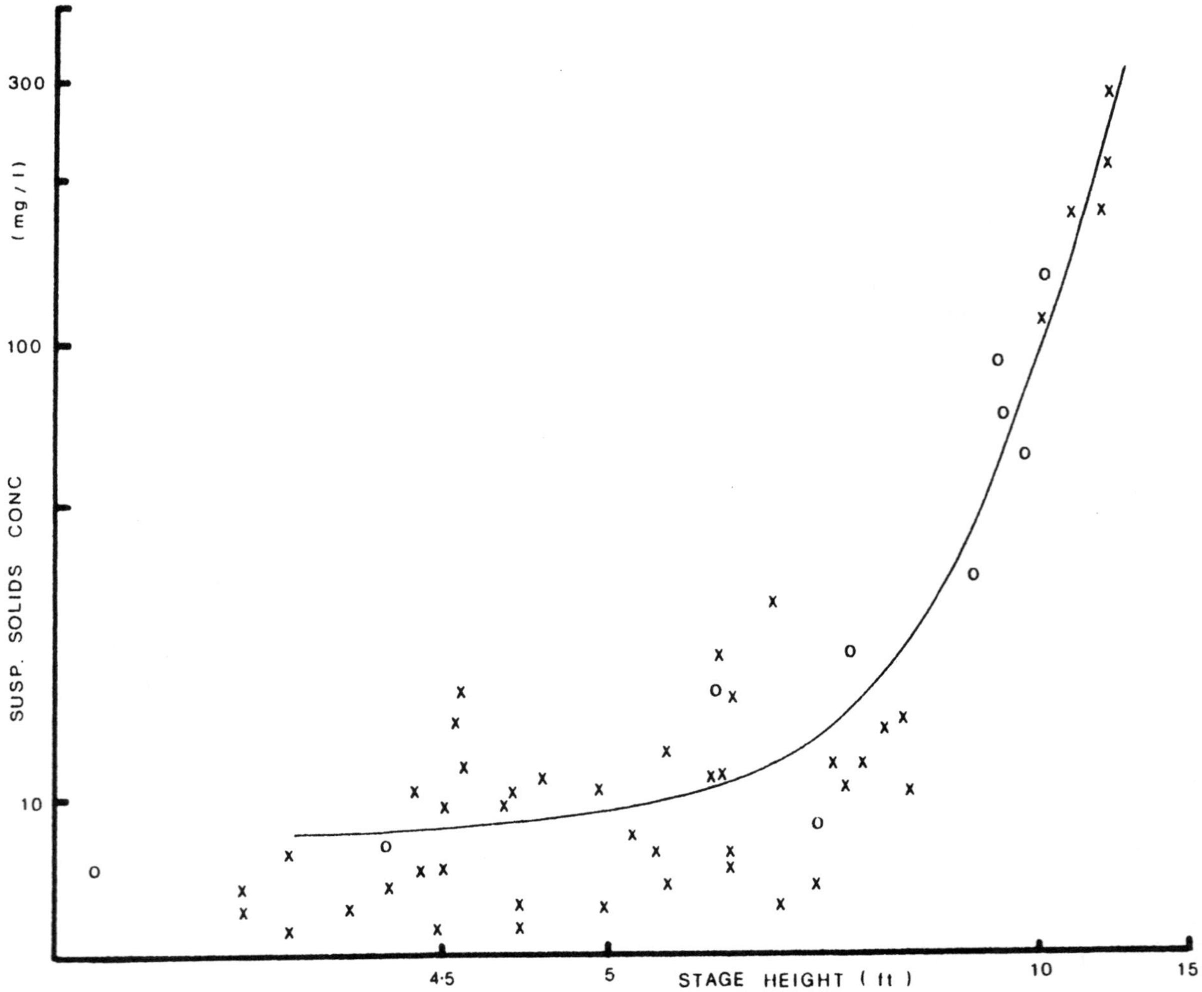

Fig. 3. Suspended solids concentration versus stage height at the Gaidrew gauging station on the Endrick Water × data before 1970; o data after 1970

ing the winter. Such uniformity implies thorough mixing between surface and bottom waters. However, in summer, and particularly in the northern part of the Loch, a notable thermocline develops at a depth of around 15 m separating overlying water at a temperature of around 14 °C from the homogeneous underlying waters which maintain their approximate winter temperature of 6 °C. Clearly the surface waters become isolated from the bottom waters, and are warmer due to direct insolation and to the higher temperature of influent runoff from the surrounding catchment. Complete mixing of the water column during summer can only be accomplished by surface wind and wave induced turbulence gradually eroding the thermocline. Energy balance equations (Fischer *et al.*, 1979) suggest that a steady wind of around 7 m s^{-1} blowing across the surface of the Loch for 24 hours, imparting momentum and energy to the water surface through frictional drag, would deepen the epilimnion by around 0.5 m and would thereby cool the surface water temperature by around 0.25 °C.

Circulation

The total freshwater flow in the narrows between Ross Point and Culag is, on average, around 17 m^3 s^{-1}. If this flow is confined to the epilimnion during summer then the resulting current speed is 0.001 m s^{-1}. It has been shown (Dooley & Steele, 1969) that near-surface currents of around 2% of wind speed are generated by frictional drag between the water surface and the wind. Clearly any wind speed above 0.05 m s^{-1} will therefore cause water movement in excess of that generated by runoff almost anywhere in the Loch except very close to river mouths; the circulation of water within the Loch must be dominated by wind action. For example,

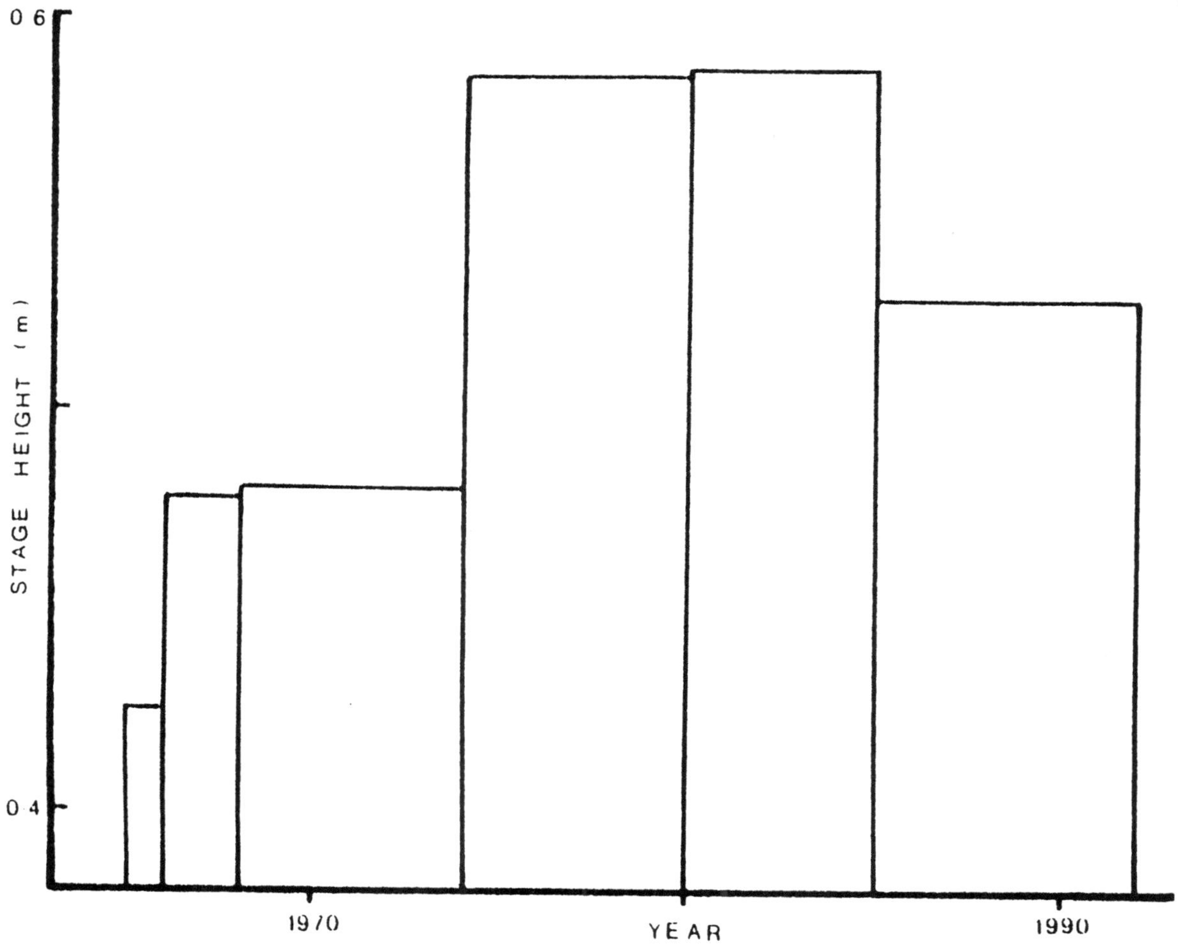

Fig. 4. Stage height required at Gaidrew gauging station in order to pass a flow of 3 $m^3 s^{-1}$

studies of littoral formations (CRPB, 1970) conclude that the near-shore drift in the south-east corner of the Loch is generally towards the north and is driven by the prevailing wind.

Specific investigations have been conducted by the CRPB using passive surface water tracers, particularly Iroquois drogues (Curran, 1986). A southwards near-shore surface drift of 0.06 m s^{-1} was recorded at Midross under the influence of a moderate easterly wind. A study conducted in the south-east corner of the Loch, during a period of light southerly wind, indicates the presence of an anticlockwise gyre with a northwards current along the eastern shore and a southerly flow near Aber Isle. It appears that the outflow from the Endrick Water generally flows northwards between Inchcailloch and the mainland even during spells of northerly wind; the drift speed in the channel may be as high as 0.13 m s^{-1}.

Turbulence

Limited studies of dilution rate within the Loch suggest that dispersion is comparable to that found within the marine waters of the Firth of Clyde where tidal energy is, admittedly, rather restricted. Turbulence scale is clearly topographically limited to gyres of the order of one kilometre which suggests that the horizontal diffusion coefficient will not exceed 1 $m^2 s^{-1}$ (Curran, 1986). Shear diffusion has been observed at Midross where a subsurface recirculation was apparent, coupled to a convergence line lying parallel to the shore (Curran, 1989). Typically, in these waters, a dilution of around 30-fold will occur within about 1.5 hours.

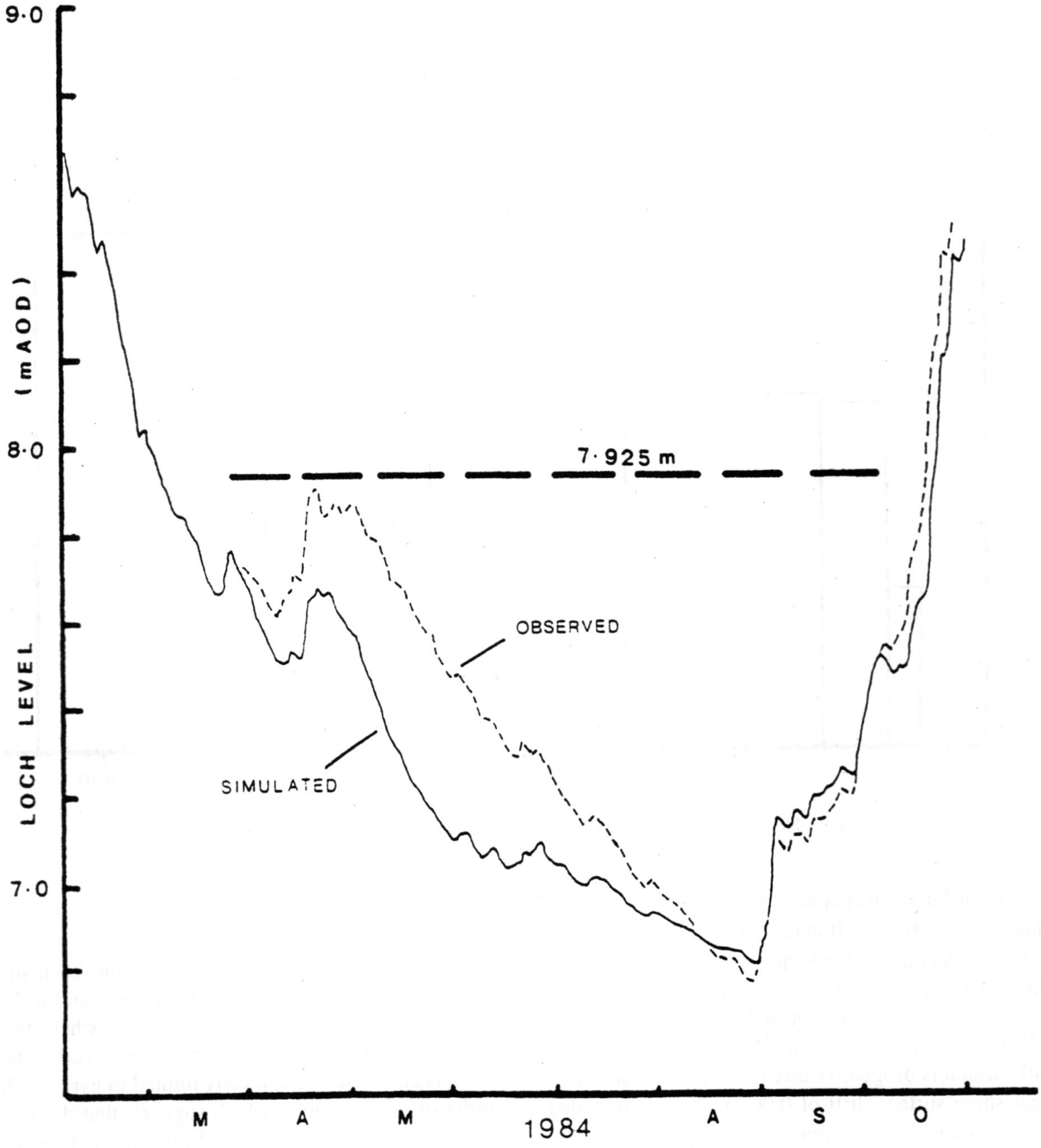

Fig. 5. Results of CRPB computer model comparing the simulated, 'natural', level of Loch Lomond, assuming no barrier across the Leven and no abstraction at Ross Priory, with the observed level at the CRPB gauging station for the period of the 1984 drought

Sediment balance

It has been shown (CRPB, 1970) that the Endrick Water comprises the main external source of suspended solids within the Loch. Data prior to 1970 reveal the expected highly non-linear relationship between river-borne suspended solids concentration and river stage height (Fig. 3). Selected data across the flow range from routine surveys since 1970 reveal no change or trend in sediment concentrations (Fig. 3). A statistical analysis of the data indicates that the total annual load of material transported to the Loch is around 14 000 tonnes, although a single high flow event may transport around 3000 tonnes in a single day (CRPB, 1970). The gauging station at Gaidrew on the Endrick is calibrated for a natural section of the river and it can be seen from

Fig. 4 that routine recalibrations since 1965 might suggest that the river bed locally has accumulated sediments up to the mid-1980s and been slightly eroded subsequently; such changes could be explained by the frequently observed phenomenon of 'pulses' of sediment being gradually translated downstream within a river system.

Studies indicate that suspended material is deposited rapidly as the current speeds and turbulence decline as the Endrick discharges into the Loch. For example, on 25/3/80 when the river flow peaked at 43 $m^3 s^{-1}$, a surface water sample at the mouth of the Endrick contained a suspended solids concentration of 180 mg l^{-1}, whereas 500 m to the north, but still within the highly visible plume, the concentration was 67 mg l^{-1}, and after 1800 m the concentration had dropped to 1.9 mg l^{-1}, only marginally above the background level of 1.3 mg l^{-1}.

On occasions, wave activity within the Loch can be sufficient not only to erode the windward shore and transport material offshore (Pender, 1992) but may also resuspend bottom sediments in areas of relatively shallow water.

Effect of leven barrage

The Central Scotland Water Development Board operates a barrage on the River Leven which maintains the mean level in the Loch approximately 0.1 m above its previous value (CRPB, 1974) (Mitchell, 1992), although this effect has been compounded by the significant increase in rainfall, runoff and consequent Loch level over the last twenty years. The barrage is operated so that it is opened completely when the Loch level exceeds 7.925 m AOD. In constructing the barrage the River Leven was both slightly widened and deepened locally so that, under these circumstances, the overall conveyance of the channel remains largely unchanged. Curran & Poodle (1992), reiterated by Pender *et al.* (1993), show that the water balance of the Loch system returns to its pre-barrage condition within about one month of the onset of such conditions and that, specifically, the 100-year peak Loch level of 10.075 m AOD, achieved on 11/3/90, was unrelated to any effect of the barrage and would also have occurred in the 'natural' system after such prolonged and intense rainfall.

In contrast, it is also instructive to review the conditions which prevailed during the severe drought of 1984. An unusual deficit in rainfall was recorded during the period April to August 1984 when, for example, the raingauge at Loch Venachar accumulated a five-month total of 193 mm compared to the next lowest comparable figure of 254 mm in 1869 (SDD, 1986). Recorded water levels in Loch Lomond declined steadily from 7.898 m AOD on 21 April 1984 to a minimum of 6.753 m AOD on 29 August 1984. The CRPB has developed a computer model which can simulate 'natural' Loch level for periods after the construction of the Leven barrage and Fig. 5 indicates that, for most of the summer period, the level was actually held higher than would otherwise have occurred, due to the barrage retaining water within the Loch. However, during August the continued abstraction at Ross Priory reduced the observed Loch levels more rapidly than the 'natural' level would have declined due solely to the Leven outflow. At the end of the drought period the simulated level reached a minimum of 6.800 m AOD, only 0.047 m above the observed level. The recorded flow in the Leven at this time was 2.83 $m^3 s^{-1}$, compared to the simulated flow of 3.29 $m^3 s^{-1}$. It is apparent that the effect of the barrage was negligible under both of the extremes of deficit and excess of rainfall in 1984 and 1990 respectively.

Acknowledgments

Central Scotland Water Development Board supplied abstraction data for input to the simulation model. This paper is published with the permission of Professor D. Hammerton, Director of the Clyde River Purification Board.

References

CRPB, 1970. Loch Lomond: suspended solids and colour. Tech. Rep. 12.

CRPB, 1974. The Loch Lomond Water Order: its effect on Loch level. Tech. Rep. 31.

Curran, J. C., 1986. Effluent disposal and the physical environment. Proc. r. Soc. Edin. 90: 97–115.

Curran, J. C., 1989. The EC bathing water directive: a sampling problem. J. Inst. Water & Env. Man. 3: 465–466.

Curran, J. C., 1991. Water quality implications of an observed trend of rainfall and runoff. J. Inst. Water & Env. Man. 5: 419–424.

Curran, J. C. & T. Poodle, 1992. Hydrological characteristics of Loch Lomond – are they changing? Proc. Symposium 'Loch Lomond 1991', Universities of Glasgow and Strathclyde.

Dooley, H. & J. H. Steele, 1969. Wind driven currents near a coast. Dt. Hydrogr. Z. 22: 213–223.

Fischer, H. B., E. J. List, R. C. Y. Koh, J. Imberger & N. H. Brooks, 1979. Mixing in inland and coastal waters. Academic Press, New York.

Met Office, 1977. Map of Average Annual Rainfall (mm). 1941–70, Northern Britain. H.M.S.O.

Met Office. Map of average annual potential evaporation in millimetres (provisional version). Met Office, Bracknell.

Mitchell, G., 1992. Water management. Proc. Symposium 'Loch Lomond 1991', Universities of Glasgow and Strathclyde.

Pender, G., 1992. Wave generated shore erosion in Loch Lomond. Proc. Symposium 'Loch Lomond 1991', Universities of Glasgow and Strathclyde.

Pender, G., G. Dickinson & J. G. Herbertson, 1993. Flooding and shore damage at Loch Lomond, January to March 1990. Weather 48: 8–15.

Reynolds, G., 1969. Rainfall, runoff and evaporation on a catchment in West Scotland. Weather 24: 90–98.

SDD, 1986. The 1984 drought in Scotland. Its effect on river systems and public water supplies, Scottish Development Department, Edinburgh.

Slack, H. D., 1957. Studies on Loch Lomond, Part I. Blackie & Son Ltd., Glasgow.

Hydrobiologia **290**: 29–37, 1994.
K. J. Murphy, M. C. M. Beveridge & R. Tippett (eds), The Ecology of Loch Lomond.

The physico-Chemical limnology of Loch Lomond

G. A. Best & I. Traill
Clyde River Purification Board, Rivers House, Murray Road, East Kilbride, Glasgow G75 1KA, UK

Key words: water chemistry, nutrients, lakes, trophic status, Loch Lomond

Abstract

Loch Lomond, the largest freshwater lake in Britain, is physically divided into a number of distinct basins by geological structures. The northern part of the loch is long, narrow and deep whilst the southern area is broad and shallow with many islands scattered throughout it.

The water chemistry of the loch is dictated by the geology of catchment and the quality of the rivers flowing into it. The two major rivers, the Falloch and the Endrick Water, are both of good quality with low nutrient content. Consequently, the loch is regarded as oligotrophic, though the southern area verges towards being mesotrophic.

Introduction

The Loch Lomond basin is of glacial origin, formed by an ice sheet moving southward from the Ben Lui area and depositing eroded material in the southern most part by Balloch. It formed the largest (by surface area) body of freshwater in Britain with a surface area of 71 km^2. The natural catchment area though is ten times greater at 781 km^2. The geology has been described by Gregory (1928) and Slack (1957) and consists principally of metamorphic schists in the northern basin, a band of slate across the centre, and sedimentary sandstones in the southern basin. There is little natural leaching of minerals into the water and it has remained as a soft, oligotrophic water.

The natural flow of water through the loch has been altered in two ways. Firstly the completion of the Leven barrage in 1971 which, as part of the water supply scheme, controls the rate of flow from the loch at times of low flow. Secondly, same water is imported into the catchment of the loch from adjacent areas for the Loch Sloy hydro-electric scheme. There are also same abstractions for water supply from different parts of the loch catchment.

The hydrology of the area has been described by Curran & Poodle (1991).

Table 1. Mean flow rates of major rivers into Loch Lomond (1989–1990)

River	Mean flow (cumecs)
Fruin	3.9 (estLmate)
Douglas	2.2 (estimate)
Luss	2.8 (estimate)
Falloch	6.8 (estimate)
Endrick	7.75 (estimate)
	23.2

Inputs into Loch Lomond

The major inputs into the Loch have been described by Curran (1992, this symposium). The data for the recorded or estimated flows are given in Table 1.

The total mean value of 23.2 cumecs should be compared with the long term average flow of the outlet river, the Leven, of 41.8 cumecs. The difference is made up by many minor streams in the catchment.

The Rivers Falloch and Endrick have markedly different catchments even though their flow rates are similar. The Falloch has a mountainous catchment area of 80 km^2 in a region of high rainfall whilst the Endrick's catchment area is over 220 km^2 with typically lowland characteristics.

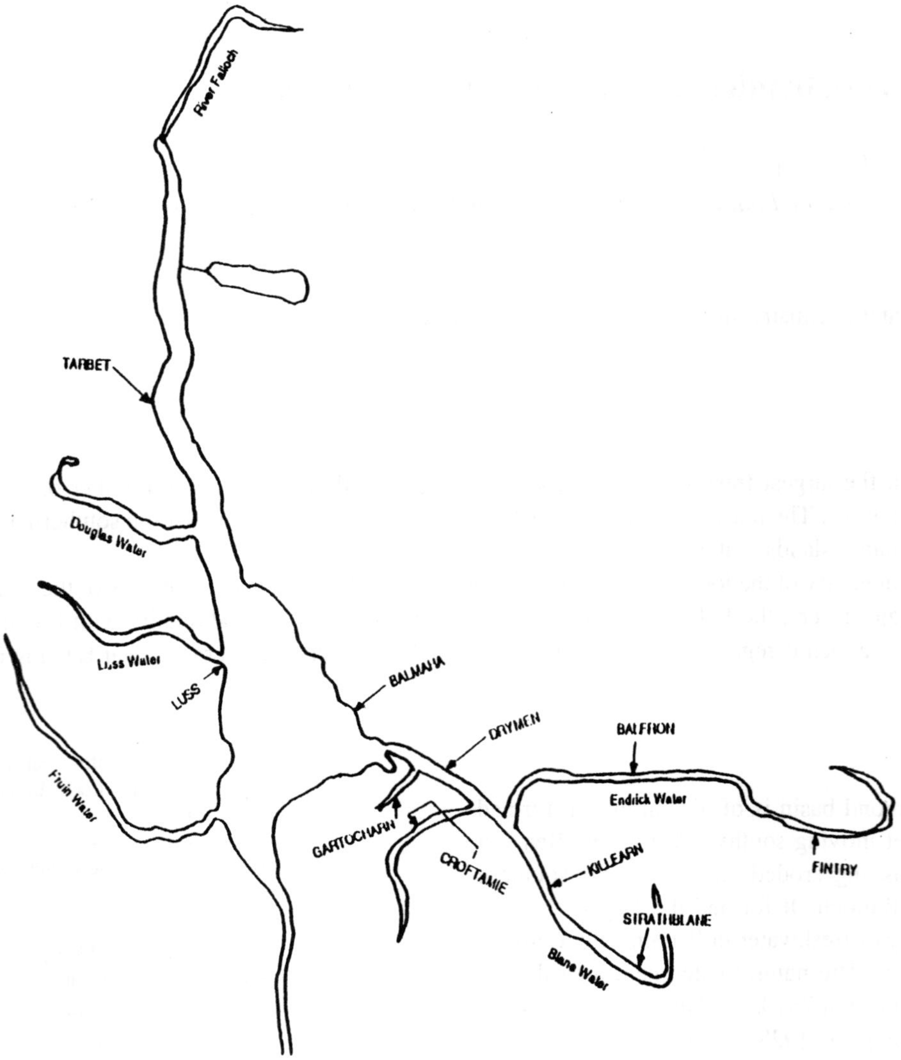

Fig. 1. Sewage treatment works in Loch Lomond catchment.

The water chemistry also reveals differences in the two rivers. The results for selected determinands for three years of monitoring, are shown in Table 2.

These differences in the water chemistry, together with the partial physical separation between the two basins of the loch fed by them, give rise to a gradation of water chemistry along the length of the loch. In Table 3, the mean results for one of the sampling points in the northern basin and for one in the centre of the Southern Basin for the period 1990–91, are shown.

A major potential influence on the chemistry of Loch Lomond, as seen from the data in Table 2, is the quality of the River Endrick.

There are about 10 000 people living in seven scattered communities in the catchment of this river.

Table 2. Waber chemistry of River Falloch and Endrick (1989–1991)

	River Falloch	Endrick water
Conductivity (μs/cm)	43	178
Total hardness (mg Ca CO_3/1)	6	66
Chloride (mg l^{-1})	6	14
pH	6.67	7.37
NH_3/N (mg l^{-1})	0.04	0.08
TON (mg l^{-1})	0.25	0.72
o.PC4 P^{-1} (μg l^{-1})	3	90

(TON: total oxidised nitrogen = nitrate + nitrage nitrogen)

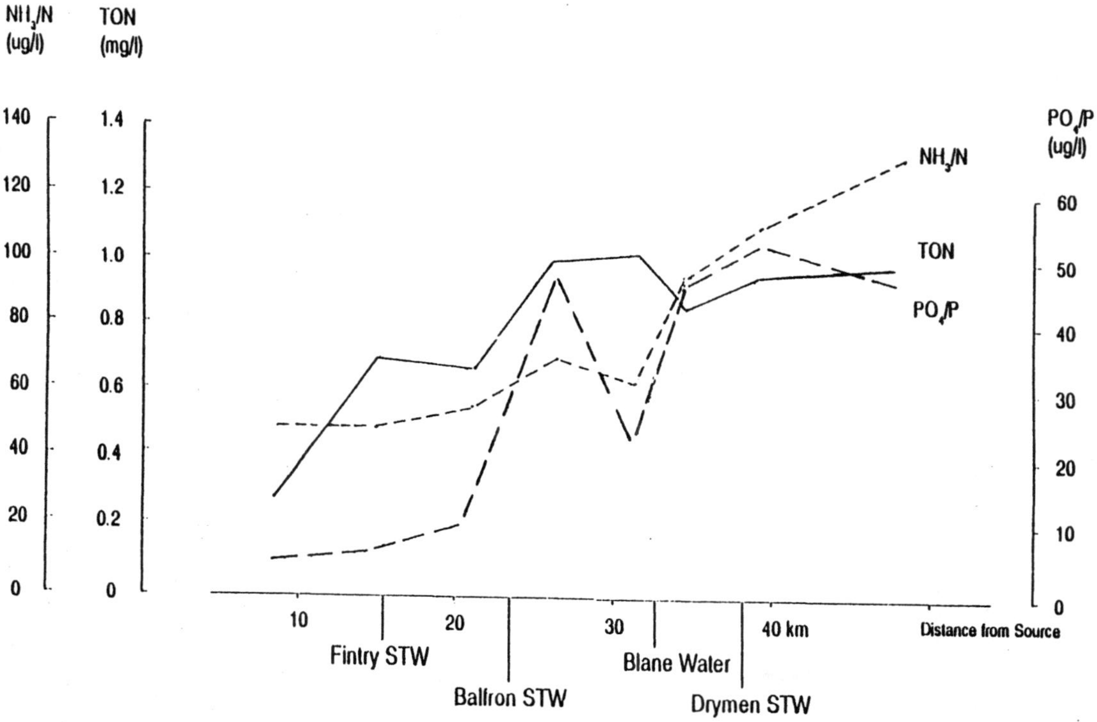

Fig. 2. Endrick water mean results for selected determinands (1991).

Table 3. Waber chemistry of Loch Lomond in Northern and Southern Basins (1990–91)

	Northern Basin at Inveruglas	Southern Basin at Creinch
Conductivity	44	62
Total hardness	11	18
Chloride	6.5	8
pH	6.9	7.0
NH_3/N	0.030	0.040
TON	0.045	0.17
PO_4/P	2	4

(TON: total oxidised nitrogen = nitrate + nitrage nitrogen)

Table 4. Compliance of STW in Endrick Catchment to CRPB standard (1989–90)

STW	Pop served	Samples taken	Compliance S	B	U
Fintry	550	15	6	3	6
Balfron	1120	12	7	1	4
Drymen	1270	12	4	5	3
Strathblane	3000	11	11	0	0
Killearn	2200	12	12	0	0
Croftamie	130	2	2	0	0
Gartocharn	300	15	14	0	1

S = Satisfactory
B = Borderline
U = Unsatisfactory

(Fig. 1) Each community is served by a sewage treatment plant and most of these purify the waste water to a satisfactory standard before discharge into the river. Table 4 summarises the performance of the works over the period 1989–1990.

The quality of the Endrick Water and its main tributary of the Blane Water, are good throughout their length. They have diverse invertebrate fauna (Doughty & Maitland, 1992, this symposium) and healthy fish populations. In Fig. 2, the increase in some water quality determinands along the length of the Endrick Water are illustrated.

These data should be compared against those for the loch water, particularly in the vicinity of the Endrick at Creinch as shown in Table 3. From this comparison, it seems that the Endrick Water does not mix well in the Loch. Work carried out by Poodle (1970) showed that there was a northward movement of Endrick water

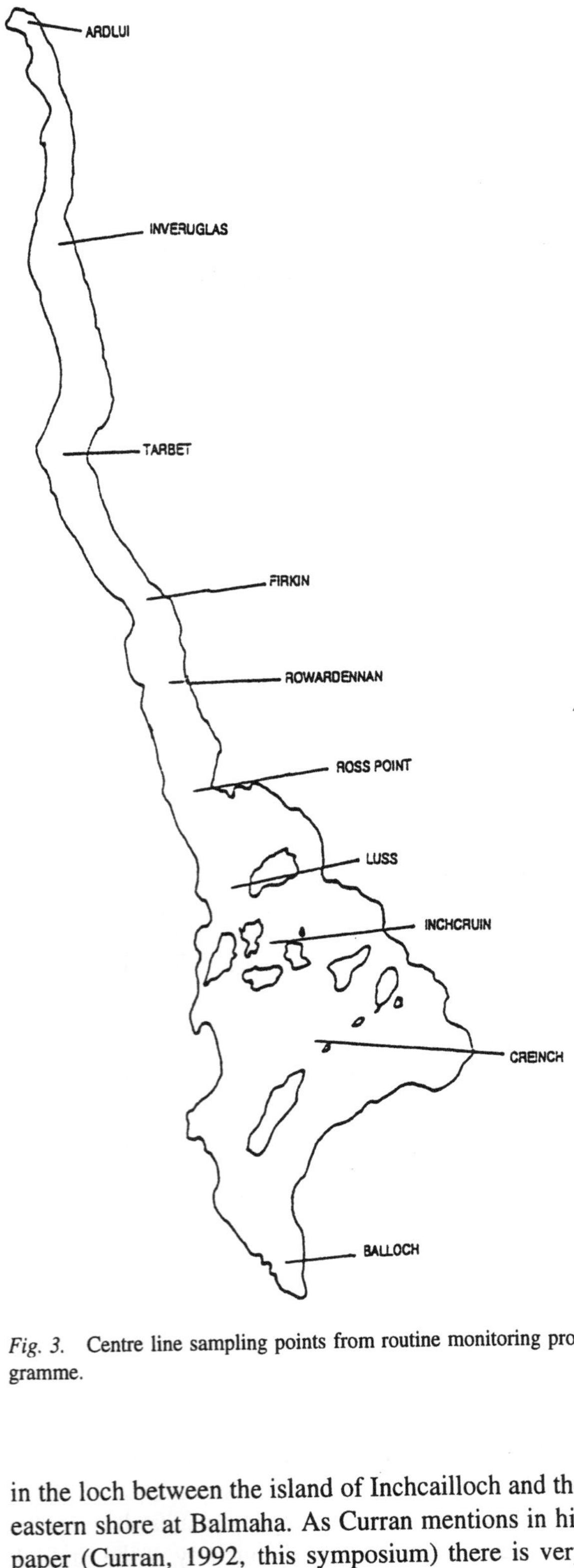

Fig. 3. Centre line sampling points from routine monitoring programme.

in the loch between the island of Inchcailloch and the eastern shore at Balmaha. As Curran mentions in his paper (Curran, 1992, this symposium) there is very little information available on the hydrography of Loch Lomond.

Loch Lomond is a great tourist attraction and around the fringes of the Loch there are camping and caravan parks, youth hostels and hotels, cafes and restaurants. Each of these premises produces waste water.

In order to maintain the high quality of the loch water, particularly because of its use for potable supply, the Clyde River Purification Board insist that all sewage must be treated before discharge to the loch. As a result, there are sewage treatment works serving the Rowardennan Caravan Park, Cashel and Milarrochy camp site, Ross Priory, Inverbeg caravan park etc. Septic tanks have been installed at some of the smaller sites such as Luss camp site and Rowardennan Youth Hostel.

Waber quality of Loch Lomond

The chemistry of the waters of Loch Lomond has been well studied over the years. Maulood & Boney (1980) concentrated on the nutrient status whilst Beveridge *et al.* (1982) reviewed various studies that have been carried out over the years including those of Clyde River Purification Board. Tippet (1988) has also assessed the nutrient status and the phytoplanktonic populations over a number of years. The Clyde River Purification Board started its survey work on the loch in the early 1960's as part of its statutory duty to assess the overall water quality of the waters in the Board's area.

Data have been gathered from 20 fixed sampling sites scattered throughout the loch which are visited six times each year. Since 1987, additional sampling and analysis has been done on seven other sites along the centre line of the loch so as to assess the nutrient status of the loch. This programme was instigated because of concern at an apparent increase in the quantity of phytoplankton in the loch water during the summer of 1986. The treatment plant at Balmore, to which Loch Lomond water is pumped from the intake at Ross Priory, includes micro-strainers which filter out planktonic organisms from the water prior to disinfection. The volume of sludge removed from the strainers in 1986 was twice that of the previous year.

The analytical data collected from the two types of surveys are shown in Table 5.

From the routine survey work, the results of surface samples taken over many years have shown that there are minimal amounts of biodegradable organic matter in the loch (with BOD values generally less than 2 mg l^{-1}), the water is circumneutral, of low total hardness (overall mean 10–20 mg Ca Co_3 l^{-1}) and contains barely measurable amounts of heavy metals. The mean results for some determinands from the cen-

Table 5. Analytical determinands carried out on Loch Lomond samples

Routine survey of 20 sites		Nutrient survey of 7 sites
Suspended solids	Zinc*	Ammoniacal nitrogen
BOD	Copper*	Nitrite nitrogen
Dissolved oxygen	Nickel*	Nitrate Nitrogen
Ammoniacal nitrogen	Chromium*	Ortho phosphate
Nitrite nitrogen	Lead*	Total Phosphorus
Nitrate nitrogen	Cadmium*	pH
Ortho phosphate	Faecal coliform*	Chorophyll *a*
pH		Silica
Alkalinity		Conductivity
Total Hardness		Secchi depth
Conductivity		

Table 6. Mean results of 18 surveys for nutrient status of Loch Lomond (1990–1991)

Sample point	Total oxidised Nitrogen (μg/l)	Total Phosphorus (μg/l)	Chlorophyll *a* (μg/l)	Secchi depth (m)
Cailness Mean	230	19	1.24	4.82
Range	(100-660)	(0-45)	(0.07-7.80)	(4-7)
Firkin	210	20	0.91	4.70
	(100-630)	(0-50)	(0.07-2.28)	(3.5-6)
Ross Point	210	10	1.15	4.02
	(100-540)	(0-50)	(0.10-2.95)	(2.5-6)
Inchcruin	220	20	1.57	3.86
	(100-510)	(0-30)	(0.09-7.03)	(2.5-5.5)
Creinch	250	20	1.79	3.34
	(100-620)	(10-30)	(0.09-4.43)	(2.25-4.5)
CSWDB	200	20	1.89	3.42
	(100-360)	(10-30)	(0.11-4.43)	(1.75-6)
Auchendennan	250	20	2.05	3.22
	(100-610)	(0-40)	(0.11-5.93)	(2.25-5)

tre line sampling points shown in Fig. 3 have been plotted in Fig. 4. These results show the slight increase in the determinands as one moves down the loch from North to South. The influence of the physical barrier at Inverbeg–Rowardennan bar can be distinguished in the water chemistry results by the increase in the determinands between the Rowardennan and Ross Point sampling sites. The concentration of nutrients in Loch Lomond is of particular importance because they control the development of phytoplankton bloom which could adversely affect the performance of the water supply treatment works. An increase in their concentration is also the first indication that human activities in the loch catchment are starting to affect water quality.

The nutrient levels have been studied extensively in the past, particularly by Maulood & Boney (1980) and Chapman (1965). The initial results of Clyde River Purification Board's nutrients survey have been reported by Traill (1991).

The sampling sites selected for the nutrient survey work are shown in Fig. 3. At each point, a 5 m long polythene tube, weighted at one end, was hung vertically over the side of the sampling boat so that the

Table 7. Variation in nutrient parameters ($\mu g\ l^{-1}$) over two year sampling programme

Date	Firkin Total P	Firkin Chlorophyll	Firkin TON	Creinch Total P	Creinch Chlorophyll *a*	Creinch TON
23.1.90	10	0.10	250	20	0.42	200
14.3.90	47	0.08	100	33	0.03	250
19.4.90	46	0.35	<100	12	3.17	100
1.5.90	5	1.19	150	13	2.98	170
19.6.90	11	0.48	110	11	0.67	140
17.7.90	9	1.70	480	19	2.62	520
15.10.90	29	1.42	100	19	1.29	200
29.11.90	22	1.41	—	20	0.76	—
23.1.91	10	0.07	250	16	0.09	330
26.2.91	13	0.16	<100	26	1.07	220
1.5.91	7	0.21	100	13	1.51	180
19.6.91	25	1.45	630	20	1.89	620
10.7.91	10	1.70	—	11	4.43	—
13.8.91	18	2.28	1.80	16	3.02	220
5.9.91	16	1.64	100	20	2.89	100
15.10.91	15	0.84	220	20	2.55	230
16.12.91	5	0.32	220	12	0.84	26

Table 8. Boundary values for Trophic States

Trophic Category	Total P ($\mu g\ l^{-1}$)	Chl *a* ($\mu g\ l^{-1}$)	Max. Chl. *a* ($\mu g\ l^{-1}$)	Secchi depth (m)	Minimum Secchi depth (m)
Ultra-oligotrophic (U-O)	<4	<1	<2.5	>12	>6
Oligotrophic (O)	<10	<2.5	<8	>6	>3
Mesotrophic (M)	10-35	2.5-8	8-25	6-3	3-1.5
Eutrophic (E)	35-100	8-25	25-75	3-1.5	1.5-0.7
Hypertrophic (H)	>100	>25	>75	<1.5	<0.7

top end was just above the water surface. This was then hauled up using a cord attached to the deepest end and emptied into a clean bucket so that a 5 m long profile of the surface layer was obtained. The water sample was then transferred to a glass half-winchester bottle and an acid washed plastic bottle. The clarity of the water was checked using a Secchi disc which is a 20 cm diameter disc with alternating black and white quadrants attached to a measured rope. This disc was lowered into the water until the markings just became visible and the depth of the disc noted.

The results of the 18 surveys carried out during 1990/91 are summarised in Table 6. More detailed analysis of these data show that there are seasonal trends in nutrient levels as have been found by earlier workers. The results of two sampling points, one from the northern basin and one from the southern are given in Table 7.

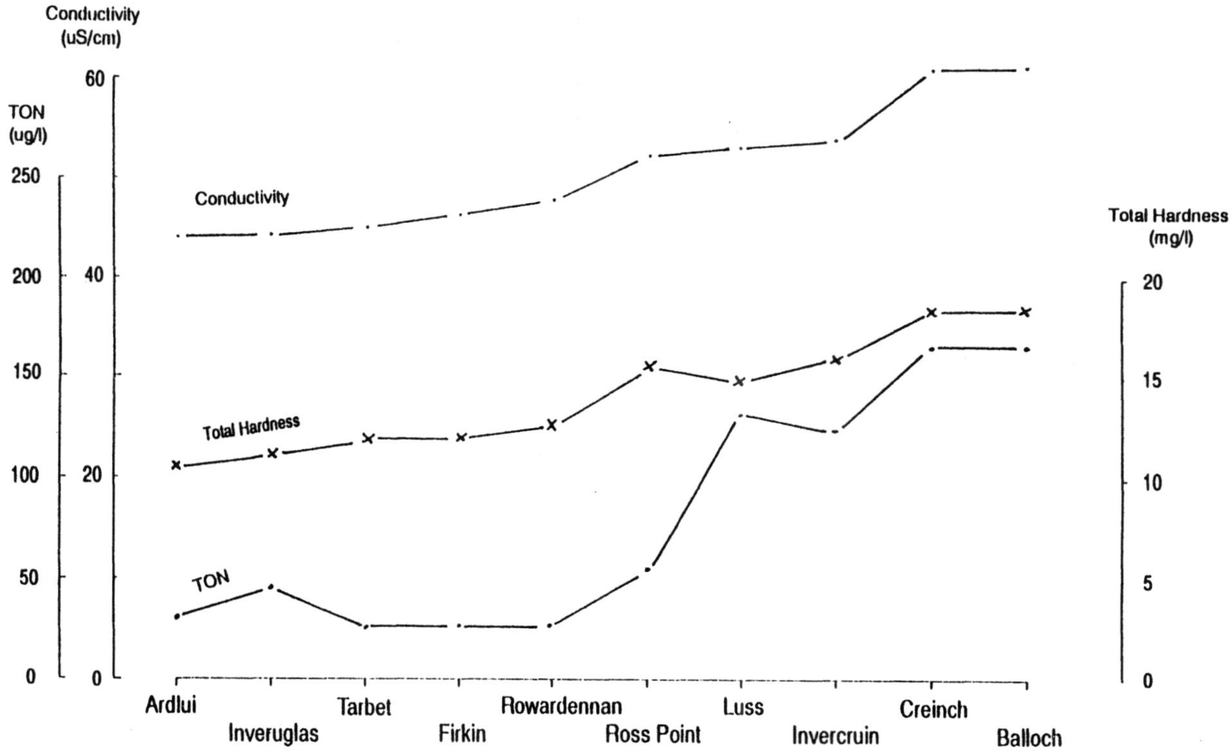

Fig. 4. Mean results of selected determinands for centre line sampling points in Loch Lomond (1990–1991).

Table 9. Trophic categories for 7 sampling sites in Loch Lomond (1990–1991)

	Total P	Chl *a*	Max Chl	Secchi	Min Secchi
Cailness	M	O	O	M	O
Firkin	M	U-O	U-O	M	O
Ross Point	O	O	O	M	M
Inchcruin	M	O	O	M	M
Creinch	M	O	O	O	M
Auchendennan	M	O	O	M	M
CSWDB	M	O	O	M	O

* on selected sites only

The results in Table 7 show the marked increase in phytoplankton growth (as measured by Chlorophyll *a*) during the summer months. For 1990 and 1991 at Firkin, the peaks occurred in July and August respectively whilst for Creinch in the southern basin they occurred in April and July in 1990 and in July 1991. Traill (1991) reported chlorophyll *a* peaks for the two sites for 1987 and 1988 as follows:

	Firkin		Creinch	
	1987	1988	1987	1988
Chlorophyll *a* Peak concentration	August	July October	March April September	June October

The magnitude of the peaks were similar from both data sets.

The results for the total phosphorus concentration are inconclusive. One would anticipate that the minimum concentration of total P would occur at the time of peak chlorophyll *a* concentration. For the Creinch site, the highest concentrations of total P occur in the winter months but this does not apply to the Firkin data except for 1990.

Although there is a broken record of the data for the analysis of silicon, the peak concentrations occur in winter whilst summer values fall to about a third of the winter values.

The important nutrients for algal growth are nitrogen, phosphorus and silicon. To determine the limiting nutrient, the total weight of nitrogen in the form

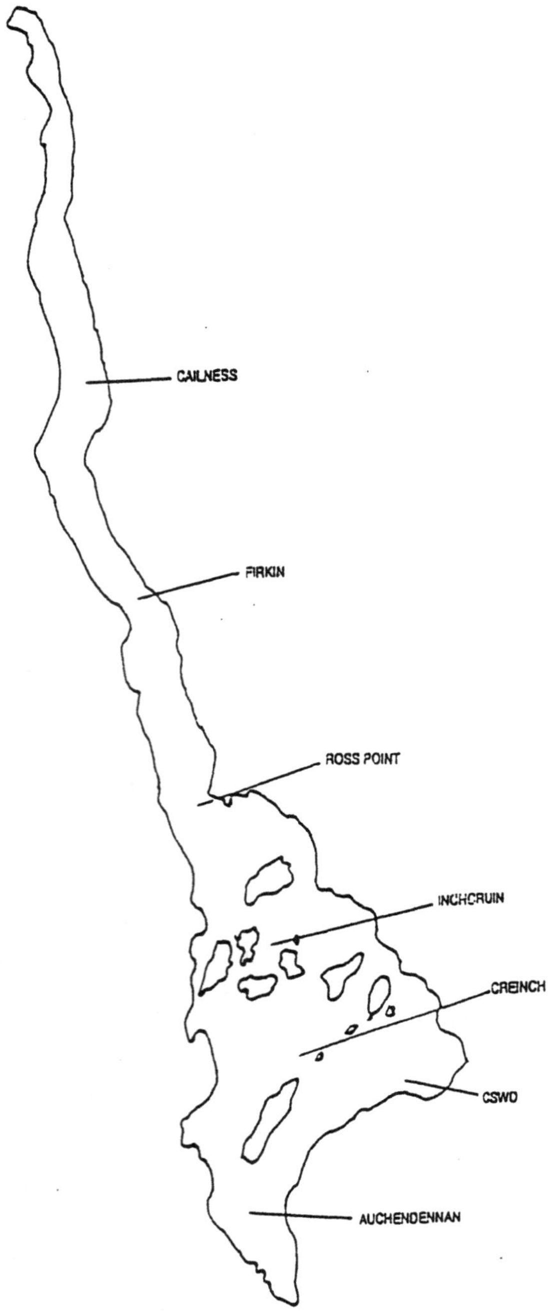

Fig. 5. Sampling points for nutrient status.

of ammonia, nitrite and nitrate is compared to that of phosphorus. If this ratio of N:P is greater than 15 then phosphorus is the limiting factor (Stumm & Morgan, 1981). From the data collected from all sampling points. this ratio is generally between 33–45.

For the phosphorus to be taken up by algae it must be in a biologically available form as simple orthophosphate. For much of the Loch Lomond data the total phosphorus values are about 10 times higher than the orthophosphate concentrations. Young *et al.* (1982) have shown however, that 72% of total phosphorus becomes biologically available within two weeks.

The results from this study of the nutrient status enable the loch water to be categorised into different trophic levels. In Table 8 the fixed boundary scheme proposed by OECD (1982) is set out. The values are arithmetic annual means or maximum values. From this scheme, the results of the monitoring from the seven sampling sites have been assessed and are summarised in Table 9.

The summary table shows that although there is generally sufficient phosphorus for algal development to classify the loch as mesotrophic water, the organisms are not present to the expected level and the loch is regarded as oligotrophic based on chlorophyll *a* levels. This could be because either the total phosphorus is not as biologically available as suggested elsewhere or else the light intensity is insufficient to allow the plankton to grow sufficiently. This latter point particularly applies to the northern basin and has been mentioned by Tippett, Maulood & Boney (1980).

The Secchi depths suggest that the loch is mesotrophic but this is likely to be because of the poor penetration of light colour rather than the presence of planktonic organisms, particularly in the northern basin. The minimum Secchi depth readings are often noted in winter months when the freshwater input of peaty water is at its highest.

Conclusions

The results of long term monitoring by Clyde River Purification Board and shorter intensive studies by other workers at different times over the last few decades suggest that the quality of Loch Lomond water is stable and of a high standard. When a good sunny spell occurs in summer months there is sufficient phosphorus available for phytoplankton to develop particularly in the southern area. However the extent of the growth is not likely to be a threat to the amenity value of the loch or to the potable supply.

References

Beveridge, M., M. Beveridge & J. Muir, 1982. Cage Fish Culture and Loch Lomond. Institute of Agriculture, University of Stirling, Stirling

Chapman, M. A., 1965. Ecological Studies on the zooplankton of Loch Lomond, Ph D Theses, University of Glasgow.

Curran, J. C. & T. Poodle, 1991. Hydrological characteristics of Loch Lomond – Are they changing? Proc. symposium 'Loch Lomond 1991'. Universities of Glasgow and Strathclyde.

Curran, J. C., 1992. Aspects of the hydrology and hydrography of Loch Lomond. Hydrobiologia 290 (Dev. Hydrobiol. 101): 21–28.

Doughty, C. R. & P. S. Maitland, 1992. The Ecology of the River Endrick: present Status and changes since 1960. Hydrobiologia 290 (Dev. Hydrobiol.101): 139–151.

Gregory, J. W., 1928. The geology of Loch Lomond. Trans. Geol. Soc. Glasgow 18 301–322.

Maulood, B. K. & A. D. Boney, 1980. A seasonal and ecological study of the phytoplankton of Loch Lomond. Hydrobiol 71: 239–259.

Organisation of Economic Co-operation and Development (OECD) (1982). Eutrophication of Waters: Monitoring Assessment and Control OECD, Paris.

Poodle, T., 1970. Loch Lomond – suspended solids and colour Technical Report No 12. Clyde River Purification Board, East Kilbride, Scotland.

Slack, H. D., 1957. Studies on Loch Lomond, Vol 1, Glasgow University Press, Glasgow.

Stumm, W. & J. J. Morgan, 1981. Aquatic Chemistry, 2nd edn. Wiley Interscience, New York.

Tippett, R. 1988. Phytoplankton of the North and South Basins of Loch Lomond. University Field Station, Rowardennan, University of Glasgow, Strathclyde.

Traill, I., 1991. Loch Lomond, An Eutrophication Study 1987–1989. Technical Report No 96. Clyde River Purification Board, East Kilbride, Scotland.

Young, T. C., J. V. De Prints, S. E. Flint, M. S. Switzenbaum & J. K. Edzwald, 1982. Algal availability of phosphorus in municipal wastewater. J. Wat. Poll, Cont., Fed. 54: 1505 01516.

Hydrobiologia **290**: 39–49, 1994.
K. J. Murphy, M. C. M. Beveridge & R. Tippett (eds), The Ecology of Loch Lomond.

Environmental change and the chemical record in Loch Lomond sediments

John G. Farmer
Department of Chemistry, University of Edinburgh, West Mains Road, Edinburgh EH9 3JJ, UK

Key words: Loch Lomond, sediments, chemical record, environmental change

Abstract

The chemical record in Loch Lomond sediments deposited since the end of the last Ice Age provides evidence of the Flandrian marine transgression some 5500–7000 ^{14}C years B.P., sedimentation rates and the influence of man's local activities, environmental pollution and its sources since the onset of the Industrial Revolution, and of elemental mobility linked with the reduction-diffusion-oxidation cycle of early sedimentary diagenesis. Information derived from vertical profiles of halogen elements bromine and iodine, radionuclides ^{14}C and ^{210}Pb, heavy metals lead, zinc, and cadmium, stable lead isotopes ^{206}Pb and ^{207}Pb, and redox-sensitive elements manganese, iron and arsenic is reviewed and assessed.

Introduction

Freshwater lake sediments can preserve physical, biological and chemical records of changes in the environment, ranging from local to global, over periods of less than one year to many thousands of years (Battarbee, 1984; Haworth & Lund, 1984; Alderton, 1985). This paper reviews the interpretation of the chemical record in Loch Lomond sediments since the end of the last Ice Age, some 10 000 years ago. It is based largely on work carried out during the last 20 years, a period which has seen enormous strides made in techniques of chemical analysis (e.g. atomic absorption spectrometry, instrumental neutron activation analysis, gamma spectrometry, inductively coupled plasma-mass spectrometry) and of radiometric dating on both short (e.g. ^{210}Pb, $t_{\frac{1}{2}} = 22.35$ years) and long (e.g. ^{14}C $t_{\frac{1}{2}} = 5730$ years) timescales.

Until the 1970's, chemical information on Loch Lomond sediments was limited to the results of the classic survey work of Slack (1954, 1965), who collected sediment cores of 12–20 cm length at 55 sampling stations throughout the loch (area 71 km^2, length 36.4 km, maximum width 8 km). The deposits were judged visually, with chemical analyses restricted to those for the content of organic carbon and nitrogen. On the basis of the nature of the deposits, hydrographic characteristics and relief morphology, Slack divided the loch into two distinct major loch regions, a 'lower' (Luss, Strathcashel, Fault basins) and an 'upper' (Ardlui, Tarbet basins), separated by the Inverbeg-Rowardennan bar. As the Luss and Strathcashel basins of the 'lower' loch resembled the rest of the 'lower' loch in the nature of bottom deposits and the 'upper' loch in terms of physical characteristics, Slack suggested they be collectively termed the 'middle' loch. In this paper, the Fault basin of the 'lower' loch is referred to as the southern basin, the Luss and Strathcashel basins of the 'lower' loch as the central basin and the Ardlui and Tarbet basins of the 'upper' loch as the northern basin (Fig. 1).

The southern basin, lying over sedimentary rocks of the Carboniferous and Devonian, is wide, relatively shallow (maximum depth 31 m) and biologically more productive than the narrow, deeper (maximum depth 200 m), oligotrophic basin to the north, excavated in metamorphosed Dalradian mica-schists, schistose grits and slates. The River Falloch, at the northern tip, drains a catchment area of metamorphic rocks overlain by peat, poor in available nutrients. The main inflow to the lower loch is the Endrick Water, draining productive farmland on beds of boulder clay and glacial and marine deposits. Outflow is via the River Leven. Even during periods of thermal stratification, more pronounced in the central and northern basins, where bottom temperatures are low all year, dissolved oxygen levels remain high (>72% saturation) throughout the loch (Maulood & Boney, 1980). The bottom sedi-

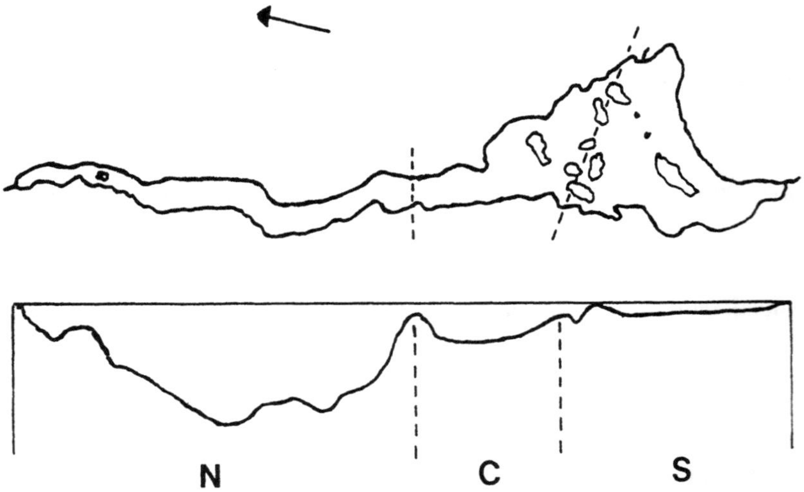

Fig. 1. Morphology of Loch Lomond, showing the three major basins - northern (N) (max. depth 200 m), separated by the Inverberg-Rowardennan bar from the central (C), in turn separated by the island chain from the southern (S) (based on Smith *et al.*, 1981) (see text).

ments range from brown muds, with a thin near-surface rust-red layer, in the south, to black micaceous oozes, overlain by thin dark-brown unconsolidated material, in the north. Porosities are high, 80–98% and 90–99% over the top 20 cm in the south and north, respectively. Surficial organic carbon content is greater in the north (≈10%) than in the south (≈5%) and decreases more slowly with depth as a consequence of the increased acidity and lower temperature of northern bottom water as well as of the greater proportion of peaty allochthonous material in the north compared with the more autochthonous planktonic debris in the south (Slack, 1954).

This review, however, is based on the much more extensive chemical and radiometric data, published between 1978 and 1991, from 18 cores collected between 1976 and 1988 (Table 1, Fig. 2). The longer cores (LLRDI, LLMA) exhibit evidence of the Flandrian marine transgression which made Loch Lomond a sea loch for a substantial period (Dickson *et al.*, 1978; Mackenzie *et al.*, 1983, 1990), while the shorter cores (LLRPMI, LL-(1-13)L, LL-lJ) provide evidence of environmental pollution during the industrial era (Farmer *et al.*, 1980; Lovell, 1985; Farmer & Lovell, 1986; Sugden *et al.*, 1991) and of the perturbatory effects of recent sediment diagenesis (Farmer & Cross, 1979; Farmer & Lovell, 1984, 1986; Johnson & Farmer, 1987). Information on sedimentation rates is provided by long (LLRDI - Dickson *et al.*, 1978) and short (LLRPMI - Baxter *et al.*, 1981; LL-10L - Lovell, 1985) cores.

Flandrian marine transgression

In LLRDI, Dickson *et al.* (1978) observed grey clay at the base of the core from 4.94 to 4.70 m and then brown silty clay from 4.70 to 3.75 m. The band of silty clay above this, from 3.75 to 3.05 m, was much darker in colour, lower in water content and higher in carbon content than sediments above or below. Furthermore, it included marine dinoflagellate cysts and marine bivalves, exclusive to this section of the core, while remains of freshwater macrophytes and algae, found elsewhere in the core, were very poorly represented. This, along with the low remanent magnetism of the darker layer and other supporting information (e.g. shoreline landforms) from the catchment area, was taken as evidence that a transient incursion of seawater from the Firth of Clyde into freshwater Loch Lomond had occurred as a consequence of the differential rates of sea level rise and isostatic recovery of land following the most recent major glaciation.

In a subsequent geochemical study, Mackenzie *et al.* (1983) showed that the well-defined band of marine deposited sediment in LLRDI was characteristically enriched in certain elements relative to the contiguous freshwater sediments. In particular, concentrations of iodine and bromine were an order of magnitude higher in the marine sediment (Fig. 3). A similar zone of enrichment was found in LLMA, collected in the general vicinity of LLRDI 12 years later, although the marine layer was found to be somewhat narrower (0.5 m) and, at 270–320 cm, about 0.5 m

Table 1. Loch Lomond Cores (1976–1988)

Core	Corer*	Date	Basin†	Water depth (m)	Core length (m)	References
LLRD1	M	/76	S	24	5.20	Dickson *et al.* (1978) Mackenzie *et al.* (1983, 1990)
LLMA	M	/8/88	S	24	5.00	Mackenzie *et al.* (1990)
LLRPM1	MM	22/8/76	S	28	0.96	Farmer & Cross (1979) Farmer *et al.* (1980) Baxter *et al.* (1981)
LL-1L	C	15/12/81	S	21	0.17	Farmer & Lovell (1984, 1986)
LL-1LA	C	26/11/82	S	20	0.15	"
LL-2L	C	3/12/81	S	20	0.155	"
LL-3L	C	4/3/82	S	31	0.16	"
LL-4L	C	3/12/81	S	24	0.17	"
LL-5L	C	15/12/81	S	19	0.19	"
LL-6L	C	19/11/81	S	18	0.11	"
LL-7L	C	18/2/82	C	37	0.09	" + Sugden *et al.* (1991)
LL-8L	C	18/2/82	C	55	0.07	"
LL-9L	C	4/3/82	C	60	0.18	"
LL-10L	C	19/5/82	N	60	0.27	" + Sugden *et al.* (1991)
LL-11L	C	9/7/82	N	140	0.30	"
LL-12L	C	2/6/82	N	175	0.44	"
LL-13L	C	2/6/82	N	55	0.23	"
LL-1J	J	9/12/85	S	31	0.15	Johnson & Farmer (1987)

* M - Mackereth; MM - Mini-Mackereth; C - Craib; J - Jenkin
† S - southern; C - central; N - northern

closer to the sediment surface (Mackenzie *et al.*, 1990). Comparison of the Loch Lomond sediment geochemical characteristics with those of the nearby smaller Dubh Loch, which was unaffected by the marine incursion as a result of its slightly higher elevation, verified that the enhanced elemental concentrations in Loch Lomond sediment resulted from the marine incursion rather than any change in the composition of the material being eroded from the catchment (Mackenzie *et al.*, 1984).

Radiocarbon dating of the LLRDl core showed that the marine incursion lasted some 1450 years from 6900 B.P. to 5450 B.P. during the middle Flandrian (Dickson *et al.*, 1978). The level the sea had to attain to enter the Loch Lomond basin was controlled by the barrier formed between Loch Lomond and the Clyde Estuary by moraine deposits from the re-advance of the Loch Lomond glacier (11 700–10 500 B.P.), which extended as far south as Alexandria in the Vale of Leven. There is evidence that this glacial re-advance was preceded by another marine incursion about 12 500 B.P., in the late Devensian (Sissons, 1974).

Loch Lomond sedimentation rates and man's activities

The radiocarbon profile for LLRDl also revealed a uniform sedimentation rate between 6300 and 1700 B.P. of 0.31 mm yr^{-1} (Dickson *et al.*, 1978). During this period much of the land surface around Loch Lomond was covered with deciduous forest upon which the early agricultural activities and settlements of man made little impact (Idle, 1974). To account for a depletion in sediment ^{14}C levels around 1700 B.P. (≡1.5 m), however, it has been suggested that man-induced inputs of fossil or old carbon through modification of run-off and erosion patterns associated with increasing agricultur-

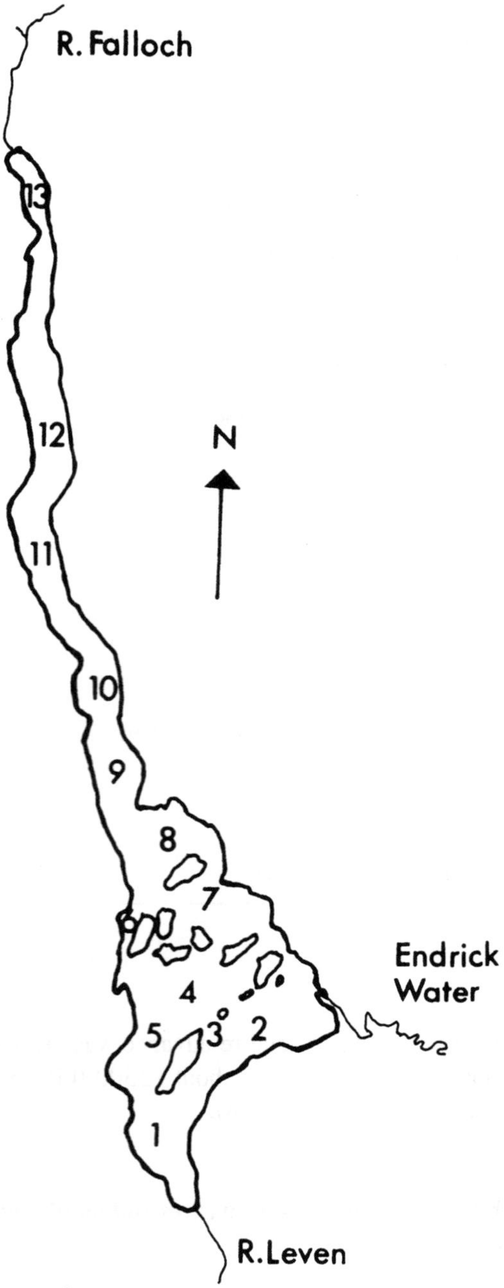

Fig. 2. Loch Lomond, showing main inflows, from the north and south-east, and outflow, to the south, and the 13 sediment core collection sites of Farmer & Lovell (1984, 1986) (1–6, southern basin; 7–9, central basin; 10–13, northern basin), designated LL-(1–13)L in Table 1. Of the other five cores listed in Table 1, LLRD1 and LLMA were collected near site 4, LLRPM1 and LL-1J near site 3, and LL-1LA between sites 1 and 5.

al activity could have occurred (Dickson *et al.*, 1978). Since then, anthropogenic inputs of fossil or old carbon and of bomb ^{14}C preclude the accurate determination of sedimentation rates via radiocarbon.

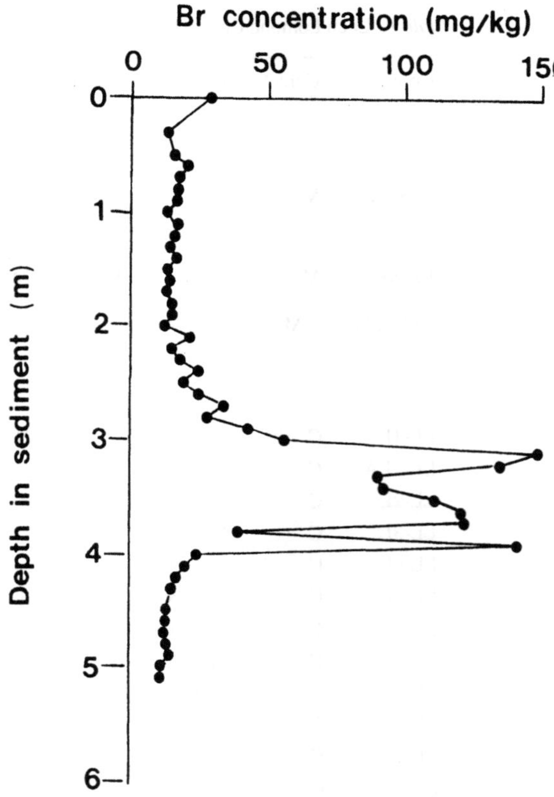

Fig. 3. Bromine concentrations in Loch Lomond sediment core LLRD1 (based on Mackenzie *et al.*, 1983, 1990).

There is magnetic susceptibility evidence in Loch Lomond sediments that the clearance of forests to provide timber for shipbuilding and especially for the preparation of charcoal, an important raw material for the then burgeoning Scottish iron-smelting industry, led to increased erosion and sediment accumulation from the late-17th to the early-19th century (Turner & Thompson, 1979). Accordingly, the rate of sediment accumulation may have been 2–3 times faster between 1700 A.D. and 1850 A.D. than the long-term historical average of 0.31 mm yr^{-1} determined by ^{14}C or the similarly uniform 0.36 mm yr^{-1} ($\equiv$220 g m^{-2} yr^{-1}) for the period from 1850 A.D. to the present, as revealed by ^{210}Pb dating of LLRPMI (Baxter *et al.*, 1981). For this core from the southern basin, palaeomagnetic dating relates a depth of 20 cm to 1790 A.D. and ^{210}Pb dating places 10 cm at 1850 A.D. For LL-10L, a core from the more porous sediments of the northern basin, ^{210}Pb yields a sedimentation rate of 210 g m^{-2} yr^{-1}, with depths of 9 cm and 17 cm corresponding to dates of 1934 A.D. and 1853 A.D. respectively (Lovell, 1985; Sugden *et al.*, 1991).

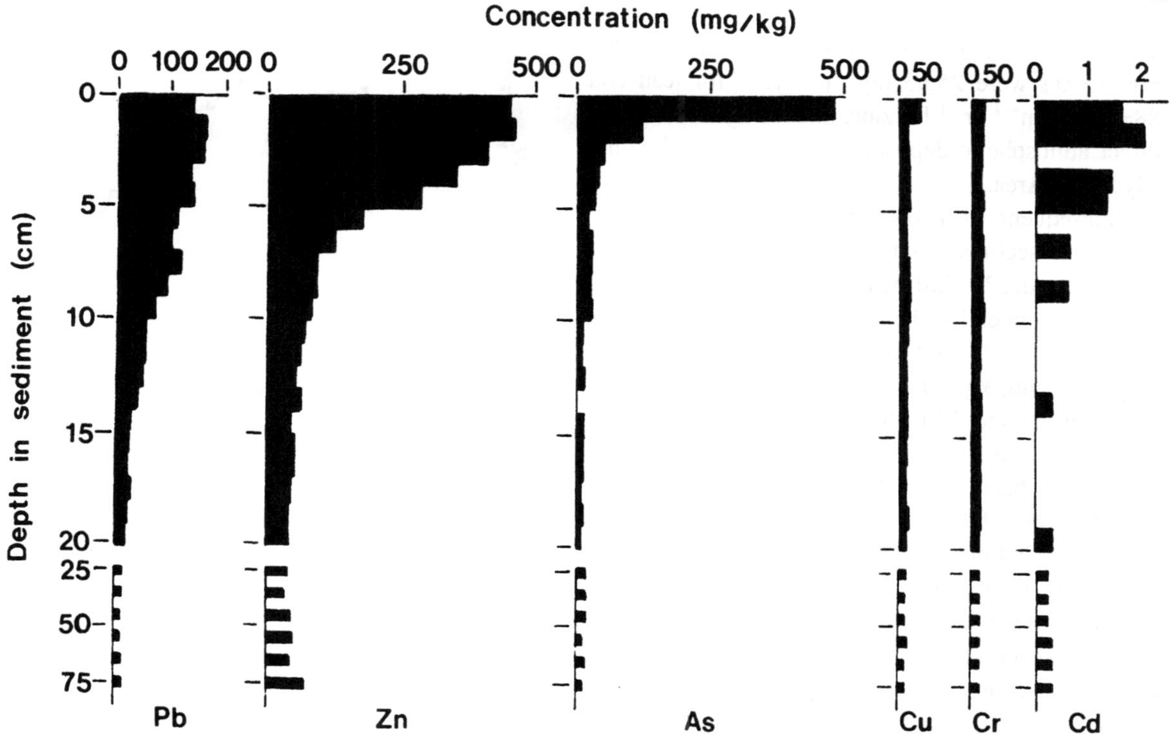

Fig. 4. Profiles of lead, zinc, copper and chromium concentrations (1M hydroxylammonium chloride/25% acetic acid extractable) and of total arsenic and cadmium concentrations with depth in Loch Lomond sediment core LLRPM1 (based on Farmer *et al.*, 1980). Note the change in scale for cadmium concentration. Not all samples from 0–20 cm were analysed for cadmium.

Environmental pollution

The attachment of a timescale to the sediment column of a freshwater lake can, under appropriate circumstances, enable reconstruction of temporal trends in the anthropogenic input to the environment of pollutants (e.g. heavy metals) which have subsequently been deposited from the atmosphere, or otherwise introduced into the lake, and then removed to the bottom sediments in association with settling particles (Alderton, 1985; Farmer, 1991).

Industrial activity in the immediate vicinity of Loch Lomond has been limited in nature and extent. Small iron-smelting bloomeries on Loch Lomondside lasted only until the mid-18th century. The Vale of Leven to the south did become a major world centre in bleaching, dyeing and printing from the early-18th until the mid-20th century, but effluents were removed by the River Leven to the Clyde Estuary. Demand for bark by the tanning industry during the 19th century did lead to the siting of a pyroligneous acid works at the south-eastern corner of Loch Lomond. In general, however, potential sources of local pollution can be considered minimal when compared with the rapid growth in industrialisation and population of the Glasgow area in west-central Scotland from the late-18th until the mid-20th century. Thereafter, while heavy industry has declined and the nature of energy generation has altered, transportation by private means has grown, leading to new patterns of environmental pollution, particularly of the atmosphere.

Thus, with the exception of arsenic, the heavy element profiles obtained for LLRPM1 (Fig. 4), have been interpreted in terms of trends in release to, and deposition from, the atmosphere (Farmer *et al.*, 1980). While copper and chromium were enriched to only a minor extent, the lead, zinc, cadmium and arsenic concentrations were markedly enhanced in upper sections of LLRPM1. Onset of an approximately ten-fold increase in lead and zinc concentrations above 'background' levels of 13±2 mg kg^{-1} and 47±13 mg kg^{-1} respectively was placed at late-18th century, in accord with the records of the commencement of significant industrial activity in the area. Taking marker levels of 3 cm (*ca* 1945), 4 cm (1933), 8 cm (1880) and 10 cm (1850), it was calculated that the percentage of total 'excess' lead (zinc) contained within these boundaries was 25% (49%), 33% (64%), 64% (90%)

and 76% (95%). Derived anthropogenic fluxes for the mid-1970's were 25±3 mg m^{-2} yr^{-1} for lead and 88±12 mg m^{-2} yr^{-1} for zinc, in good agreement with direct atmospheric deposition data for the adjoining Clyde Sea Area.

Subsequent work on cores LL-(1-13)L revealed maximum sectional concentrations of 100–200 mg kg^{-1} for lead, 270–450 mg kg^{-1} for zinc and 13–21 mg kg^{-1} for copper for the southern basin (LL(1-6)L), 160–220 mg kg^{-1} for lead, 330–390 mg kg^{-1} for zinc and 16–19 mg kg^{-1} for copper for the central basin (LL-(7-9)L), and 120–350 mg kg^{-1} for lead, 250–450 mg kg^{-1} for zinc and 9–22 mg kg^{-1} for copper for the northern basin (LL-(10-13)L) (Lovell, 1985). Sediment from one of these cores, LL-10L, which exhibited lead concentrations in excess of 100 mg kg^{-1} over the top 14 cm, with a maximum of 210 mg kg^{-1} at 10–11 cm, was later subjected to stable lead isotopic analysis by Sugden *et al.* (1991). After correcting for an intrinsic baseline lead contribution with a $^{206}Pb/^{207}Pb$ ratio of 1.175 and employing a ^{210}Pb-based sedimentation rate of 210 g m^{-2} yr^{-1}, post-1850 trends in the flux (directly related to the concentration) and $^{206}Pb/^{207}Pb$ ratio of 'excess' (i.e. anthropogenic) lead deposited in Loch Lomond sediments were derived as shown in Fig. 5. Interpretation of the isotopic data showed that lead emissions from heavy industry and the combustion of coal ($^{206}Pb/^{207}Pb \approx 1.17$) predominated until the early 1930s, by which time more than 50% of the post-1850 anthropogenic lead in the sediment column had been deposited. Since then, the introduction of alkyllead additives ($^{206}Pb/^{207}Pb \approx 1.09$) to petrol, the growth in car-exhaust emissions of this comparatively ^{206}Pb-depleted lead and the decline in emissions from other sources have strongly influenced the reduction in sediment $^{206}Pb/^{207}Pb$ from 1.17 to 1.11. Calculated fluxes of lead to the sediments since the early 1950's, however, were 40% lower than for the first half of the century.

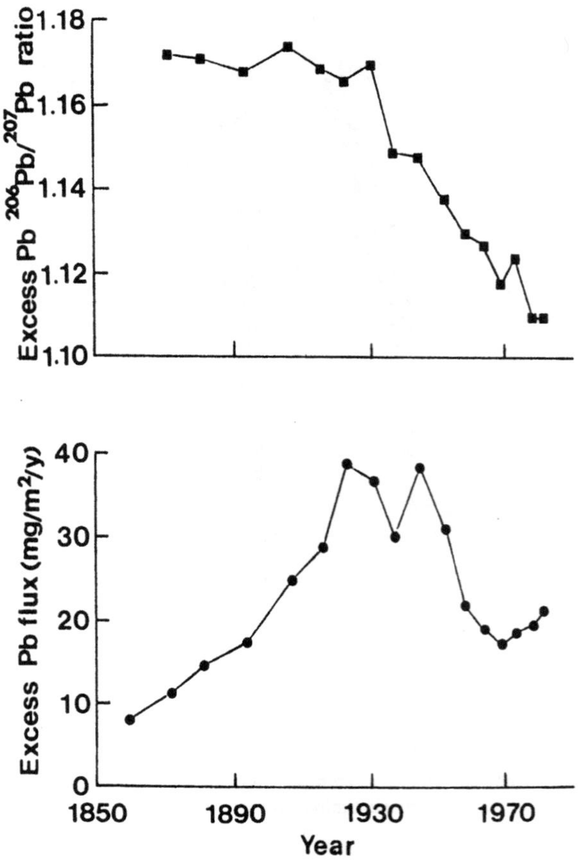

Fig. 5. Post-1850 trends in fluxes (•) and $^{206}Pb/^{207}Pb$ (■) ratios of 'excess' lead deposited in Loch Lomond sediments (LL-10L) (based on Sugden *et al.*, 1991).

Sediment diagenesis and element mobility

The use of vertical concentration profiles of heavy elements in dated lake sediments as historical records is dependent upon lack of post-depositional disturbance of sediment (e.g. via mixing) and of diagenetic release and mobility of elements in the sediment column (Farmer, 1991). While any mixing at sites LLRPM1 and LL-10L has been insignificant and lead can be considered immobile, the same cannot be said of arsenic. This element, despite a lack of any known significant local source of arsenic contamination, exhibited a *ca* 25-fold enhancement in the surface sediment of LLRPM1 relative to a baseline arsenic level at depth in the sediment column of 18±5 mg kg^{-1} (Fig. 4). In view of its association with an enhanced content of iron, a redox-active element of known mobility, Farmer & Cross (1979) postulated that the arsenic profile was controlled by similar post-depositional diagenetic remobilisation processes.

This hypothesis was tested on a loch-wide basis in the detailed 13-core study (LL-(1-13)L, Table 1, Fig. 2) of Farmer & Lovell (1986). Concentrations of arsenic in surface or near-surface sediment in 12 of the 13 cores from the three major basins were indeed substantially enhanced, up to 675 mg kg^{-1} compared with background values of 15–50 mg kg^{-1} (Fig. 6). Peaks of arsenic were again found to be closely associated with those of iron but were typically displaced downwards relative to manganese, another element whose vertical distribution is strongly influenced by redox cycling (Fig. 7). The relationship between these three elements and direct confirmation, via porewater analysis, of the mechanism controlling the natural enrichment of arsenic in Loch Lomond sediments are shown for core LL-1J in Fig. 8 (Johnson & Farmer, 1987). Total arsenic concentrations were substantially enhanced in the near-surface (2–5 cm) sections, the peak of 283 mg kg^{-1} (2–3 cm) coinciding with that for iron, 1 cm below the manganese peak. Porewater arsenic rapidly declined above and below the 5–11 cm zone where maximum concentrations (76–103 $\mu g\ l^{-1}$) were recorded. Both solid phase and porewater arsenic profiles (as found also for core LL-1LA in Farmer & Lovell (1986)) are consistent with arsenic release into solution from an association with solid phase ferric oxyhydroxides, which dissolve under sufficiently reducing conditions, followed by upward migration, oxidation and co-precipitation or readsorption on iron oxyhydroxides in the aerobic layers of sediment. Figure 8 shows that As(III) is the predominant species in the reduction zone and, with the exception of the uppermost section, even in the oxidised zone, due to preferential adsorption of As(V) on iron oxyhydroxides. The observed northwards increase in the length of displacement between the manganese and iron-arsenic peaks and in the thickness of the zones of enhancement for each element is probably due to a more gradual transition from surface oxic to deeper anoxic conditions in the northern basin sediments.

The enriched manganese levels (Fig. 7), with maximum sectional concentrations of 2.6–6.8% in the south (LL-(1-6)L), 4.0–5.9% in the middle (LL-(7-9)L) and 1.3–9.1% in the north (LL-(10-13)L), contrast with typical background levels of 0.15–0.5% and are amongst the highest recorded for freshwater lake sediments (Farmer & Lovell, 1984). A significant distinguishing feature of Loch Lomond sediments, however, is the apparent absence of ferromanganese nodules, in line with Slack's survey of the loch bottom some 40 years ago. The accumulation of flocculent, organic-rich, fine-grained sediments is not conducive to the formation of nodules in the central areas of the major basins of Loch Lomond. The clear distinction between the depths of manganese and iron peaks in the Loch Lomond sediment column, attributable to the thermodynamic and kinetic factors controlling elemental oxidation along the pronounced redoxcline, in conjunction with probable minimal release of the two elements to the oxygenated overlying waters, also restricts opportunity for nodule formation.

Nevertheless, in some shallower, littoral locations in the loch, black coatings of manganese oxides were observed on rocks and pebbles and a 4-cm thick crust composed of a black manganese-rich (11%) layer overlying an orange iron-rich (22%) layer was found over a 0.25 km^2 barren area of glacial clay in shallow water west of the LL-4L site (Farmer & Lovell, 1984). This probably corresponds to one of the areas qualitatively described as 'firm black iron pan' by Slack (1954). Iron phosphate hydroxides were identified by X-ray diffraction in the iron-rich layer. No other mineral species of iron or manganese could be so identified either in this material or in surface layers of lake sediment, where it is presumed that amorphous hydrous oxides of manganese and iron predominate.

It can be concluded that this comparatively oligotrophic loch affords an excellent opportunity for investigation of recent diagenetic effects on the distribution and mobility of manganese, iron and trace elements in freshwater sediments characterised by transition from oxic to progressively anoxic layers.

Current studies

In the light of the foregoing, there is currently a collaborative research programme on the loch, involving personnel from the Chemistry Department, University of Edinburgh, the Scottish Universities Research and Reactor Centre, East Kilbride, and the N.E.R.C. Institute of Freshwater Ecology, Penicuik, and focusing on the study areas listed below.

(i) The potential mobility of heavy metals in sediments, via investigation of metal partitioning and speciation in different sedimentary phases and porewaters.
(ii) Sedimentary distribution and behaviour of fallout radiocaesium from nuclear weapons testing and the Chernobyl reactor accident.
(iii) High-resolution profiling (0.2 cm intervals) of manganese, iron, trace elements and radionuclides in sediment cores.
(iv) Determination of sedimentation rates and investigation of mixing via ^{210}Pb and radiocaesium.
(v) Characterisation of lead inputs via stable lead isotopic analysis of sediments.

The University Field Station at Rowardennan continues to provide excellent facilities and support for sampling on the loch and preliminary sample handling onshore.

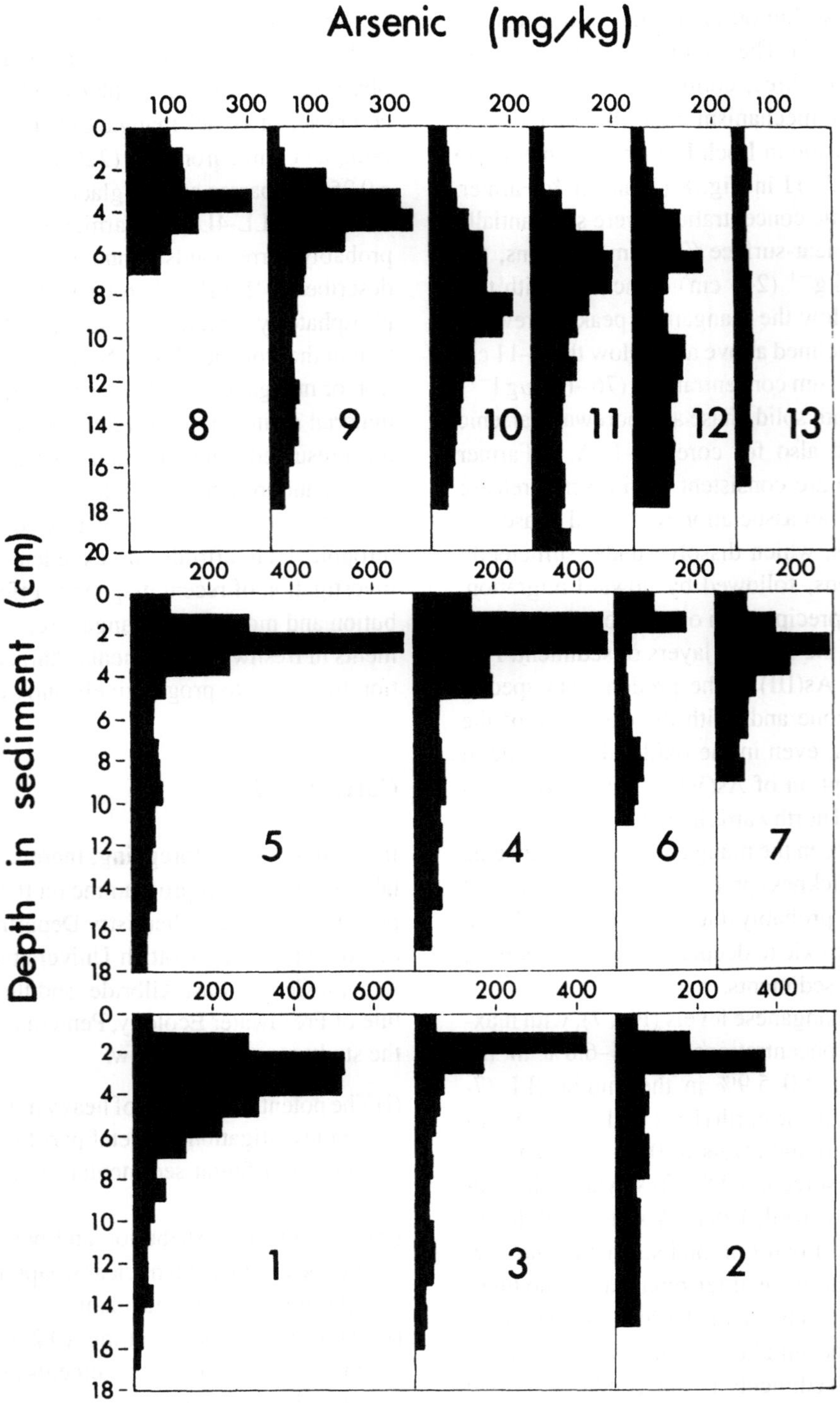

Fig. 6. Profiles of total arsenic concentration with depth, over collected core lengths, up to 20 cm, in 13 sediment cores (LL-(1–13)L) from Loch Lomond: 1–6 (southern basin), 7–9 (central basin), 10–13 (northern basin) (based on Farmer & Lovell (1986)).

Acknowledgements

In addition to those workers whose publications on Loch Lomond have been cited in the text, I should like to thank A. E. Bailey-Watts, M. S. Baxter, C. L. Bryant, R. W. Crawford, J. D. Cross, L. R. Johnson, A. Kirika, M. A. Lovell, A. B. Mackenzie, R. McMath, C. L. Sugden, D. S. Swan and R. Tippett, with whom I have collaborated directly, for their greatly appreciated efforts and expertise in joint studies of the loch during the past

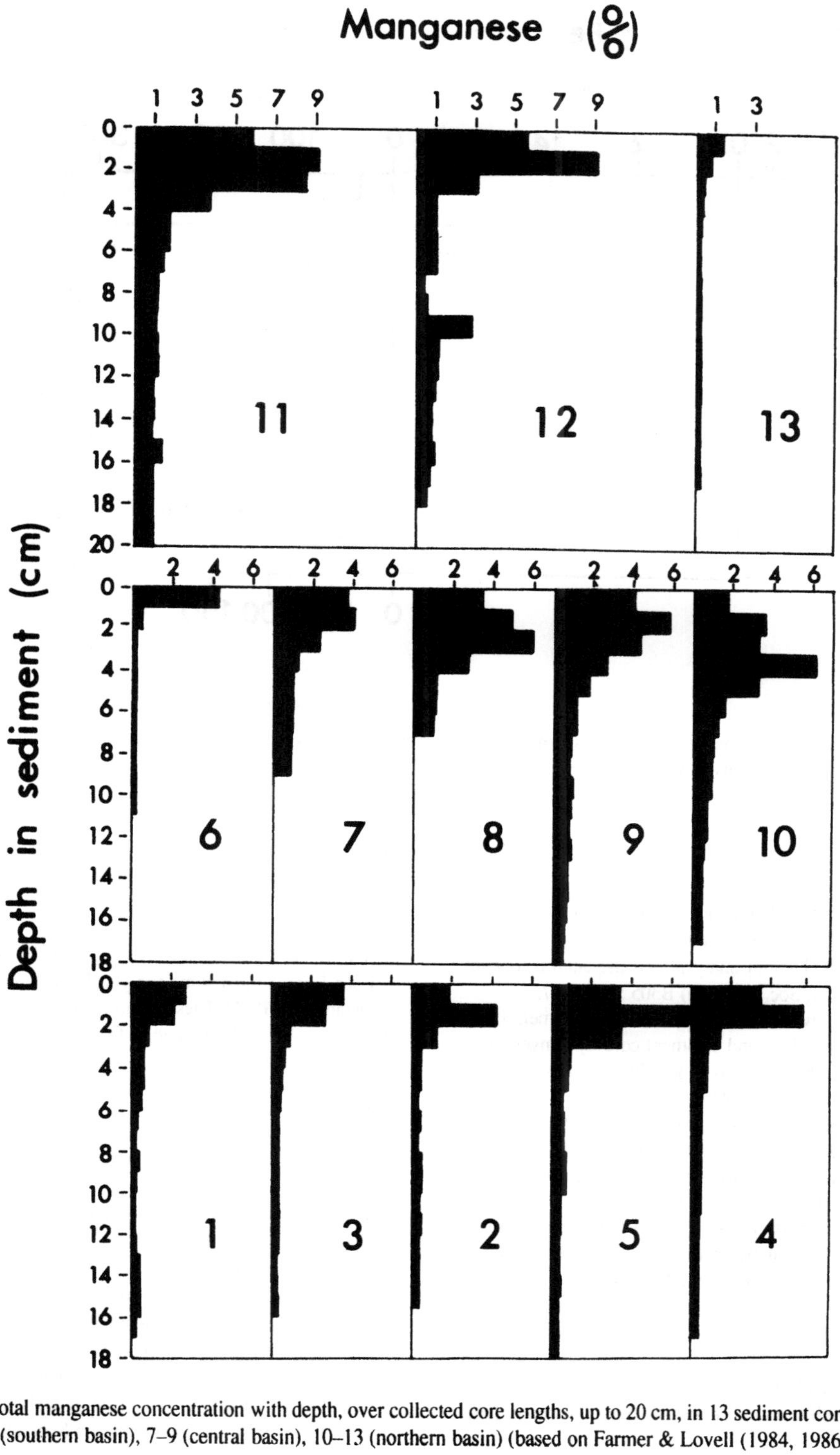

Fig. 7. Profiles of total manganese concentration with depth, over collected core lengths, up to 20 cm, in 13 sediment cores (LL-(1–13)L) from Loch Lomond: 1–6 (southern basin), 7–9 (central basin), 10–13 (northern basin) (based on Farmer & Lovell (1984, 1986)).

15 years or so. P. R. O. Barnett, J. H. Dickson, K. McKay, I. G. McKinley, F. Oldfield and R. Thompson also provided invaluable assistance at various stages. The research studentship support of C. L. Bryant, M. A. Lovell and L. R. Johnson by N. E. R. C. and of C. L. Sugden by S. E. R. C. is gratefully acknowledged.

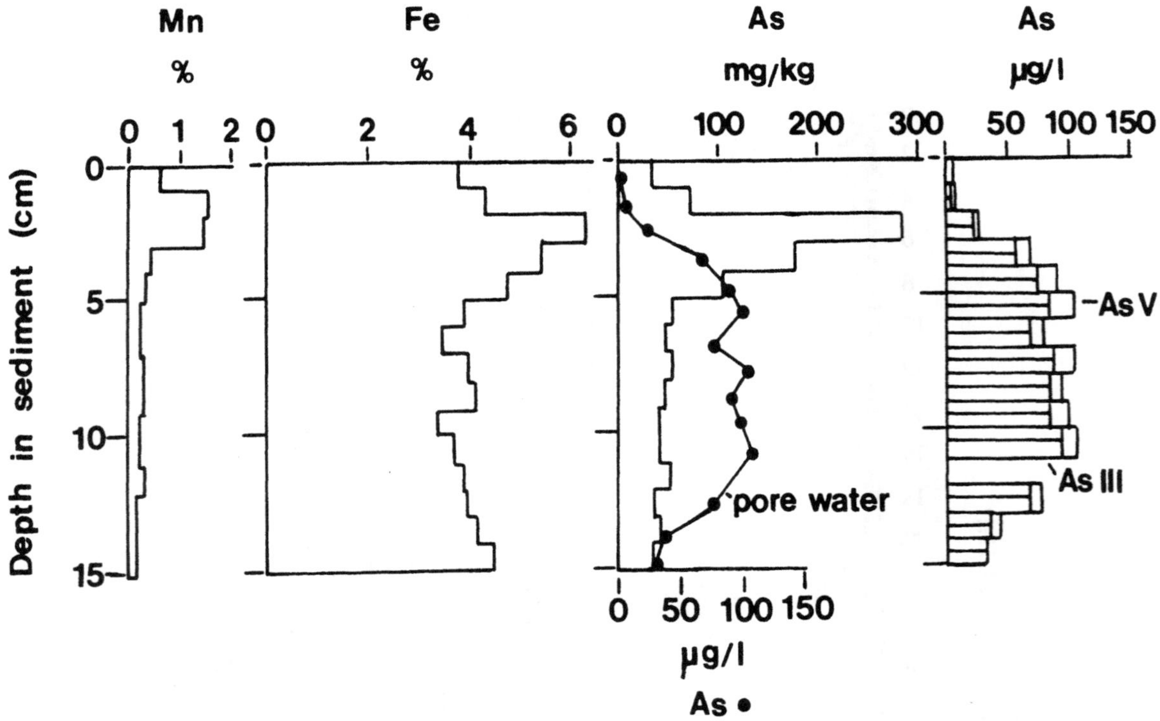

Fig. 8. Profiles of manganese, iron and arsenic in sediment and of arsenic (and its species AsIII, AsV) in pore water from Loch Lomond sediment core LL-1J (based on Johnson & Farmer, 1987).

References

Alderton, D. H. M., 1985. Sediments. In Historical Monitoring, MARC Report No. 31. MARC, London, 1–95.

Battarbee, R. W., 1984. Diatom analysis and the acidification of lakes. Phil. Trans. Roy. Soc. (London) B305 : 451–477.

Baxter, M. S., R. W. Crawford, D. S. Swan & J.G. Farmer, 1981. ^{210}Pb dating of a Loch Lomond sediment core by conventional and particle track methods and some geochemical observations. Earth Planet. Sci. Lett. 53: 434–444.

Dickson, J. H., D. A. Stewart, R. Thompson, G. Turner, M. S. Baxter, N. D. Drndarski & J. Rose, 1978. Palynology, palaeomagnetism and radiometric dating of Flandrian marine and freshwater sediments of Loch Lomond. Nature 274: 548–553.

Farmer, J. G. and J. D. Cross, 1979. The determination of arsenic in Loch Lomond sediment by instrumental neutron activation analysis. Radiochem. Radioanal. Lett. 39: 429–440.

Farmer, J. G., D. S. Swan & M. S. Baxter, 1980. Records and sources of metal pollutants in a dated Loch Lomond sediment core. Sci. Total Envir. 16: 131–147.

Farmer, J. G. & M. A. Lovell, 1984. Massive diagenetic enhancement of manganese in Loch Lomond sediments. Envir. Technol. Lett. 5: 257–262.

Farmer, J. G. & M. A. Lovell, 1986. Natural enrichment of arsenic in Loch Lomond sediments. Geochim. Cosmochim. Acta 50: 2059–2067.

Farmer, J. G., 1991. The perturbation of historical pollution records in aquatic sediments. Envir. Geochem. Health 13: 76–83.

Haworth, E. Y. & J. W. G. Lund (eds), 1984. Lake Sediments and Environmental History. Leicester Univ. Press.

Idle, E. T., 1974. Land use. In Tippett, R. (ed.) A Natural History of Loch Lomond. Univ. of Glasgow Press, 96–105.

Johnson, L. R. & J. G. Farmer, 1987. Arsenic mobility and speciation in the sediments of Scottish inland and coastal waters. In Proc. Sixth Intern. Conf. Heavy Metals Environ. CEP, Edinburgh, 2: 218–222.

Lovell, M. A., 1985. Arsenic cycling in the freshwater sediments of Loch Lomond and some analytical speciation studies of arsenic metabolism. Ph.D. Thesis, Univ. of Glasgow.

Mackenzie, A. B., R. D. Scott, I. G. McKinley & J. M. West, 1983. A study of long term (10^3-10^4 y) elemental migration in saturated clays and sediments. Rep. Fluid Processes Unit Inst. Geol. Sci. FLPU 83-6. Institute of Geological Sciences, N.E.R.C., Harwell.

Mackenzie, A. B., R. D. Scott, I. M. Ridgway, I. G. McKinley & J. M. West, 1984. A study of long term (10^3-10^4 yr) elemental migration in saturated clays and sediments (Part II). Rep. Fluid Processes Res. Gp., Brit. Geol. Surv. FLPU 84-11. British Geological Survey, N.E.R.C., Keyworth.

Mackenzie, A. B., T. M. Shimmield, R. D. Scott, C. M. Davidson & P. J. Hooker, 1990. Chloride, bromide and iodide distributions in Loch Lomond sediment interstitial water. Brit. Geol. Surv. Tech. Rep. WE/90/2. British Geological Survey, N.E.R.C., Keyworth.

Maulood, K. & A. D. Boney, 1980. A seasonal and ecological study of the phytoplankton of Loch Lomond. Hydrobiol. 71: 239–259.

Sissons, J. B., 1974. The Quaternary in Scotland: a review. Scott. J. Geol. 10: 311–337.

Slack, H. D., 1954. The bottom deposits of Loch Lomond. Proc. Roy. Soc. Edin. B65: 213–238.

Slack, H. D., 1965. The profundal fauna of Loch Lomond, Scotland. Proc. Roy. Soc. Edin. B69: 272–297.

Smith, I. R., A. A. Lyle & A. J. Rosie, 1981. Comparative physical limnology. In Maitland, P.S. (ed.) The Ecology of Scotland's largest lochs : Lomond, Awe, Ness, Morar and Shiel. Monogr. Biol. 44: 29–65.

Sugden C. L., J. G. Farmer & A. B. Mackenzie, 1991. Isotopic characterisation of lead inputs and behaviour in recent Scottish freshwater loch sediments. In Proc. Eighth Intern. Conf. Heavy Metals Environ. CEP, Edinburgh, 1: 511–514.

Turner, G. M. & R. Thompson, 1979. Behaviour of the earth's magnetic field as recorded in sediment of Loch Lomond. Earth Planet. Sci. Lett. 42: 412–426.

Hydrobiologia **290**: 51–52, 1994.
K. J. Murphy, M. C. M. Beveridge & R. Tippett (eds), The Ecology of Loch Lomond.

Assessment of eutrophication in Loch Lomond by desk analysis

I. D. M. Gunn, A. E. Bailey-Watts & A. A. Lyle
Institute of Freshwater Ecology, Edinburgh Laboratory, UK

Key words: Loch Lomond, Scotland, eutrophication, phosphorus, loss coefficients, chlorophyll

Abstract

Phosphorus and chlorophyll *a* levels were predicted for Loch Lomond from desk-derived P loss coefficients and flushing rates related according to existing eutrophication models. These were in close keeping with measured values.

Eutrophication is, literally, the supply of nutrients to an ecosystem. However, it is of increasing global concern in aquatic environments due to problems stemming from the biological manifestations of accelerated nutrient enrichment, rather than from the elevated concentrations of phosphorus (P) and of nitrogen (N) themselves.

Long term, intensive sampling programmes are essential if nutrient budgets, trophic status and chemical and biological interactions are to be fully understood (Bailey-Watts *et al.*, 1987; Bailey-Watts *et al.*, 1990). However, pressures on the aquatic environment are such that planners, conservationists and managers need more rapid assessments. The Institute of Freshwater Ecology has applied such an approach to the Loch Lomond system. It involves desk analysis of land use and human occupancy of the catchment (see e.g. Harper, 1992) and the incorporation of research results on nutrient inputs and the impacts on plankton dynamics into existing eutrophication models (Bailey-Watts & Kirika, 1993). The 3 main elements of the work are: catchment pressures, loch sensitivity factors and loch responses. These can be represented by the following relationship:

$$\text{catchment pressures} \times \text{loch sensitivity factors} = \text{loch responses}$$

Catchment pressures by way of nutrient burdens are expressed as P loadings, estimated by combining map information on land use and human population, with P loss coefficients appropriate to, for example, areas of arable land, forest, improved pasture, sewage treatment works and septic tanks.

The predicted losses of P from the Loch Lomond catchment, increase from north to south. This reflects the general increase in the proportion of agricultural land towards the south, but also the much higher population in the Endrick sub-catchment.

Loch sensitivity factors determine the responses of a waterbody to its nutrient income, for example, the numbers of algal cells produced per unit of nutrient supply. Basic physical features such as area, depth and volume are very important as are other factors such as flushing rate and thermal stratification. The contrasting morphometry of the two main basins in Loch Lomond (broad and shallow in the south, narrow and deep in the north) must also be taken into account when assessing eutrophication impacts.

The loch responses are the outcome of the interactions between catchment pressures and the physical and chemical characteristics of the loch. Nutrient and chlorophyll levels and the types of algae present are the responses of major concern. Phytoplankton in the shallow southern basin are likely to respond the more rapidly and more in line with eutrophication trends because, (a) it is the more favourable environment for photosynthesis and (b) since the summer (stratified) flushing rates of the two basins are similar, the larger P loading will give higher P concentrations.

The results of the assessment of the north and south basins of Loch Lomond clearly show the south basin to be more productive (Table 1). However, in practice, the strength of wind-driven mixing may at times reduce the differences in phytoplankton abundance.

Comparison with recently published data (Traill, 1991) indicates that this desk-based method is accurate

Table 1. Comparison of predicted P loadings, and, P and chlorophyll *a* concentrations in the north and south basins of Loch Lomond

North Basin	Predicted responses	South Basin
0.18	* Specific areal loading of P (g P $m^{-2}y^{-1}$)	0.15
4.90	** Annual mean P concentration - corrected for flushing rate (μg tot. P l^{-1})	12.2
<1.0	*** Annual mean chlorophyll *a* concentration (μg l^{-1})	1.0–2.0?

* The predicted annual loads divided by the surface areas of the basins.
** The predicted annual loads divided by the flushing volumes of the basins (i.e. the product of the physical volumes and flushing rates).
*** These values assume (i) that only a third at most of the total P is likely to be in the form of biologically available soluble reactive P and (ii) that losses incurred by phytoplankton populations due to sedimentation and zooplankton grazing, render a large proportion of the algae produced, 'unseen'.

in assessing the trophic status of both basins of Loch Lomond although Traill did record total P values 2–3 times higher than predicted for the north basin.

This paper was written in consultation with the following: G. Best, K. Johnson and I. Trail of the Clyde River Purification Board; R. Tippett of Glasgow University; A. Kirika and L. May of the Institute of Freshwater Ecology.

References and Literature Consulted

Bailey-Watts, A. E. & P. Duncan, 1981a. The ecology of Scotland's largest lochs: Lomond, Awe, Ness, Morar and Shiel. 3. Chemical characterisation – a one-year comparative study. Monographiae biol. 44: 67–89.

Bailey-Watts, A. E. & P. Duncan, 1981b. The ecology of Scotland's largest lochs: Lomond, Awe, Ness, Morar and Shiel. 4. The phytoplankton. Monographiae biol. 44: 91–118.

Bailey-Watts, A. E. & A. Kirika, 1993. Phytoplankton and controlling factors in the rapidly flushed, upland Loch Dee (Galloway, Scotland). Proceedings of the Loch Dee Symposium: Foundation for Water Research, Medmenham 83–96.

Bailey-Watts, A. E., A. Kirika, L. May & D. H. Jones, 1990. Changes in phytoplankton over various time scales in a shallow eutrophic loch: the Loch Leven experience with special reference to the influence of flushing rate. Freshwat. Biol. 23: 85–111.

Bailey-Watts, A. E., R. Sargent, A. Kirika M. Smith, 1987. Loch Leven phosphorus loading. Report to the Department of Agriculture and Fisheries for Scotland, the Nature Conservancy Council, Scottish Development Department and Tayside Regional Council. Institute of Terrestrial Ecology, Edinburgh.

Harper, D. M., 1992. Eutrophication of Freshwaters: principles, problems and restoration. London, Chapman and Hall.

Maitland, P. S., 1981. The ecology of Scotland's largest lochs: Lomond, Awe, Ness, Morar and Shiel. 1. Introduction and catchment analysis. Monographiae biol. 44: 1–27.

Maitland, P. S., I. R. Smith, A. E. Bailey–Watts, B. D. Smith & A. A. Lyle, 1981. The ecology of Scotland's largest lochs: Lomond, Awe, Ness, Morar, Shiel. 10. Comparisons and synthesis. Monographiae biol. 44: 253–282.

Maulood, B. K. & A. D. Boney, 1980. A seasonal and ecological study of the phytoplankton of Loch Lomond. Hydrobiologia 71: 239–259.

Maulood, B. K., G. C. F. Hinton & A. D. Boney, 1978. Diurnal variation of phytoplankton in Loch Lomond. Hydrobiologia 58: 99–117.

Murray, J. & L. Pullar, 1910. Bathymetrical Survey of The Freshwater Lochs of Scotland. Edinburgh: Challenger. Vols. 1–6.

Smith, I. R., A. A. Lyle & A. J. Rosie, 1981. The ecology of Scotland's largest lochs: Lomond, Awe, Ness, Morar, Shiel. 2. Comparative physical limnology. Monographiae biol. 44: 29–65.

Tippett, R., 1988. The Ecology of The Phytoplankton of the North and South Basins of Loch Lomond. Report to Central Scotland Water Development Board.

Traill, I., 1991. Loch Lomond: A Eutrophication Study 1987–1989. Clyde River Purification Board. Technical Report 96.

Hydrobiologia **290**: 53–61, 1994.
K. J. Murphy, M. C. M. Beveridge & R. Tippett (eds), The Ecology of Loch Lomond.

Vegetation and land use in the Loch Lomond catchment

Gordon Dickinson
Department of Geography and Topographic Science, University of Glasgow, UK

Abstract

The Loch Lomond catchment lies across the boundary between the Lowlands and Highlands of Scotland, and its vegetation and land-use are influenced by the environmental contrasts which occur on either side of the Highland Boundary Fault. In the northern part of the catchment area elevations exceed 800 metres, and thus there is an altitudinal zonation component to vegetation and land-use patterns.

The role of man has been profound in modifying natural vegetation, and recent land-utilisation is reviewed. Changing human and environmental circumstances have a significant effect on vegetation and land-use, and these factors are evaluated. Though natural and semi-natural vegetation occupies only a small fraction of the catchment, its conservation importance is considerable. In terms of scenic and heritage resources, the whole area has international significance and the Loch Lomond catchment requires special protection and management.

Introduction

The Loch Lomond catchment basin includes some of the most attractive, easily accessible scenery in Western Europe. A major element in this scenic resource is the vegetation of the area. This vegetation cover, the majority of which to a greater or lesser extent has been modified by human activities, has considerable scientific interest in its own right. This is related to the location of the catchment which lies across the most profound environmental and cultural divide in Scotland, the Highland Boundary fault. Consequently a wide range of types of vegetation and land use activities are found within the catchment basin. The objects of this paper are to summarise the patterns of vegetation and land use in the basin, to analyse the main factors which influence contemporary vegetation patterns and land use, and to discuss some of the major problems related to management of vegetation and to land use.

The environmental setting

All the major elements of the physical environment exhibit a very considerable range of variation over the catchment. The Highland Boundary Fault (Fig. 1) marks a fundamental divide which has climatological, geological, physiographic and biogeographical dimensions. Jones (1991) states that the loch is divided into northern and southern components across the middle part of the basin, by the boundary between the Koppen Cfs and Cfc types of climate. Birse (1971) showed that much of the southern part of the catchment was in the H2B2 (Very humid southern boreal) bioclimatic sub-region, whilst the northern and higher areas fell into the PB3 (Perhumid hemiboreal) sub-region. The character of the change across this boundary can be summarised by a significant increase in precipitation, decrease in accumulated and average temperatures and an increase in exposure in moving from south to north, and from lower to higher ground. These climatic changes have an important effect on vegetation patterns. This is accentuated by similar variations in other environmental attributes.

The Highland Boundary fault marks a profound contrast in solid geology. More significantly from the ecological perspective it delimits a fundamental boundary in geomorphological conditions. The hard metamorphic rocks to the north of the fault zone form high rugged areas, reaching 974 m at the summit of Ben Lomond on the east side of the Loch, and 943 m at the summit of Ben Vorlich on the west side. Glacial erosion has been the dominant process in the past million years leaving an area characterised by steep slopes, immature drainage and thin soils. To the south the solid geology is predominantly sediments of Devonian age. Here lower ground predominates. Rocks include

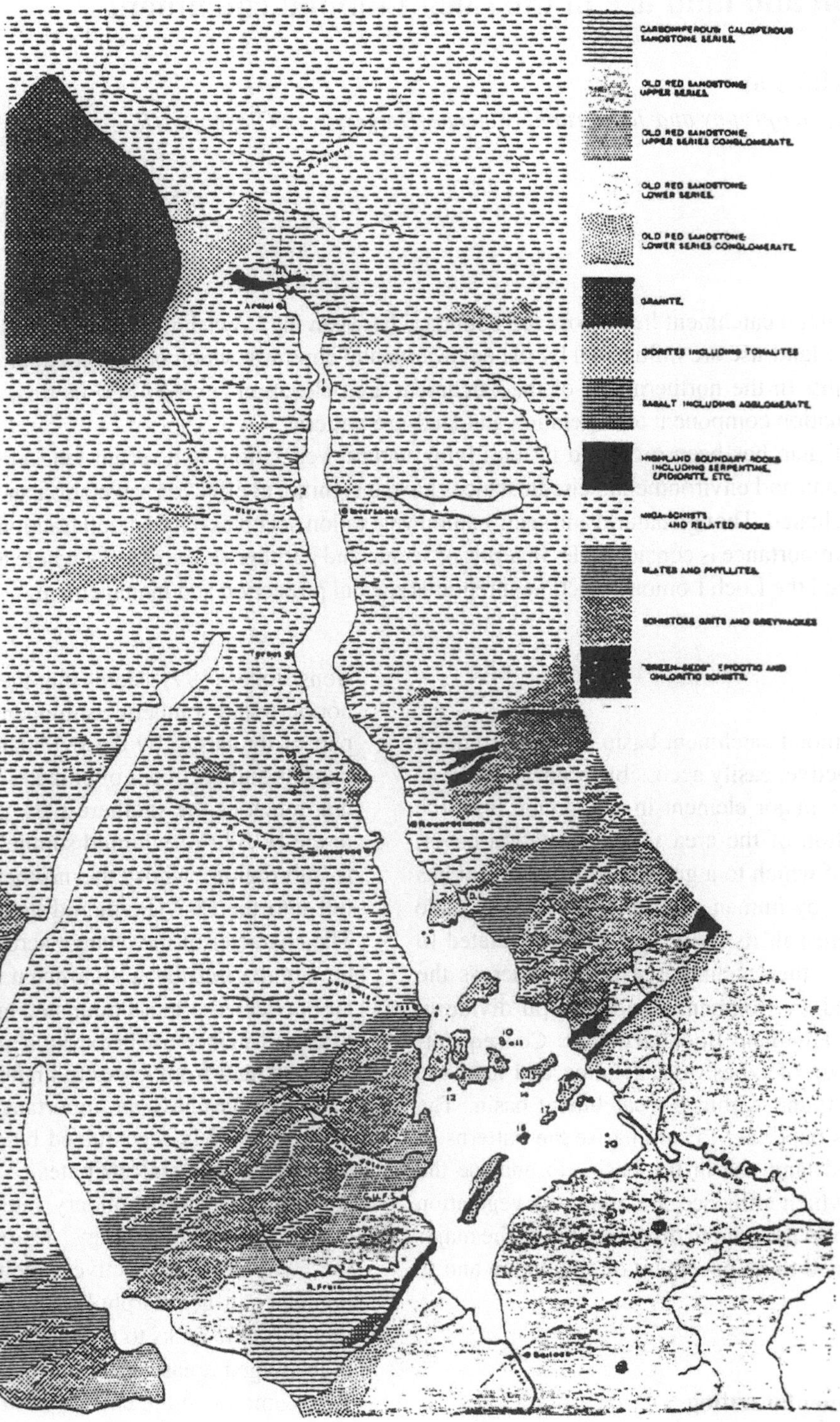

Fig. 1. Geology of the Loch Lomond Basin. Source: Slack, H. D. (1957) *Studies on Loch Lomond I*, Blackie, London, pp 18–19. Reproduced by permission of the University of Glasgow.

a wide range of conglomerates and sandstones, and are often overlain by glacial and fluvio-glacial deposits. Local drainage conditions vary considerably according to the detail of relief and the characteristics of the superficial deposits. For example in the south-eastern section of the catchment there are extensive areas of outwash sands and gravels. These not only form a valuable mineral resource in their own right, but also produce freely drained soils which are a good base for agriculture in an area with significant annual water budget surpluses.

Soil conditions are adversely affected by the climate, which is cool and wet even by Scottish standards. Impeded drainage is common and there are very extensive areas of peaty soils. Acidic parent materials do little to offset the poor inherent quality of much of the soil, and virtually all of the land in the northern part of the catchment is given over to rough grazings. In the south where lower ground, a somewhat drier climate and a mantle of superficial deposits combine to give areas of better soils, there are significant areas of improved grassland. The range of soil types which occur in the Loch Lomond N.N.R. exemplifies profiles in lower ground locations for the basin as a whole. These are described by Gauld and Bell (1986) who recognise four major soil forming factors. These are accumulation of organic matter, weathering, leaching and gleying. On this basis 7 types of soils are classified, the key diagnostic element of which is that all are developed on specific types of parent material. These are: parent material derived from Lower Old Red Sandstone (L.O.R.S.), parent materials derived from Dalradian series rocks and L.O.R.S. sediments, parent material derived from Highland Schists, L.O.R.S. sediments and lavas, parent material derived from ultrabasic rock, parent material derived from L.O.R.S. conglomerate, alluvium and peat.

As far as land use capability is concerned severe restrictions apply very widely, with the only significant areas of cultivable land occurring in the south. In the north the land is predominantly in classes 5 and 6, with some class 7 on the higher ground. In the south most of the land is in classes 4 and 5. Without improvement much of the potentially better land in the area is poorly drained and rather acidic. A long history of improvements in the Endrick valley and a few other more favoured areas in the south has led to substantial gains in the productivity of the soils of these areas. Nevertheless soil conditions in the catchment as a whole impose considerable constraints on both the development of natural and semi-natural vegetation, and on the utilisation of these soils for agriculture, forestry and outdoor recreation.

The pattern of vegetation

The history of vegetation in this area is analysed by Dickson (1992). Virtually all of the catchment is covered by vegetation which to a greater or lesser extent has been modified by man. Indeed, over most of the area the degree of modification is such that human activities are the single most important factor in shaping the current pattern of vegetation. The present situation is the result of an evolution over millennia, during which the activities of man and changing environmental conditions have played a continuous role. The broad pattern of environmental influences have been discussed in the previous section and the impact of human activities is considered in the following sections dealing with land use.

McVean in Burnett (1964) derived a regional pattern of potential natural vegetation for Scotland. Given the range of variation in environmental conditions in the basin it is not surprising that, in terms of vegetation patterns, the Loch Lomond area lies athwart a major boundary zone. To the south of the Highland boundary fault lies the area of cultivated land, whilst to the north, the western part of the basin is in the predominantly oak forest with birch region, whilst the eastern area lies in the predominantly pine forest with birch and oak region. (ibid., p. 569). These are described as woodland tendencies and current woodland cover is largely human in its origin. The tendencies identified by McVean are related mainly to abiotic environmental conditions, particularly climate.

There is a marked altitudinal dimension to the pattern of vegetation. This operates at a more detailed level of scale than the tendencies described above. In the West-central Highland region delimited by McVean (ibid., p. 575), four major associations are recognised. These are well represented in the northern part of the catchment. The associations are *Calluna – Molinia* heath, *Molinia – Myrica* bog, *Molinia*, *Agrostis – Festuca* grassland, and sub-alpine *Nardus* and *Juncus squarrosus* grasslands. The latter association tends to dominate at the higher altitudes, typically above 600 m, where there are significant micro-scale niches, such as crags and corries, with distinctive stress-tolerant plant communities. At altitude *Nardus* is less associated with snow cover than in the east, and generally excessive soil moisture and severe exposure are dominant factors shaping the detail of the vegetation cover. Paton (1954) gives an account of the main botanical habitats of the area, largely by altitudinal arrangement, distinguishing a mountain flora of upper grasslands, crags and corries, a lochside flora, a flora of river valleys and finally the heaths and moorlands.

At lower levels utilisation of the vegetation cover intensifies. Regeneration of natural tree cover has been prevented by centuries of burning and grazing. In some parts of the catchment, particularly in the east, bracken (*Pteridium aquilinum*) infestation is a serious problem. This may be symptomatic of heavy grazing pressures, and in a number of locations the spread of the plant seems to be on the increase, but in general animal stocking rates are not particularly high. The current pressures on hill farming may lead to major changes in hill land management and with it the development of semi-natural tree cover in those locations in which the pastoral agricultural system withers. This is one of the most interesting issues for the future of vegetation in the area.

There are many important small scale habitats, with resultant distinctive vegetation cover. Two examples from the wide range of these small scale habitats which occur in the catchment illustrate this feature of the vegetation. The mainland core of the Loch Lomond N.N.R., the Endrick marshes, comprises an area of former water meadow, with a relatively rich flora. The abandonment of cropping for hay has led to encroachment by tree species, though this is now controlled by active conservation management. This area is an internationally significant habitat for wild-fowl, particularly for over-wintering It is thus is of particularly high conservation importance within the catchment as a whole. Raised bogs, such as that located at Blairbeich S.S.S.I. near Balloch, are of importance. Largely the creation of pastoral agriculture which arrested succession, the bog contains a wide range of plant species including several members of the moss genus *Sphagnum* which subtly indicate micro-scale variations in moisture conditions. Recent drainage and encroachment of tree species threaten this interesting habitat. Active management involving blockage of some of the drains, together with elimination of the colonising trees are appropriate conservation strategies currently under consideration.

The woodlands, which form such a distinctive and valued element in the landscapes of Loch Lomond also owe their existence to man. The history of woodlands and their management which is described by Cameron (1986) for the area of the Loch Lomond N.N.R., gives an understanding of the present extent of mixed oak woodlands on the eastern shores of the Loch. The recent coniferous plantations are discussed more fully later in this paper. When the area was settled by Neolithic man about 5000 B.P. there was an extensive tree cover around the loch shores which probably extended to about 400 m elevation. The woodland was mixed in species composition, comprising birch, alder, oak, elm and hazel, with pine in some areas. A long history of clearance as a consequence of agriculture followed, so that by the seventeenth century this woodland cover had been very greatly reduced.

From the early eighteenth century until about 1850 there was a great demand for oak bark, which led to the re-establishment of tree cover, and the origin of the present day woods. The main species planted was imported *Quercus robor* which subsequently hybridised with the native *Quercus petraea*. The woodland was managed in a coppice with standards system of between 21 and 24 years rotation. Bark was harvested and used in the chemical industry to produce tannin for the treatment of leather. In the early stages of the industrial revolution there was a large demand for heavy duty leather belts to drive machinery from power shafts. By the middle of the nineteenth century, however, more sophisticated machinery resulted in a collapse in the demand for leather, and thus the end of the bark industry. Though active management is now limited, there remains a rich legacy of oak woodland, with alder in the wetter areas close to the loch, together with a rich under-storey vegetation. This is one of the most important scenic resources in the area, and furthermore has considerable conservation significance due to the wide range of habitats which are provided by the broad-leaf woodland cover.

Table 1. Crops, grass and rough grazing in selected areas of Scotland (thousand hectares)

	Rough grazings	Grass	Wheat	Barley
Scotland	3552.7	1041.7	88.8	418.2
Strathclyde Region	674.9	229.9	0.5	28.3
Central Region	138.7	39.8	1.7	11.5

(Source: The Scottish Environment - Statistics, HMSO, 1987, p20)

Land use – The overall pattern

The general patterns of land use in the catchment are complex. Many specific activities are undertaken. These include agriculture and forestry, outdoor recreation (including tourist activities), wildlife and landscape conservation, field sports, water catchment and utilisation, hydro-electric power generation, mineral extraction, settlement and, in the south in the Vale of Leven, some industry. Not only is there a long list of individual land uses, but the spatial patterns are complex. In many cases individual tracts of land are used for several activities simultaneously. Multiple purpose land use is attractive because it allows optimum development of the natural resource base, but it requires very careful and integrated management if environmental problems are to be avoided. This discussion is focussed on the first four of the land uses cited above. These categories are at the core of the land use system, and the present land utilisation problems in the catchment.

Agriculture

Current land use patterns are well illustrated by data from the 1986 agricultural census, shown in Table 1.

The table shows that grass and rough grazing predominate. Though the Loch Lomond catchment forms only a small part of the two Regions, the pattern for the basin is very similar. Only in the southern part of the area is there any significant amount of arable land. The vast majority of this is given over to grass, which is harvested for conservation chiefly as silage. The only cereal production is in the middle part of the Endrick valley. For the purposes of analysis of contemporary agricultural activities, the area is thus divided into two regions; the low ground of the south and the hill and mountain country of the north.

In the south until very recently dairying has been dominant. This is changing as a result of reduced E.C. quotas, with a corresponding increase in lowland sheep. This has implications for the hill farm sector. The extension of grass conservation represents efforts to reduce costs in these more favoured lower ground areas. The implications of wildlife and landscape conservation are becoming more significant for agriculture in the area. The catchment, less the Sloy and middle to upper Endrick basins, is designated an Environmentally Sensitive Area (ESA). In the ESA grants may be awarded for the maintenance of faming systems sympathetic to conservation. As Selman (1988) has pointed out, this designation has been effective in promoting landscape conservation in Scotland. In the Loch Lomond area there has been an approximately 90% take up of the scheme, a further demonstration of its success in the Loch Lomond area. About 70% of the lower Endrick valley has S.S S.I. status, and thus intensification of agriculture is not possible. Future trends in the south are likely to be towards further extension of lowland sheep production, larger individual dairy units and a significant number of farmsteads being owned by commuters, the land sub-let and limited areas kept in paddocks for equestrian activities.

Hill sheep farming constitutes the backbone of the industry in the northern part of the area, and covers by far the largest extent in the basin. Here again there are significant changes taking place. It has been estimated that there has been a 10% decline in heather moorland over the past decade. Heather moorland does not flourish in the wetter west of Scotland, and the recent decline can be traced to problems in getting sufficient labour at an economic cost, to continue the muirburn management regime. As a consequence sheep have tended to concentrate on the limited areas of better ground on the hills, with consequent detrimental impact on these areas. Lack of low ground for overwintering, as lowland sheep production has grown in importance, has led to static or declining flocks of hill ewes, although lambing percentages have risen from 60–65% to 80–85% for sheep raised 'at home', *i.e.* on the hill.

Grants are now available for natural woodland regeneration. This scheme promotes regeneration by fencing, without further action. The maintenance of ancient semi-natural woodland is also supported. These uses are likely to become more significant as economic pressures on traditional hill farming increase. On the Montrose estates on the east side of the Loch, returns on sporting land repay about 75% of the costs incurred. Most of the shortfall is attributable to 'wardening' costs which are necessary given the level of recreational use in the area. Overpopulation of red deer is a problem throughout the catchment. A significant cull of hinds is needed to reduce the population to a sustainable level, and the Red Deer Commission supports a deer management group, to which most land owners on the east side of the Loch belong. Though 300 foxes are killed annually in the eastern catchment, land owners maintain that this makes no difference to their overall numbers, and foxes are regarded as a significant cause of losses of lambs.

Table 2. Species composition in East Loch Lomond Forestry Commission Areas

Species	Per cent of total productive area (*c* 1300ha)
Spruce	47
Larch	16
Pine	5
Other conifers	5
Oak	15
Other broadleaves	12

Source: Forestry Commission, personal communication.

Forestry

Contemporary trends in forestry in Scotland have been examined by Mather (1988). Yet again change is the order of the day. The evolution of broadleaf woodlands in the catchment has been examined briefly earlier in this paper. The pattern of species planting on the east side of the Loch by the Forestry Commission is shown in Table 2. Most of the conifers were planted between 1952 and 1970, with some of the older stands in the Balmaha area dating back to the period 1936 to 1942. In common with Scotland as a whole, the single most common species planted in the Loch Lomond area is Sitka spruce (*Picea sitchensis*).

On the east side of Loch Lomond, recreation and landscape conservation have increasingly high priorities in Forestry Commission activities. Current work is focussed on conservation projects in the recently designated Forest Nature Reserve, in accordance with a management plan agreed with Scottish Natural Heritage (SNH). There is also considerable maintenance work associated with the heavy recreational pressure that the western part of the Queen Elizabeth Forest Park receives. It is the intention of the Commission that more operational emphasis will be given only after the landscape plan is fully completed, in what is acknowledged by the Commission, to be a scenically sensitive area.

As has been discussed previously, forest management largely directed to amenity and landscape enhancement is likely to become an increasing element in the hill land in coming decades. The highly welcome indicative forest strategies prepared by the Regional Councils, will help to ensure that commercial forestry is harmoniously integrated into the overall rural development strategy of the Loch Lomond catchment. However it is likely that the scenic and conservation value of the forests of this beautiful area, will have priority in development and management work over strictly commercial forestry.

Outdoor recreation

Land utilisation for outdoor recreation, including tourism, is characterised by the wide range of specific activities undertaken and by changing activity patterns. These have been analysed by Dickinson (1988) for Scotland as a whole. The growth of outdoor recreation activity since the 1960's can be traced directly to growth in personal access to a motor car. Even on Loch Lomondside, which is less than 50 km from Glasgow, and served in its southernmost part by adequate public transport, more than 90% of visitors come by car. Access is restricted to a relatively limited area in which car parking and suitable sites for recreation are available. This concentration causes considerable problems for resource management. Such issues in the Loch Lomond area have been discussed more fully by Dickinson (1991).

A further problem for recreational management lies in the peakiness of activities with respect to time. About 75% of all visits are made on Sundays, with most of the rest occurring on Saturdays. This is compounded by weather, summer season and public holidays. The result is that peak use occurs on only a few days of the year, and average use levels are very much lower. However, averages in this context are meaningless, as management must cope with peak levels of demand. On both sides of the Loch use levels are close to maximum in terms of both ecological and recreational carrying capacities. The continuing major road improvements on the west side enable significantly larger volumes of traffic to reach the Loch, and in shorter journey times. However, whether this will result in greatly increased pressures on the popular recreational sites around Luss, Inverbeg, Tarbet and Ardlui remains to be seen. It is possible that much of the increased recreational traffic will go to destinations to the north of Loch Lomond. However in the longer term it seems highly likely that increased use of the west side will occur. Consideration should be given to the impact of road improvements on recreational activity, when any future developments are contemplated.

The situation on the east side illustrates this latter point. The narrow road north of Balmaha is the Lochside's prime recreational corridor along its entire length. Presently the limited vehicular capacity of this road helps to maintain a precarious balance between visitor use and carrying capacity. On a few peak days in the summer in recent years, the road has had to be closed to all but emergency vehicles as congestion has become unmanageable. At several of the most popular recreational sites along this part of the eastern banks of the Loch there is striking evidence of impact problems. Beaches have been severely eroded, parking areas stripped of vegetation and left with seriously compacted soils and the fragile shore-line tree cover of alder (*Alnus glutinosa*) undercut by bank erosion and scarred by picnic fires. However as is shown by Dickinson (1993) there is clear evidence that most of the erosion problem is the result of wave action at periods of abnormally high loch level, such as occurred during the winters of 1990 and 1993.

There are numerous other contentious issues relating to outdoor recreation in the catchment. The most serious include pressures on the angling resources of the water body, the use of the Loch by various types of pleasure craft and the provision of suitable types and amounts of amenities for visitors. Problem issues have been addressed through two policies to date. The first has been the adoption of a local structure plan, by the two Regions and two Districts with responsibility for the area. This is a plan 'for recreation, tourism and conservation'. (Loch Lomond Planning Group, 1982). The plan proposes general principles and strategy, which are translated into policies and which have then been used to define five main policy areas around the Loch. This is an effective and progressive plan, and has been a catalyst for further constructive appraisal and action in resource management and planning around the Loch.

The second action has been the long-awaited designation of the Loch Lomond Regional Park. This action reflects the landscape, wildlife and recreational qualities of the catchment, which previously had status as a National Park Direction Area (1948), and a National Scenic Area (1980). Though the Park Authority has done sterling work in the context of recreational management, it recognises itself that its remit is too narrow. (Pennington, 1991). As is indicated by the Loch Lomond plan of 1982, recreation is bound in with conservation and indeed with all of the major land uses of the area, in an interactive system. Effective management of the natural resource base requires a similarly integrated approach. At the time of writing, there is much appraisal of the area's needs. These include the proposals for designation as a national park or similar structure put forward by the Countryside Commission for Scotland (now a partner in the newly constituted body Scottish National Heritage) (C.C.S., 1990), and the review working party set up by the Secretary of State for Scotland. The outcome of the current debate is not yet clear, but further changes in the organisational status of the Loch Lomond region to cater for its conservation priority and recreational resource base seem highly likely.

Conservation

The theme of conservation has been manifest in many parts of this review. It is clear that the water, land and biological resources of the catchment area are of national and international importance, and that conservation in its widest sense should have priority in future land use policies. This is in part related to specific elements in the area's environment and biota, and in part to the range of habitats and environments which occur in a relatively small area. For example, the large population in the loch of powan (*Coregonus clupeoides*), and the Endrick Marshes with its wildfowl are of high scientific conservation value. These two examples also illustrate the type of problems facing conservation management in the area. In the first case, the fish species ruffe (*Gymnocephalus cernus*) was accidentally introduced to the Loch in the 1970's, most probably as a release of live bait being used by a angler. Ruffe numbers have increased exponentially so that they are now the commonest fish in the Loch. This is causing concern for the well-being of other fish and fish eating animals in Loch Lomond (Adams & Tippett, 1990). In the second case, active conservation management of the former water meadows which form a particularly important habitat in the Endrick Marshes, has proven to be demanding of time and money, though the results which have been achieved are highly encouraging.

Moreover, it should not be overlooked that conservation in Britain has two antecedents. One is scientific ecology, and the need to sustain fragile ecosystems. The other is appreciation of the aesthetics of landscape in the broadest sense of that term. In the Loch Lomond area, landscape conservation, for the enjoyment of present and future visitors also has a high priority. Therefore many of the problems relating to resource use, which have been discussed in preceding sections, are significant factors in protecting the landscape of the area. As the landscape is an important resource for recreational activities and tourism, its conservation can be justified on both economic and aesthetic criteria.

There are significant areas of land specifically dedicated to conservation management in the study area. These include the Loch Lomond National Nature Reserve, which comprises the Endrick marshes, the island of Inchcailloch and adjacent islets. Lying on the Highland boundary fault zone, there is a very wide range of ecological conditions represented within the reserve, as well as areas of geological and geomorphological significance. There are 19 S.S.S.I.s in the catchment, and the whole Loch itself qualifies as such, though has not been notified for practical reasons. Protected areas in the voluntary sector are well represented in the Loch Lomond basin. Examples of these include the R.S P.B. reserve at Inversnaid, located on a grade 1 S.S.S.I., and the heritage site of Ben Lomond under the control of the National Trust for Scotland. Much important scientific and practical conservation work is carried on at these locations and elsewhere in the Loch Lomond region.

Whilst these dedicated areas form the backbone of conservation in the area there is a need for a comprehensive overall approach, in which conservation, in its widest sense is integrated into all land use activities. As has been pointed out by Mowle (1989), this is the exception rather than the rule in Scotland. The adoption of effective land use policy for all activities is hindered by the absence of an integrated approach. The vast majority of landowners and farmers are deeply concerned about long-term resource conservation. However current economic pressures, stemming from a sectorial approach to land use policy may force them to give conservation management a lower priority than they would wish. The urban-dwelling majority, who constitute the bulk of the transient visitor population to the Loch provide much of the area's income, either directly through purchase of services or indirectly through taxation. They too wish Loch Lomond to be conserved for the long-term future. Prompt and decisive action on conservation, to integrate conservation management into all spheres of land use activity, is essential, if the aspirations of both local people and visitors are to be satisfied.

Conclusion

This review of the nature of vegetation and land use in the Loch Lomond catchment basin inevitably has been selective but has given an indication of the diversity and importance of the area as a whole. It is worth remembering that this unique area is located literally on the doorstep of one of the larger cities in western Europe, yet retains important biological and environmental characteristics. It also provides high quality recreational experiences for many thousands of visitors from the local area and beyond. To maintain the high conservation value of the area, whilst continuing to support its local populace, progress in land management policies required urgently. The arguments for National Park, or similar designation have been put forward by Dickinson (1991). This is the best way to conserve the landscape and biota, and to maintain the local economy. Experience elsewhere in Europe shows that these are not irreconcilable aims. The Loch Lomond catchment with its beautiful landscapes, rich diversity of environments and important wildlife deserves European priority in its management of land use.

Acknowledgments

I am grateful to His Grace, the Duke of Montrose for information on farming and estate management in the area. Mr J. P. McLean, District Forest Manager for the Forestry Commission kindly supplied data on forestry on the east side of the Loch. Over the years colleagues, in what is now Scottish Natural Heritage, at Balloch Castle and at Battleby, Perth, have been highly supportive. I should also like to thank all those others, too numerous to name individually, both Loch Lomondsiders and colleagues at Glasgow University who have directly and indirectly contributed to this study.

References

Adams, C. E & R. Tippett, 1990. Status of the fish community of Loch Lomond, Report to the Scottish Development Agency, Glasgow.

Birse, E. L., 1971. The Assessment of Climatic Conditions In Scotland 3. Bio-climatic Sub-regions, The Macaulay Institute for Soil Research, Craigiebuckler, Aberdeen.

Burnett, J. H., 1964. The Vegetation of Scotland, Oliver and Boyd, Edinburgh.

C. C. S., 1990. The Mountain Areas of Scotland: Conservation and Management, C.C.S., Battleby, Perth.

Cameron, J., 1986. Woodland History and Management, in Placido, C. (ed.), Loch Lomond N N R Reserve Record Part 1 (Preliminary), Nature Conservancy Council, Balloch Castle Country Park, Balloch, Dunbartonshire.

Dickinson, G., 1988. Countryside Recreation, in Selman, P. (ed.), Countryside Planning in Practice: the Scottish experience, S.U.P., Stirling.

Dickinson, G., 1991. National Parks – Scottish Need and Spanish Experience, Scottish Geographical Magazine, 107: 124–129.

Dickinson, G., 1991. Recreational Impacts, in Dickinson, G., Jones, G. and Pender, G. (eds.), Loch Lomond 1991 – Proceedings of the Symposium on Environmental Conditions in the Loch Lomond Basin, University of Glasgow and the University of Strathclyde, Glasgow.

Dickinson, G., 1993. Environmental Degradation in the Loch Lomond Area: A Case Study of the Roles of Human Impacts and Environmental Changes, in Dawson, A. H., Jones, H. R., Small, A. & Soulsby, J. A. (eds.), Scottish Geographical Studies, University of Dundee and the University of St Andrews, Dundee and St Andrews.

Dickson, J. H., 1992. Vegetation history of the catchment, (The 1992 limnology of Loch Lomond symposium paper).

Gauld, J. H. & J. S. Bell, 1986. The Soils of the Loch Lomond Nature Reserve, in Placido, C (ed.) Loch Lomond N N R Reserve Record Part 1 (Preliminary), Nature Conservancy Council, Balloch Castle Country Park, Balloch, Dunbartonshire.

Jones, G., 1991. Local Climate and Climatic Change, in Dickinson, G., Jones, G. and Pender, G. Loch Lomond 1991 – Proceedings of the Symposium on Environmental Conditions in the Loch Lomond Basin, The University of Glasgow and the University of Strathclyde, Glasgow.

Loch Lomond Planning Group, 1982. The Loch Lomond Local (subject) Plan for tourism, recreation and conservation, Loch Lomond Planning Group, c/o Department of Physical Planning, Strathclyde Regional Council, Glasgow.

Mather, A. S., 1988. Agriculture and Forestry, in Selman, P. (ed.), Countryside Planning in Practice: the Scottish experience S.U.P., Stirling.

Mowle, A., 1989. Conservation Problems and the Need for an Integrated Rural Land Use Policy, in Dickinson, G. (ed.), The Scottish Countryside in Conflict. The Private Use of a Public Resource, C.U.R.R. Discussion Paper No. 40, University of Glasgow.

Paton, D., 1954. The Botany of The Area, in Walton, J. (ed.), The Queen Elizabeth Forest Park Guide, H.M.S.O., Edinburgh.

Pennington, N. (1991), Shoreline Management in the Loch Lomond Regional Park, in Dickinson, G., Jones, G. and Pender, G., Loch Lomond 1991 – Proceedings of the Symposium on Environmental Conditions in the Loch Lomond Basin, Symposium Proceedings, The University of Glasgow and the University of Strathclyde, Glasgow.

Selman, P., 1988. Landscape and Wildlife, in Selman, P. (ed.), Countryside Planning in Practice: the Scottish experience, S.U.P., Stirling.

Hydrobiologia **290**: 63–74, 1994.
K. J. Murphy, M. C. M. Beveridge & R. Tippett (eds), The Ecology of Loch Lomond.

Freshwater and wetland plant communities of Loch Lomond

K. J. Murphy, K. D. Hudson & J. Mitchell[1]
Department of Botany, University of Glasgow, Glasgow G12 8QQ, UK
[1]*Scottish Natural Heritage, The Castle, Balloch Castle Country Park, Balloch, Dunbartonshire G83 8LX, UK*

Key words: aquatic vegetation, wetland vegetation, Loch Lomond, lake ecology

Abstract

Information on the aquatic and wetland vegetation of Loch Lomond, Scotland, is summarised for the period 1957–90. Aquatic macrophyte growth is estimated to occupy about 1% of the loch's total surface area, being limited to the 0–10 m euphotic zone, and probably excluded from much of this area by wave disturbance and unsuitable substrates. Aquatic vegetation is however abundant in sheltered bays and less-exposed shorelines, particularly in the South Basin of the loch. Although *Littorella uniflora* (L.) Aschers is the commonest, ubiquitous, submerged plant of the loch, three separate euhydrophyte communities have been identified within the loch. One of these is characterised by abundant stands of *Elodea canadensis* Michx., which appears to have invaded Loch Lomond some time between 1967–88, and has rapidly spread throughout the loch.

Introduction

The aim of this paper is to summarise the current state of knowledge of the aquatic macrophyte and wetland plant communities of Loch Lomond. Remarkably little attention has been paid to this aspect of the ecology of Britain's largest lake by botanists: there is information going back only some 35 years on the distribution and composition of the principal submerged and floating-leaved (euhydrophyte), emergent, and wetland plant communities of Loch Lomond and its surrounds. Data can be found in Slack (1957); Spence (1964); Idle (1978), Mitchell & Stirling (1980); and E. T. Idle's unpublished 1967 survey, summarised in Bailey-Watts & Duncan (1981).

In terms of macrophyte ecology, Loch Lomond has much in common with oligotrophic - mesotrophic Scandinavian lakes, such as Tyrifjord and Steinsfjord (Norway), which are dominated by *Myriophyllum alterniflorum* DC and isoetid communities, and in which invasion by *Elodea canadensis* may also have occurred (Murphy *et al.*, 1990; Rørslett, 1987).

Aquatic vegetation of Loch Lomond

Early studies

Slack (1957) seems to have been the first to describe the zonation of euhydrophytes in relation to substrate and depth in Loch Lomond: in Chapter 3 of 'Studies on Loch Lomond' there is a brief outline of plant zonation in Auchentullich Bay, near the University Field Station at Rowardennan (Fig. 1). The wave-lap zone, with coarse gravels and cobblestones, between the high and low water marks supported only epilithic diatoms (notably *Gomphonema*), filamentous algae and occasional clumps of the moss *Fontinalis antipyretica*. Between approximately the low water mark and the 1–2 m isobath zone, finer sand substrates supported a sward of *Littorella uniflora* (L.) Aschers and *Lobelia dortmanna* L., replaced by *Isoetes lacustris* L. at the downward edge of the isoetid zone. In deeper water, on fine silts and muds, *Myriophyllum alterniflorum* was dominant, together with charophytes (*Nitella opaca*). Sometimes also present in the 2–3 m depth zone were tall stands of *Potamogeton perfoliatus* L.. There was little or no macrophyte growth below 4 m. This zonation was modified on steep and/or exposed shorelines by the loss of all substrates except rock/coarse gravel

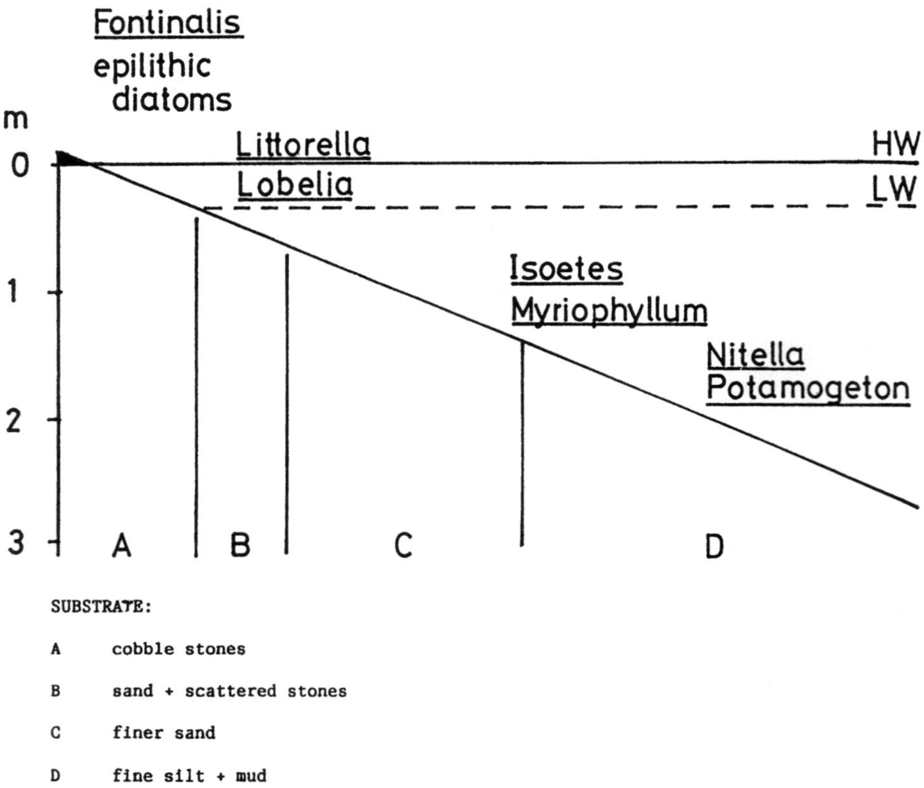

Fig. 1. Euhydrophyte and algal community distribution related to depth and substrate type: Auchentullich Bay, Loch Lomond (drawn from information given in Slack, 1957).

and stones, with only occasional clumps of *Myriophyllum alterniflorum.* At the other end of the exposure gradient, well-sheltered localities with 'fine, richly-organic muds' supported marginal zones of *Phragmites australis* (Cav.) Trin. ex Steudel, and *Scirpus lacustris* (L.) Palla, with stands in deeper water of *Nuphar lutea* (L.) Sm., *Potamogeton lucens* L. and *Potamogeton praelongus* Wulfen.

Some information can be gleaned from the chapter by Spence (1964), on macrophyte vegetation, in Burnett's 'The Vegetation of Scotland' (1964). Spence calculated that macrophyte colonisation of Loch Lomond probably occupied no more than 1% of the total loch area, within the area (perhaps up to 20% of the total surface area of water) which was potentially open to submerged plant growth (Spence stated that this equated approximately to the 10 m depth contour: see Fig. 2). This limited success of macrophyte colonisation is probably due to the prevalence of unsuitable substrates, especially rippled sand, which in turn is a function of the high exposure ratings of large areas of the loch. For example, in the extensive areas of shallow sandy hydrosoil in the south-west corner of the loch, below Balmaha, which are less than 9.15 m deep and therefore unlikely to be light-limited, Spence estimated submerged macrophyte cover at <1%. Using phytosociological terminology, Spence listed two submerged macrophyte community types for the loch: respectively an *Isoetes* society and a *Juncus fluitans* (=*J. bulbosus.* var. *fluitans*) – *Lobelia dortmanna* association. Two floating-leaved community types were recorded: a *Nuphar lutea* and a *Polygonum amphibium* L. (=*Persicaria amphibia* (L.) Gray) society. Swamp and fen community-types were:

Carex vesicaria L./*Carex vesicaria* – *Veronica scutellata* L.
Equisetum fluviatile L. – *Littorella uniflora*
Juncus articulatus L. – *Angelica* L.
Phalaris arundinacea L./*Phalaris* – *Filipendula ulmaria*
Scirpus lacustris – *Juncus bulbosus*

Spence had this to say about the sedge community of the loch: 'On the southeastern shore of L. Lomond, above the entry of the Endrick Water, there is some

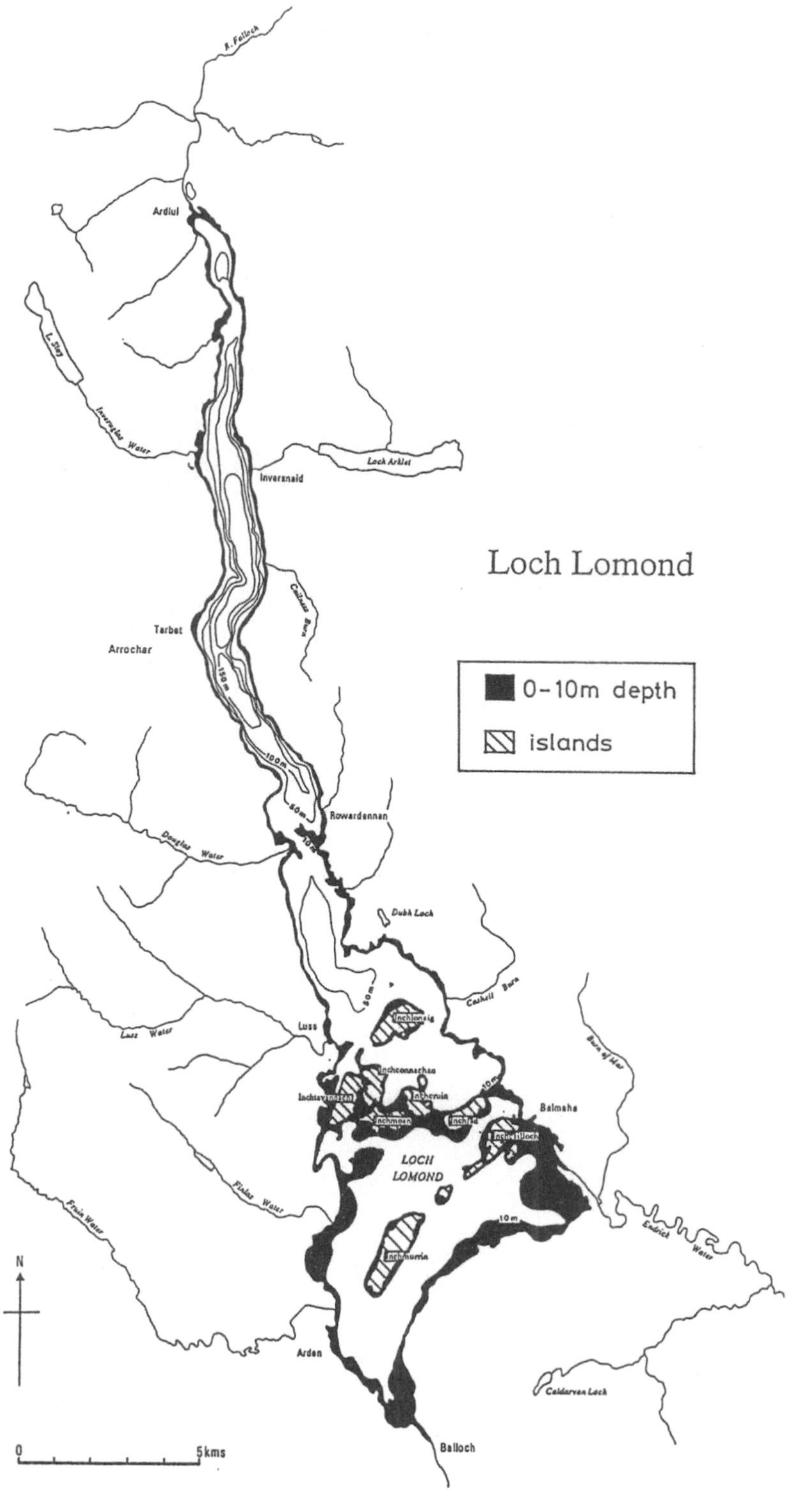

Fig. 2. Distribution of shallow water (approximate euphotic zone: 0–10 m) in Loch Lomond.

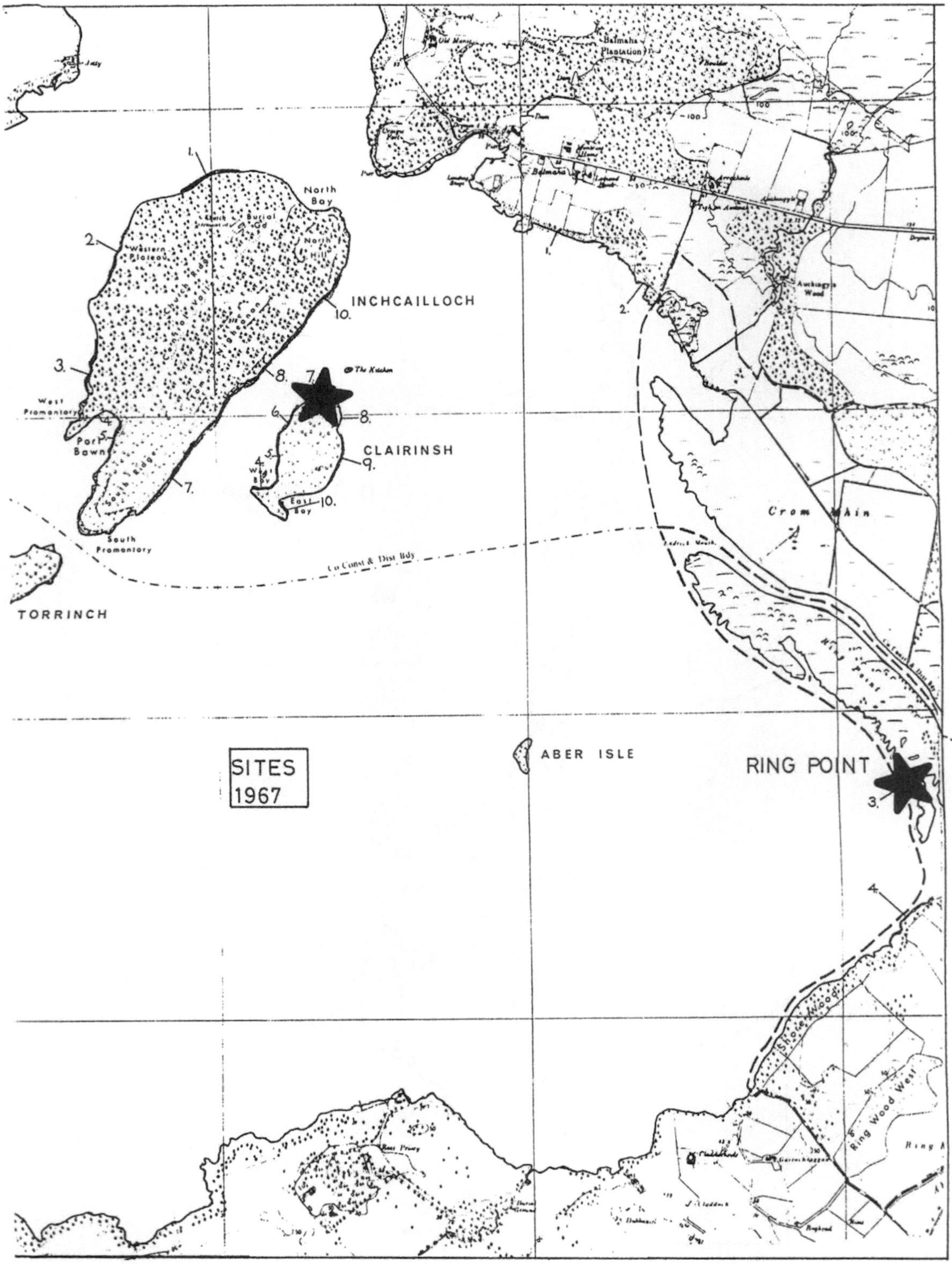

Fig. 3. Location of macrophyte survey transects in south-east corner of Loch Lomond studied by E. T. Idle in 1967. Transects depicted in Figs 4 and 5 are starred (*).

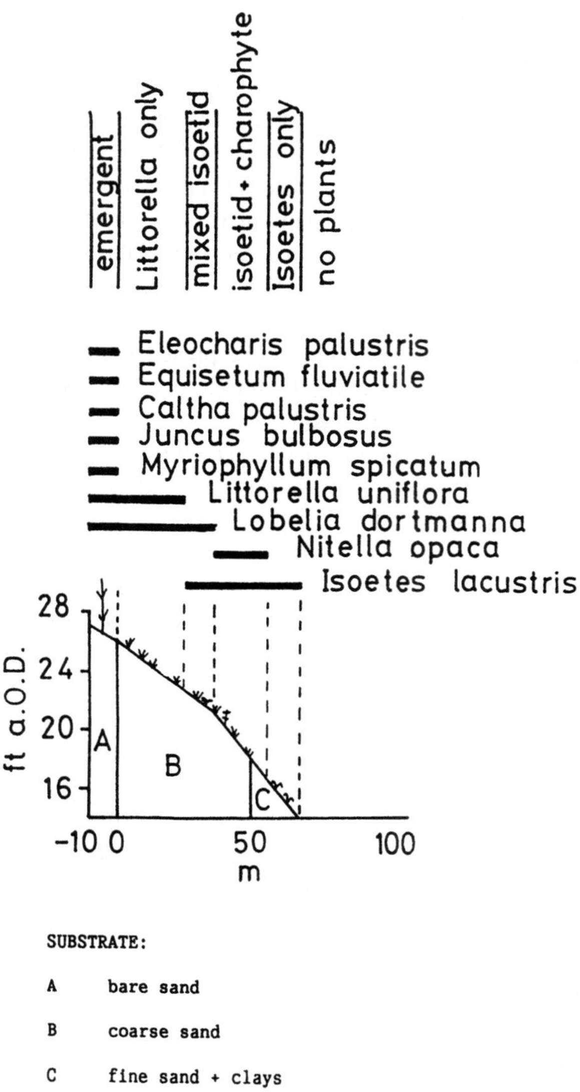

Fig. 4. Depth distribution of aquatic plants in relation to substrate on steep-sloping shore of Transect 7, Clairinsh, Loch Lomond (drawn from original unpublished data of E. T. Idle's 1967 survey).

development of a *Carex vesicaria* zone between open water and the grazed zone of *Agrostis canina*, *Juncus effusus* or *Juncus articulatus* communities. In bays the sedges give way to *Polygonum amphibium* or *Nuphar*'.

In 1967 E. T. Idle carried out a survey (summarised by Bailey-Watts & Duncan in Maitland, 1981) of a series of transects on shorelines within the National Nature Reserve area of the south-eastern corner of the loch. Each transect was located along a line stretched out to a buoy located 'beyond the photic limit', with species recorded at 5 m intervals along the transect (Fig. 3). *Littorella uniflora* and *Isoetes lacustris* were recorded at all sites around the islands of Clairinch and Inchcailloch, and at Crom Mhin and Ring Point on the mainland. *Nitella opaca* and *Myriophyllum alterniflorum* were also common at most sites. Two contrasting transect profiles, constructed from Idle's description of vegetation and substrate along 'Clairinsh 7' and 'Ring Point' transects are shown in Figs 4 and 5. Although the depth profiles are very different for the shallow transect there was relatively little difference in plant community composition or zonation pattern, at least in the submerged zone.

Aquatic vegetation of Loch Lomond in 1990

During 1990, we carried out a pilot survey of submerged macrophyte distribution in Loch Lomond, in relation to chemical and physical characteristics of the loch environment. In total 11 sites were sampled, ranging from Ardlui in the north, to Ring Point in the south basin of the loch (Table 1, Fig. 6). The small size of the sample base meant that only limited conclusions can be drawn, but the study nevertheless seems to be the largest of its kind yet carried out on he loch. Frequency of occurrence (%F) of submerged macrophytes was assessed by means a timed-run towed-grapnel method, from a boat, along transects located parallel to the shore in differing depths of water. In addition biomass samples were collected using an Ekman grab in deep water, and a Lambourn sampler in waderdepth water. We measured water chlorophyll $a+b$ content, dissolved oxygen, pH, temperature, downwelling light (PAR) attenuation and conductivity. Particle size analysis was undertaken for sediment samples from each location. Wave exposure at each site was calculated from fetch and prevailing wind direction data (using meteorological data from Rowardennan field station), following the method of Keddy (1982). We analyzed the data using TWINSPAN (Hill, 1979b) to produce a classification of macrophyte communities within the euhydrophyte zone of the loch (Fig. 7).

Within the euhydrophyte zone the evidence suggested that three main community types were present, with strong evidence for gradients of vegetation change related both to geographical location from south to north, and to exposure to wind and wave action. These three community types are clearly separated when the samples were ordinated using Detrended Correspondence Analysis (Hill, 1979a; Fig. 8). Following DCA ordination of the sites in terms of species present, we correlated site ordinate score on the first axis (which accounted for 62% of the variation) with environmental variables for the sites. There were significant neg-

Table 1. Locations of submerged macrophyte survey sites in Loch Lomond 1990

Site		Site code	National Grid reference
North Basin:	Ardlui (Marina)	AM01	NN 318156
	Ardlui (E shore)	AES02	NN 318157
Mid Basin:	Camas an Laosgainn	CAL11	NS 373956
	Sallochy Bay	SB10	NS 360968
	Inchcailloch (N)	IN03	NS 413907
	Luss Narrows	LN07	NS 373918
	Luss Bay I	LBI08	NS 362921
	Luss Bay II	LBII09	NS 359917
South Basin:	Inchcailloch (Marina)	INM04	NS 411908
	Crom Mhin	CM05	NS 424899
	Ring Point	RP06	NS 430891

ative correlations with exposure and conductivity, and a significant positive correlation with distance northwards. Disturbance due to wind and wave action (Tivy, 1980; Pender, 1991) seems likely to be an important potential regulator of macrophyte vegetation in Loch Lomond. No work has as yet been done to ascertain the possible effects of other sources of stress and disturbance potentially affecting the survival of aquatic macrophytes in Loch Lomond, despite the fact that there have been substantial changes in human influences on the loch in recent years (Hamilton, 1988). One such has been the major increase in powered pleasure boat traffic in Loch Lomond, especially within the southern basin (Adams *et al.*, 1993). Boats may be a cause of disturbance and damage to aquatic plant communities in shallow waters (Murphy & Eaton, 1983; Liddle & Scorgie, 1980), and their effects need to be monitored in those areas of the loch where usage is most intense.

Throughout the loch the commonest euhydrophyte species currently appears to be *Littorella uniflora*, which is ubiquitous (Marrs, 1994). In the south basin, during 1992–93, *Littorella* and other isoetid plants were observed to be covered by dense growths of epiphytic algae. Such growths appear to be a new phenomenon in Loch Lomond: Marrs (1994) found little evidence of epiphytic algal growth on *Littorella*, growing in the same area, up to 1991. The three main euhydrophyte communities identified by TWINSPAN analysis of the 1990 vegetation data from the loch were:

(i) a community indicated by *Elodea canadensis* Michx., together with *Myriophyllum alterniflorum* and *Isoetes lacustris*. This community is present in deeper waters throughout the loch, and is probably a fairly recent development, in response to *Elodea* invasion (Idle's 1967 survey of 22 sites within the south basinof the loch did not record *Elodea canadensis*), and perhaps increasing eutrophication. By 1991 *E. canadensis* had spread throughout Loch Lomond, from the South Basin to Ardlui in the extreme north end of the loch;
(ii) a community indicated by the presence of charophytes, especially *Nitella flexilis*, occurring in the Central and South Basins;
(iii) a diverse community in which *Callitriche hamulata* is prominent, occurring in sheltered bays in the North Basin.

The principal difference noted between the 1990 survey and previous work is the invasion of *Elodea canadensis*. This appears to have occurred sometime between Idle's survey in 1967 and 1988, when *Elodea* was one of the common species recorded by a Glasgow University field course based at Rowardennan. It is known that *Elodea canadensis* was present in a tributary of the loch, the Endrick Water from the 1970s, and this is a likely source of propagules for the plant's subsequent invasion of Loch Lomond.

Emergent and wetland vegetation

Some previous information is available from studies on the flora of the national Nature Reserve area of Loch Lomond (Idle, 1978; Mitchell & Stirling, 1980a; Eccles, 1989), but little work has been done to characterise the emergent and wetland vegetation of the remainder of the loch, other than studies carried out in relation to problems of bank erosion around the loch shore (Tivy, 1980).

Six major habitat types are occupied by emergent and wetland plants, between the low and high water levels of the loch:

Alluvial silt and mud flats.
Inundated for the best part of the year, this habitat is well represented around the mouth of the River Endrick (NS 425896) in the south-east corner of the loch. Being subject to a high degree of exposure to wind and wave action, the plant cover on this disturbed, unstable substrate comprises small low-growing plants, dominated

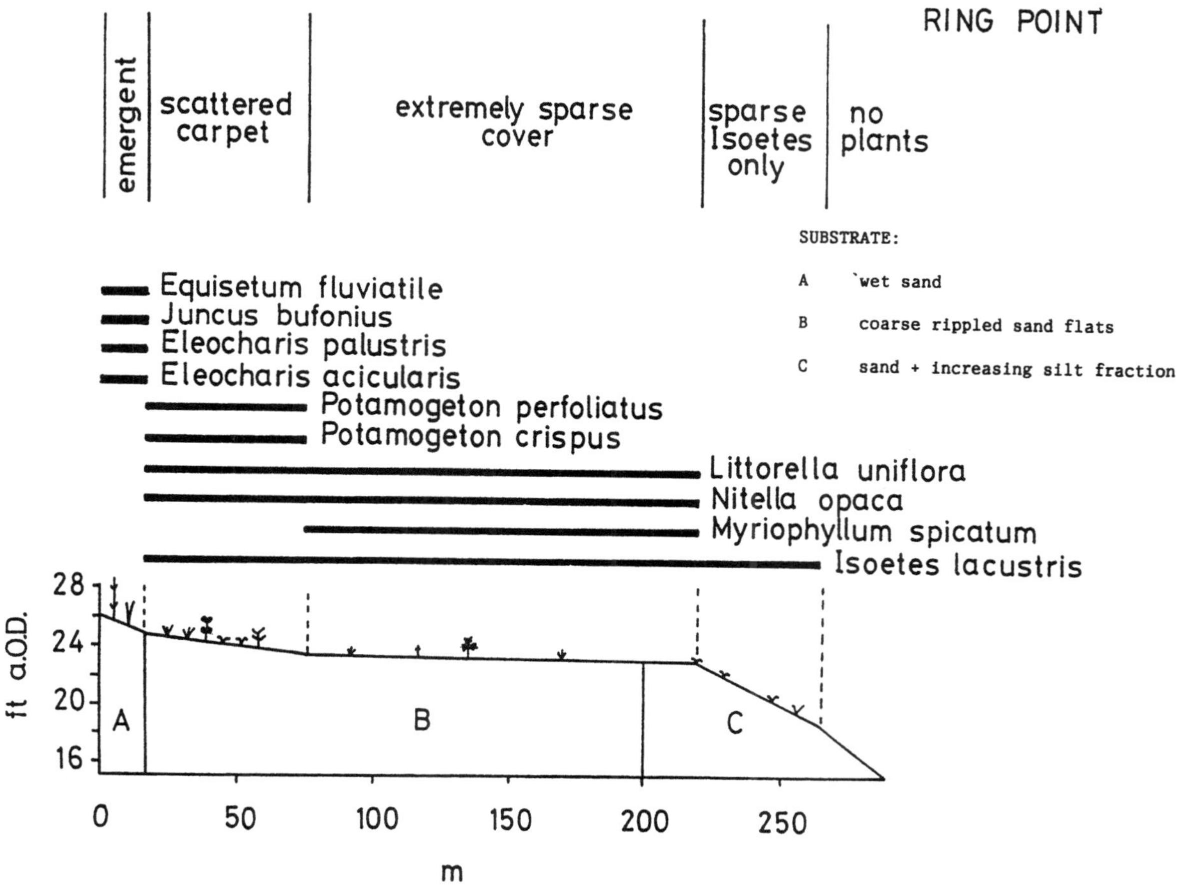

Fig. 5. Depth distribution of aquatic plants in relation to substrate on shallow-sloping shore of Transect 3, Ring Point Loch Lomond (drawn from original unpublished data of E. T. Idle's 1967 survey).

Table 2. Correlations between sample scores on DCA axis 1 and environmental and vegetation variables.

	EXP	BIO	CON	D(N)	P(L)	P(F)	OM	P(M)
DCA AXIS 1 SCORE	–	+	–	–	NS	NS	NS	NS
EXPOSURE		–	+	NS	+	NS	–	NS
BIOMASS			NS	NS	NS	NS	NS	NS
CONDUCTIVITY				–	NS	NS	NS	NS
DISTANCE (N)					NS	NS	NS	NS
PARTICLES (LARGE)						NS	–	–
PARTICLES (FINE)							+	NS
ORGANIC MATTER								NS

by amphibious forms of *Littorella uniflora*. The rest of the community is made up of small perennials such as *Callitriche stagnalis* Scop. and *Eleocharis acicularis* (L.) Roem. & Scultes, which are unable to compete successfully with larger aquatic macrophytes in more sheltered areas, and opportunist annuals capable of completing their life cycle during a brief period of exposure or beneath a shallow cover of water (*Subularia aquatica* L., *Lythrum portula* (L.) D. Webb, *Elatine*

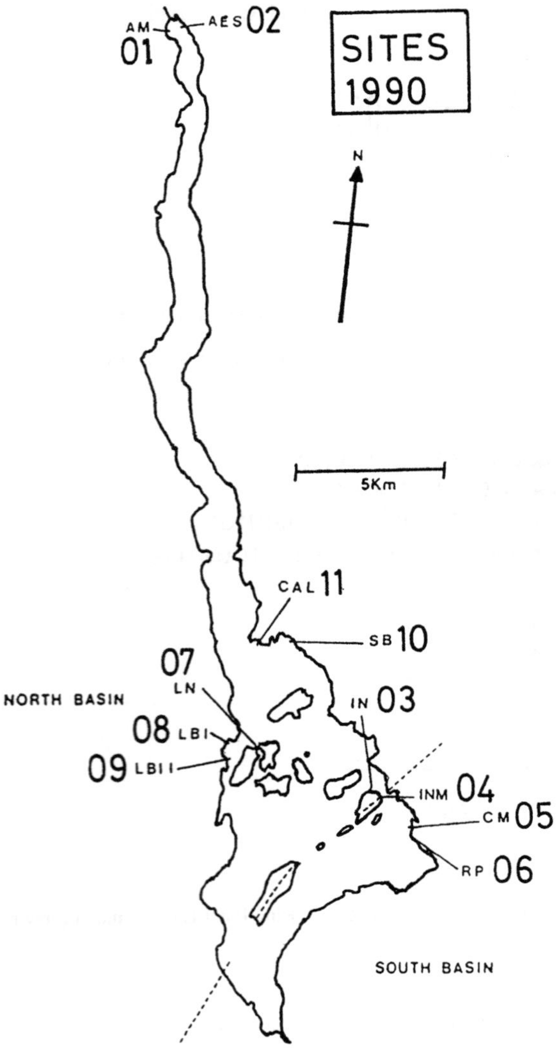

Fig. 6. Location of 1990 macrophyte survey sites in Loch Lomond (see Table 1 for site codes and National Grid references).

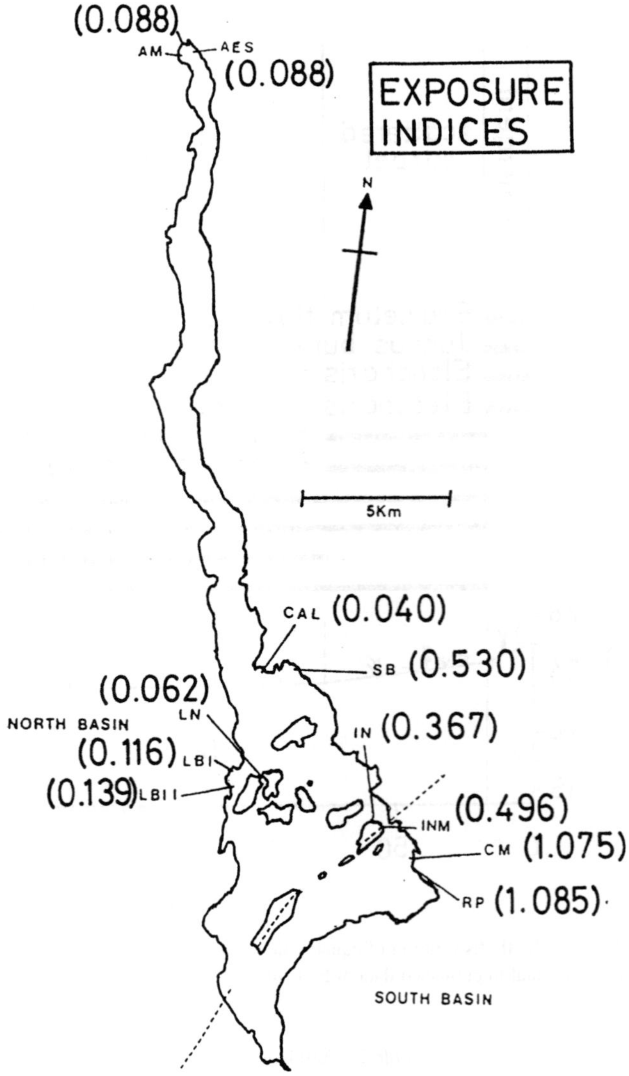

Fig. 7. Wave exposure indices calculated for 11 sites surveyed in 1990 in Loch Lomond (see also Fig. 6 and Table 1).

hexandra (Lapierre) DC., and the nationally-rare *Elatine hydropiper* L. (Idle *et al.*, 1970; Idle, 1979).

Given shelter from wave action, as is the case in Crom Mhin Bay (NS 426900), a dense surface cover of floating-leaved macrophytes may be present, including *Nuphar lutea* (L.) Smith and *Persicaria amphibia* (L.) Gray. Crom Mhin Bay is one of the few Scottish localities for *Juncus filiformis* L., which grows in open turf right by the waters-edge (Blackstock 1981).

Periodically-inundated boulder and gravel shores

This rather poor habitat for emergent plant growth is extensive around the shores of the loch. At the northern end the usually-steep gradient of the shore profile produces only a very narrow band of nutrient-poor, stoney substrate between the open water and adjoining woodland. The best development of the plant community typifying these shores is seen on the more-sheltered western shore, e.g. opposite Island I Vow (NN 326128). A tussocky flora, dominated by *Myrica gale* L., *Carex nigra* (L.) Reichard, and *Molinia caerulea* (L.) Moench is present, together with occasional few and far between populations of *Caltha palustris* L., *Ranunculus flammula* L., *Lysimachia vulgaris* L. and the less common *Carum verticillatum* (L.) Koch.

In comparison with the north basin shores, the stoney shores of the southern end of the loch are more gentle in gradient and more nutrient-rich. Again, exposure is important, with optimum protection from

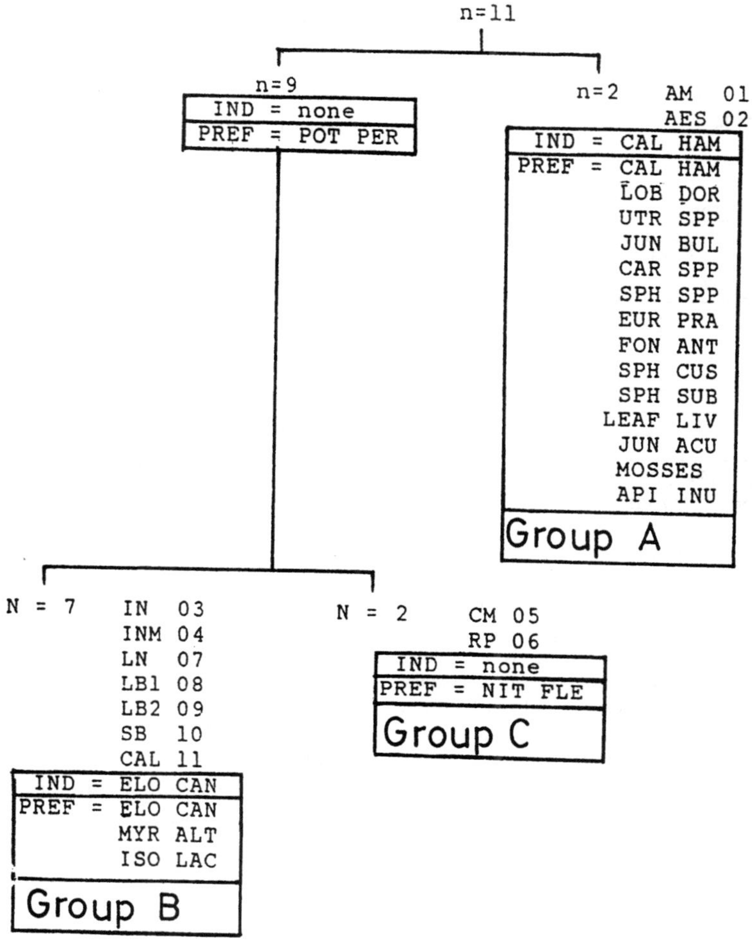

Fig. 8. TWINSPAN classification of Loch Lomond freshwater vegetation, for 11 sites surveyed in 1990. Ind = indicator species; Pref = preferential species for sample-groups. For site-codes see Table 1. Species codes: API INU *Apium inundatum*; CAL HAM *Callitriche hamulata*; CAR SPP *Carex* spp.; ELO CAN *Elodea canadensis*; EUR PRA *Eurhynchium praelongum* [= *Rynchostegium riparioides*]; FON ANT *Fontinalis antipyretica*; ISO LAC *Isoetes lacustris*; JUN ACU *Juncus articulatus*; JUN BUL *Juncus bulbosus*; LEAF LIV leafy liverworts; LOB DOR *Lobelia dortmanna*; MOSSES submerged moss species (other than *Sphagnum* spp., *Eurhynchium praelongum* [=*Rynchostegium riparioides*] or *Fontinalis antipyretica*); MYR ALT *Myriophyllum alterniflorum*; NIT FLE *Nitella flexilis*; POT PER *Potamogeton perfoliatus*; SPH CUS *Sphagnum cuspidatum*; SPH SUB *Sphagnum subsecundum*; SPH SPP *Sphagnum* spp.(other than *subsecundum* or *cuspidatum*); UTR SPP *Utricularia* spp.

wind and wave action occurring on lee shores of the loch's larger islands, such as Clairinsh (NS 413899). The enriched shore-zone community of Clairinsh supports a well-developed tall herb community, including *Trollius europaeus* L., *Oenanthe crocata* L., *Angelica sylvestris* L., *Persicaria bistorta* L., and occasional localised occurrences of *Aquilegia vulgaris* L. and *Listera ovata* (L.) R. Br. (Idle, 1979).

Low-lying valley bog

Now almost entirely confined to the lower valley of Glen Falloch as a result of 19th century land drainage elsewhere in the catchment, this habitat type is subject to occasional inundation from both river and loch. The best remaining remnant of lowlying valley bog occurs in the vicinity of Geal Loch (NN 318164), comprising *Molinia-Myrica-Sphagnum* mire influenced by localised flushing along its eastern edge from base-rich schists on the hill slopes above. Characteristic bog plants of this site include *Drosera longifolia* L., *Vaccinium oxycoccos* L., *Rhynchospora alba* (L.) Vahl. and *Carex pauciflora* Light, with submerged species including *Utricularia intermedia* Hayne and *Eleogiton fluitans* (L.) Link in pools and old drainage ditches.

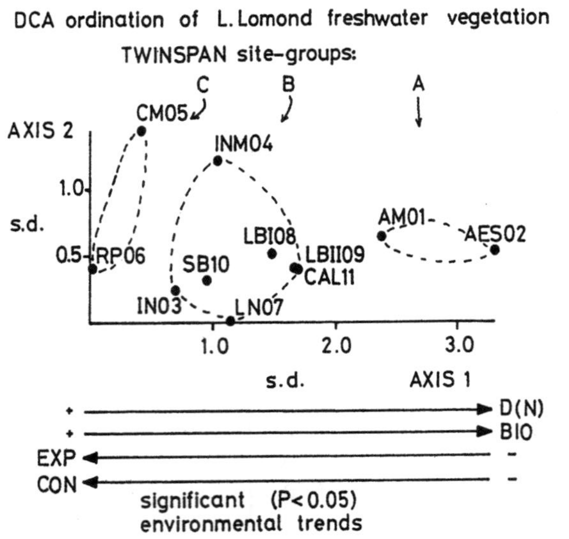

Fig. 9. Detrended Correspondence Analysis (DCA) of 11 sites from Loch Lomond, showing clear separation of three TWINSPAN sample-groups (see Fig. 8) along Axis 1. For site-codes see Table 1; axis units: standard deviations of species turnover. Significant ($P<0.05$; + = positive; − = negative correlation) correlative trends occur along Axis 1 between sample ordinate-score and environmental variables measured at individual sites: D(N) = distance (north) in km from Balloch (see Fig. 2); BIO = biomass of submerged plants (g m^{-2} dry weight); EXP = wave exposure index; CON = conductivity μS cm^{-1}). See Table 2 for full correlation matrix.

Sheltered hinterland waters

The Geal Loch is an 11 ha nutrient-poor natural waterbody near the northern end of the main loch, which has shown little change in appearance during this century on the evidence of old photographs. It supports a very sparse emergent fringe of *Menyanthes trifoliata* L., *Lobelia dortmanna*, *Carex rostrata* Stokes, *Carex lasiocarpa* Ehrh. and *Phragmites australis* (Cav.) Trin. ex Steudel, with some *Nymphaea alba* L. in deeper water.

In complete contrast to the typical highland lochan community of the Geal Loch is the nutrient-rich man-made waterbody known as Wards Ponds near the southern end of Loch Lomond (NS 444878). Formerly arable fields, which were turned over to rough grazing in the 1920's, the land was also deliberately flooded in winter, to form flight ponds for hunting wild duck. The ponds, which vary in size seasonally, are noteworthy for their extensive sedge beds, which comprise mainly *Carex rostrata*, *Carex vesicaria* L. and *Carex aquatilis* Wahlenb., with scattered tussocks of an uncommon hybrid, *Carex* × *hibernica* A. Bennett (*Carex aquatilis* × *Carex nigra*). The rate of encroachment of sedge-dominated vegetation into the open water of the ponds has noticeably accelerated in recent years, and at some time in the future this vegetation will probably extend over the entire site. The most likely cause of this increased growth of sedges is fertliser run-off from adjoining agricultural land.

Fen and fen meadow

Mainly in the lower flood plain of the River Endrick, these wetland communities, again *Carex*-dominated have been greatly altered by changing management over the years. The 27 ha Aber Bogs (NS 436876) in the lower flood plain of the R. Endrick is the largest example of fen vegetation within the loch catchment. Due to modification through past drainage and agricultural practice, followed by total abandonment in the 1950's, the fen communities appear at present to be in a transitional state. A number of communities can however be identified with reasonable certainty. These include a set of *Carex*-dominated communities, similar to those of Wards Ponds, and communities dominated by *Filipendula ulmaria* (L.) Maxim., *Juncus acutiflorus* Ehrh ex. Hoffm., *Juncus effusus* L., and *Phalaris arundinacea* L. Two species characteristic of northern fens are present: *Cicuta virosa* L. and *Lysimachia thyrsiflora* L. (Eccles, 1989). In an adjoining 8 ha fen meadow (NS 438881), where annual cutting was resumed in 1981, over 100 marsh and meadow plant species have been recorded to date (Mitchell, 1984).

Flood plain alluvial woodland

Alder and willow carr, with birch and oak on drier ground, are present in some areas of regularly-flooded woodland, notably the 35 ha Gartfairn Wood (NS 434894) in the lower Endrick flood plain. Without the check of regular cutting of fen vegetation, successional processes would ensure that alluvial woodland would rapidly replace the fen communities. *Leucojum aestivum* L., and *Carex elongata* L. are notable species of these regularly flooded low-lying woodlands (Stirling & Mitchell, 1978; Mitchell & Stirling, 1980). The British distribution of a third species, *Rumex aquaticus* is confined to this habitat-type in the southern half of Loch Lomondside (Idle, 1968).

Conclusions

1. Loch Lomond and its associated catchment waterbodies and wetlands constitute a habitat of national sig-

nificance for freshwater and wetland plants, supporting a wide range of plant communities located along uninterrupted gradients from oligotrophic to eutrophic conditions, and along a hydrosere running from deep water through to fen and alluvial woodland.

2. Much of the interest of the loch's plants in scientific and conservation terms is associated with the unique position of Loch Lomond as a major freshwater habitat straddling the highland - lowland boundary. The low intensity of anthropogenic disturbance and pollution up till now has contributed to the maintenance of the freshwater and wetland plant community diversity of the loch and its catchment.

3. There is an increasing threat to the aquatic plant habitats of Loch Lomond. If eutrophication increases (Best & Traill, 1994; in press), there is a possibility that increased growth of invading nuisance species adapted to richer nutrient conditions, such as *Elodea canadensis*, may outcompete and exclude those existing submerged species which are adapted for growth in an oligotrophic- mesotrophic waterbody. The advent of this plant has already had a substantial effect on the submerged flora of the loch (Marrs, 1994). The South Basin is particularly at risk not just from aggressive macrophyte invasion, but also from changes in algal growth. The fairly localised phytoplankton blooms which have so far been recorded (e.g. Habib, 1993) could, if current trends continue, become a threat to submerged macrophyte growth in the future. In the south basin, during 1992–93, *Littorella* and other isoetid plants were observed to be covered by dense growths of epiphytic algae. Such growths appear to be a new phenomenon in Loch Lomond: Marrs (1994) found little sign of epiphytic algal growth on *Littorella*, growing in the same area, up to 1991. Given that increases in epiphytic algae are a well-known warning sign of cultural eutrophication (Phillips *et al.*, 1978), this suggests that there is a need for further work to investigate the potential for eutrophication-related damage to the macrophytes of Loch Lomond. Increasing boat traffic (Adams *et al.*, 1992; Adams, 1993) also has the potential to cause localised damage to macrophyte beds, both from propeller and wash disturbance, and via increased turbidity caused by stirring up silt in the shallows (Murphy & Eaton, 1983).

4. In the loch's emergent zone, and in wetlands within the catchment, changing land use and growing visitor pressure, coupled with pollution (especially nutrient runoff) all have the capacity to cause habitat damage. We are fortunate that the most significant area of wetland (the Endrick Marshes), and parts of the loch littoral zone along the Endrick foreshore and around Inchcailloch and Clairinch all have the highest available level of protection, in the form of National Nature Reserve status. However the point has to be made that for the plants as much as for birds and other organisms, these parts of the catchment cannot be preserved in isolation from the rest of it. Some form of catchment-wide protection is vital if the current value of Loch Lomond's waters as a plant habitat is to be maintained for much longer.

References

Adams C. E., R. Tippett, S. Nunn & G. Archibald, 1992. The utilization of a large inland waterway (Loch Lomond, Scotland) by recreational craft. Scottish Geogr. Mag. 108: 113–118.

Adams, C., 1993. Environmentally-sensitive predictors of boat traffic loading on inland waterways. Leisure Studies 12: 71–79.

Bailey-Watts, A. E. & P. Duncan, 1981. A review of macrophyte studies. Ch. 5 in The ecology of Scotland's largest lochs: Lomond, Awe, Ness, Morar and Shiel (ed. P.S. Maitland), 119–134. Monogr. Biol. Vol. 44, W. Junk, The Hague.

Best, G. & I. Traill, 1994. The physico-chemical limnology of Loch Lomond. Hydrobiologia 290: 29–37.

Blackstock T. H., 1981. The distribution of *Juncus filiformis* L. in Britain. Watsonia 13: 209–214.

Burnett, J. H. (ed.), 1964. The vegetation of Scotland. Oliver & Boyd, Edinburgh.

Eccles, C. R. J., 1989. Vegetation survey of the Aber Bogs. May–June 1988. Unpublished report, Nature Conservancy Council, Balloch.

Habib, O. A., 1993. Environmental and seasonal influences on lake phytoplankton community structure. M.Sc. Thesis, University of Glasgow.

Hamilton, J. D., 1988. Recent human influences on the ecology of Loch Lomond, Scotland. Verh. Internat. Verein. Limnol. 23: 403–413.

Hill, M. O., 1979a. DECORANA – a FORTRAN program for Detrended Correspondence Analysis and Reciprocal Averaging. Cornell University, Ithaca, NY., 52 pp.

Hill, M. O., 1979b). TWINSPAN – a FORTRAN program for arranging multivariate data in an ordered two-way table by classification of the individuals and attributes. Cornell University, Ithaca, NY., 48 pp.

Idle, E. T., 1968. *Rumex aquaticus* at Loch Lomondside. Trans. Bot. Soc. Edinb. 60: 445–449.

Idle, E. T., 1978. The flora of the Loch Lomond National Nature Reserve. Glasg. Nat. 19: 403–421.

Idle, E. T., 1979. Ecological considerations relating to changes in shore-line. In: Shore Erosion around Loch Lomond (ed. T. Huxley), Countryside Commission for Scotland. Perth: 27–31.

Idle E. T., J. Mitchell & A. McG. Stirling, 1970. *Elatine hydropiper* L. – new to Scotland. Watsonia 8: 45–46.

Keddy, P. A., 1982. Quantifying within-lake gradients of wave energy: interrelationships of wave energy, substrate, particle size and shoreline plants in Axe Lake, Ontario. Aquat. Bot. 14: 41–58.

Liddle, M. J. & H. R. A. Scorgie, 1980. The effects of recreation on freshwater plants and animals: a review. Biol. Conserv. 17: 183–206.

Marrs, S., 1994. Influence of algae on *Littorella uniflora* in Scottish lochs Ph. D. Thesis, University of Glasgow.

Mitchell, J., 1984. A Scottish bog hay meadow. Scottish Wildlife 20: 15–17.

Mitchell, J. & A. McG. Stirling, 1980a. The flora of the Loch Lomond National Nature Reserve: a supplement. Glasg. Nat. 20: 77–81.

Mitchell, J. & A. McG Stirling, 1980b. *Carex elongata* in Scotland. Glasgow Naturalist 20: 77–81.

Murphy, K. J. & J. W. Eaton, 1983. Effects of pleasure-boat traffic on macrophyte growth in canals. J. appl. Ecol. 20: 713–729.

Pender, G., 1991. Wave generated shore erosion in Loch Lomond. In: Loch Lomond 1991: A symposium on environmental conditions in the Loch Lomond basin and their implications for resource management and development. University of Glasgow and University of Strathclyde: 31–38.

Phillips, G. L., D. Eminson & B. Moss, 1978. A mechanism to account for macrophyte decline in progressively eutrophicated freshwaters. Aquat. Bot. 4: 103–126.

Rørslett, B., 1987. Niche statistics of submerged macrophytes in Tyrifjord, a large oligotrophic Norwegian lake. Arch. Hydrobiol. 111: 283–308.

Slack, H. D., 1957. Studies on Loch Lomond. Blackie, Glasgow.

Spence, D. H. N., 1964. The macrophytic vegetation of freshwater lochs, swamps and associated fens. Ch. 9 in The Vegetation of Scotland (ed. J. H. Burnett), Edinburgh: 306–425.

Stirling, A. McG. & J. Mitchell, 1978. Summer snowflake. Glasgow Naturalist 19: 429–430.

Tivy, J., 1980. The effect of recreation on freshwater lochs and reservoirs in Scotland. Report to Countryside Commission for Scotland. University of Glasgow.

Hydrobiologia **290**: 75–90, 1994.
K. J. Murphy, M. C. M. Beveridge & R. Tippett (eds), The Ecology of Loch Lomond.

Zooplankton in Loch Lomond: perspectives, predation and powan

P. P. Pomeroy
Zoology Department, University of St. Andrews; Present address: NERC Sea Mammal Research Unit, High Cross, Madingley Road, Cambridge CB3 0ET, UK

Key words: Zooplankton, Loch Lomond, seasonal cycles, predation

Abstract

Published and unpublished accounts of the zooplankton community of Loch Lomond are reviewed. Loch Lomond's zooplankton community is unique amongst those of Scotland's other large lochs. *Holopedium gibberum* appeared in Loch Lomond in the years between the studies of Slack & Hamilton in the 1950s and zooplankton surveys in the late 1970s.

The annual cycles of *Eudiaptomus gracilis*, *Cyclops strenuus abyssorum* and *Mesocyclops leuckarti* are described for the mid basin of Loch Lomond. *E. gracilis* overwinter mainly as adults with variable egg production during their lifespan of 5–8 months. The cyclopoid copepods have variable life history characteristics. *C. str. abyssorum* produce a single generation per year, overwinter by means of resting eggs and a small population of planktonic adults, while *M. leuckarti* is present in the plankton throughout the year, but in low numbers during winter.

Zooplankton populations in Loch Lomond are predated by the diverse fish community which includes a facultative planktivore in the powan, *Coregonus lavaretus*. Powan feed heavily on zooplankton from late spring until late autumn. Powan display selectivity in the prey species taken, in the size distribution and morph types taken within prey species. Zooplankton species, morphs and forms in Loch Lomond tend to reflect this predation.

Introduction

Loch Lomond lies 30 km north-west of Glasgow with its long axis running north to south. Its maximum length is 36.4 km and with a surface area of 71.1 km^2 it is the largest expanse of freshwater in mainland Britain. The northern loch has a narrow steep sided basin, reaching a depth of 190 m, but south of Ross Point opens out to form a wider basin to 66 m depth. South of this middle basin the loch reaches a maximum breadth of 8.8 km and the depth rarely exceeds 20 m. There are numerous islands in the southern loch. Physical differences between the basins of Loch Lomond and their catchments are geological in origin (Slack, 1954). These differences are indicated by the water chemistry of the loch. The highland loch shows typical oligotrophic base-poor characteristics while the lower loch's water reflects a mainly lowland base-rich agricultural catchment. Geological and hydrological information are given by Slack (1957) and in Maitland (1981). Seasonal variations in the temperature regimes in the three basins of Loch Lomond have been described by Slack (1957). Stratification typically persists from May to December in the upper loch and from May to November in the middle basin, reaching 30 m depth in September. In the shallow southern basin calm weather in the summer months allows temporary stratification.

The extent of current knowledge on Loch Lomond's zooplankton community will be reviewed briefly here. I will compare the results of zooplankton surveys carried out by Slack & Hamilton between 1955 and 1957, Chapman (1965, 1969, 1972), Maitland *et al.* (1981) and Pomeroy (1987). Finally, the nature of the zooplankton community will be examined with special reference to predation by fish, particularly that of a facultative planktivore, the powan, *Coregonus lavaretus* (L.).

Table 1. Species found in zooplankton samples from Loch Lomond, 1982-1985, classified by limnological origin. Nomenclature of Cladocera after Scourfield & Harding (1966).

Pelagic	Littoral	Profundal/Other
Eudiaptomus gracilis (Sars)		
Cyclops strenuus abyssorum Sars	*Macrocyclops* spp.	*Acanthocyclops viridis* (Jurine)
Mesocyclops leuckarti (Claus)		
Bosmina coregoni var. *obtusirostris* (Sars)	*Sida crystallina* (O.F. Müller)	*Alona* spp.
Daphnia hyalina var. *galeata* Sars	*Diaphanosoma brachyurum* Liéven	*Chydorus* spp.
Holopedium gibberum Zaddach	*Ceriodaphnia quadrangula* O.F. Müller	
Bythotrephes longimanus (Leydig)	*Polyphemus pediculus* (L.)	
Leptodora kindti (Focke)		
Asplanchna spp.		*Chaoborus flavicans* (Meigen)
Notholca spp.		
		Chironomidae
		Hydracarina

Table 2. Comparison of clutch size and proportion of ovigerous *Eudiaptomus gracilis* in the plankton samples from Loch Lomond: data from Chapman (1969) and Pomeroy (1987). Bimodal clutch sizes occur where overwintering and spring females occur together.

Sample	Modal clutch size		Ovigerous females %	
date	Chapman	Pomeroy	Chapman	Pomeroy
Late February	6	6	21	11
Early March	6	6	20	12
Late March	6	7	31	18
Mid April	6	6	34	24
Early May	8	11	72	22
Late May	–	8,16	–	12
Early June	6,18	–	–	–
Mid June	4,14	–	24	–
Late July	8	10	35	12

The zooplankton community of Loch Lomond

A summary of the species which occurred in zooplankton samples between 1982 and 1984 is given in Table 1. Nomenclature of Cladocera follows that of Scourfield & Harding (1966).

The small calanoid copepod *Eudiaptomus gracilis* (Sars) is the most numerous species in the zooplankton for most of the year. *Cyclops strenuus abyssorum* Sars and *Mesocyclops leuckarti* (Claus) are common members of the plankton community particularly in late spring and summer. *Acanthocyclops viridis* (Jurine) is normally regarded as a benthic species but has been found in plankton samples in moderate numbers, particularly towards dusk.

Daphnia hyalina var. *galeata* Sars and *Bosmina coregoni* var. obtusirostris (Sars) are the principal phytoplankton-grazing cladocerans. Although it was absent from samples taken by Slack and Hamilton, *Holopedium gibberum* Zaddach has appeared in the loch within the last few decades. *Daphnia* and *Bosmina* are most abundant during the autumn and spring respectively. The predatory limnetic Cladocera comprise *Bythotrephes longimanus* (Leydig) and *Leptodora kindti* (Focke). Both have a limited season from spring to autumn and occur at low densities, disappearing from the plankton during winter. Other cladoceran species are found frequently but can be considered as vagrants from other habitats (see Table 1).

Rotifers are numerous during spring and summer in Loch Lomond. Predatory *Asplanchna* spp. are common, but the group as a whole has received little attention. Larvae of the phantom midge *Chaoborus flavicans* (Meigen) make regular vertical migrations in a nearby lochan (Goldspink & Scott, 1971) and are taken regularly in zooplankton hauls. Other insect larvae are found occasionally, particularly chironomid larvae. Although Hydracarina are taken in very low numbers, they occur regularly in plankton samples.

I. Zooplankton studies of Loch Lomond

Scott (1899) provided the first brief description of the zooplankton of Loch Lomond, and this was followed by the faunal lists of Murray & Pullar (1910), in their bathymetrical survey of the Scottish lochs. Murray & Pullar (1910) described the distribution of zooplankton species by the number of lochs in which they were found, thus species such as *E. gracilis* were described as having a general distribution, while *Bosmina lon-*

girostris was of local occurrence. Murray & Pullar (1910) recognised that many of the species taken by plankton nets were not truly planktonic, and that some of the species had clearly limited seasons in the plankton community.

The proximity of the loch to Glasgow made it an obvious location for field studies and in 1938 a field course in limnology was started for students of the Zoology Department of the University of Glasgow. When the University established a field station by the side of the loch in 1946 more extensive limnological studies became possible. Dr Harry Slack was instrumental in priming many of the studies on the loch and his interest and enthusiasm initiated much of the work described here. Unpublished data collected by Slack & Hamilton in the period 1955-7 are discussed below. The work of Chapman (1965, 1969, 1972) gives the most detailed information to date on the main copepod species of the loch. Chapman was particularly interested in the processes of production and loss within population cycles and her results are considered below in parallel with other relevant data. Loch Lomond was one of the subjects of a comprehensive ecological survey of Scotland's largest lochs carried out in the late 1970's (Maitland *et al.*, 1981). Pomeroy (1987) investigated the diet of powan and their predation on zooplankton.

Methods

Slack & Hamilton collected zooplankton samples using a Hardy plankton sampler towed at the surface, obtaining samples from all three basins. Their results from the mid basin are discussed here. Chapman used a series of vertical hauls from a Nansen plankton net, of 26 cm mouth diameter and mesh size 235 μm. Duplicate hauls were taken at 2–6 sample stations in the Strathcashel Basin within the mid basin of the loch.

Maitland *et al.*, (1981) and Pomeroy (1987) used exactly the same sampling gear and regime. A standard plankton net of 1 m length and 28 cm mouth diameter with a 250 μm mesh was used for all vertical hauls. A detachable phosphor-bronze filter (mesh size 100 μ m) was fixed at the weighted end of the net. Samples were obtained from the mid basin of Loch Lomond opposite Luss, in approximately 60 m depth. Samples were taken closer inshore towards Sallochy Bay occasionally, but never from less than 20 m depth. The net was lowered to the bottom (or 50 m depth) and recovered at a constant rate of approximately 1 m per second. The filter was removed after washing down the sides of the net and placed in a labelled container of 5% formaldehyde solution. Five vertical hauls were taken on each sample date.

Results

The following summary is based on Chapman (1965, 1969, 1972) unless otherwise indicated. A degree of caution should be exercised in comparing median counts from Maitland *et al.*'s (1981), Pomeroy's (1987) and Chapman's (1965, 1969, 1972) mean values for vertical hauls against single estimations using a different sampling method for Slack & Hamilton's data. *E. gracilis* overwinters in the plankton largely as adults with some older copepodites, though few of these animals survive until June. The maximum longevity of overwintering animals was 5–8 months. Egg bearing females appear at the end of February and the numbers of ovigerous females and their clutch sizes increase rapidly from this time. Clutch sizes reach a maximum in early June but decline thereafter (Table 2.). Up to 6 clutches were produced in the spring.

Young copepodites first occurred in plankton samples in April. These animals grew quickly, reaching a larger size than overwintering adults by late spring and early summer (Fig. 1). Longevity of these animals was approximately 3 months, although breeding occurred throughout this period. Clutch sizes of overwintering females were smaller than those of the faster growing spring females (Fig. 1). By mid June the growth of copepodites was reduced greatly. Stage III copepodites from mid June did not mature until October. These adults then formed the overwintering generation. Most of the eggs laid by summer adults did not reach maturity, therefore it is likely that most of the overwintering generation is derived from eggs laid by the preceding overwintering generation.

Pomeroy (1987) recorded low counts of *Eudiaptomus* in many samples but confirms the temporal trend of changes in clutch size over the spring, both in modal clutch size and in the range of clutch sizes being greatest before mid June (Table 2). However, the much lower occurrence of ovigerous females in samples recorded by Pomeroy (1987) is unexpected, and may be a sampling artifact.

The two limnetic cyclopoid copepods in Loch Lomond form an important part of the zooplankton community, particularly in the spring. However, their seasonal cycles are subject to some variation. Chapman's (1972) description of *C. str. abyssorum* states that the species was absent from the winter plankton.

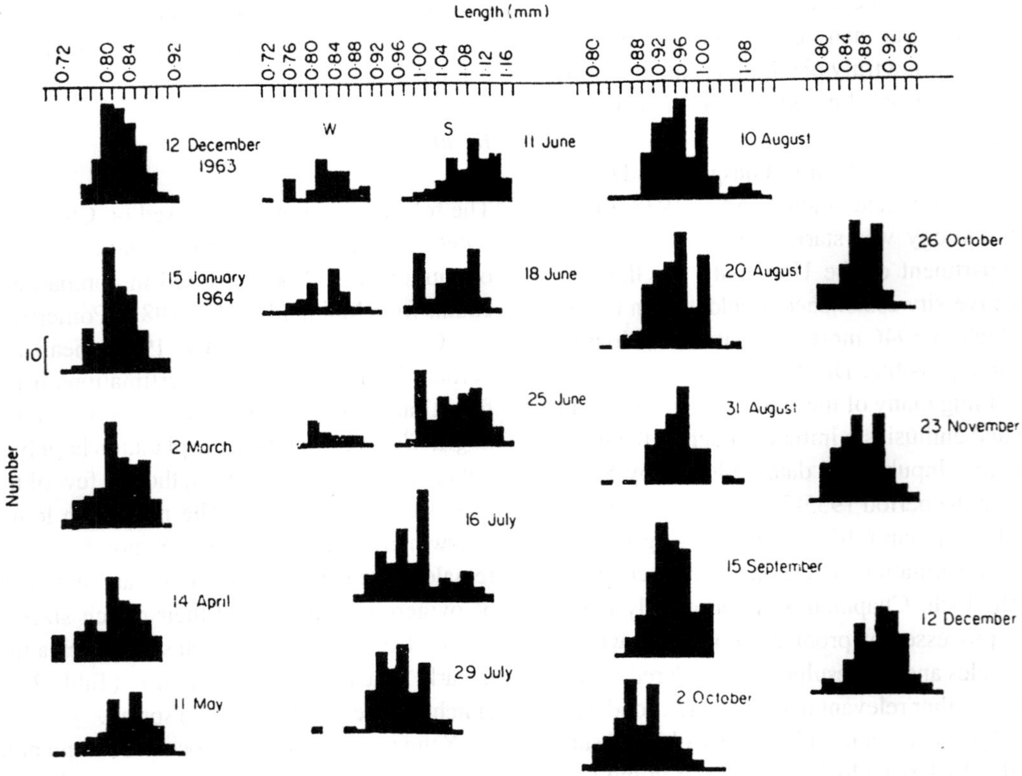

Fig. 1. Variations in metasomal lengths of adult female *Euiaptomus gracilis*. W, Overwintering animals; S, summer animals. (Reproduced after Chapman, 1969.)

Table 3. Seasonal pattern of powan electivities for zooplankton prey from pooled powan diet samples, Loch Lomond.

Date	Preferred prey species (by sample)
Oct 82	*Bythotrephes, Chaoborus*
Nov 82	*Bosmina, Cyclops, Bythotrephes*
Dec 82	*Chaoborus, Bosmina*
Feb 83	*Chaoborus*
Mar 83	*Chaoborus*
Apr 83	*Chaoborus*
May 83	*Chaoborus, Bosmina*
Jul 83	*Bythotrephes, Daphnia, Leptodora*
Aug 83	*Bythotrephes, Bosmina*
Oct 83	*Chaoborus*
Nov 83	*Daphnia, Cyclops, Chaoborus*
Jan 84	*Chaoborus*
May 84	*Bosmina, Holopedium, Leptodora*

C. str. abyssorum only appeared as very occasional unripe females in Chapman's (1972) samples from Strathcashel Basin. In mid April the first three copepodite stages appeared, but a month later few stage V's were present, indicating slow development. Adults appeared in June and subsequently copepodites formed a declining portion of the population until August when few remained. Males predominated in samples before mid August with females more numerous after August. A large proportion of the females were carrying eggs throughout the summer months. The greater part of these adults disappeared from the plankton by November. Overwintering by means of resting eggs produced during the summer breeding period was suggested by Chapman as the most likely method of sustaining the seasonal cycle. Chapman noted that previous records had not mentioned a winter absence from the plankton in Scottish lochs, and that a small planktonic population overwinters in the lakes of the English Lake District as stage V copepodites as well as a profundal population (Smyly, 1968). By contrast, *C. str. abyssorum* was recorded by Maitland *et al.* (1981) and Pomeroy (1987) throughout the winter in the plankton, although in low numbers (Fig. 2). Counts of *C. abyssorum* in

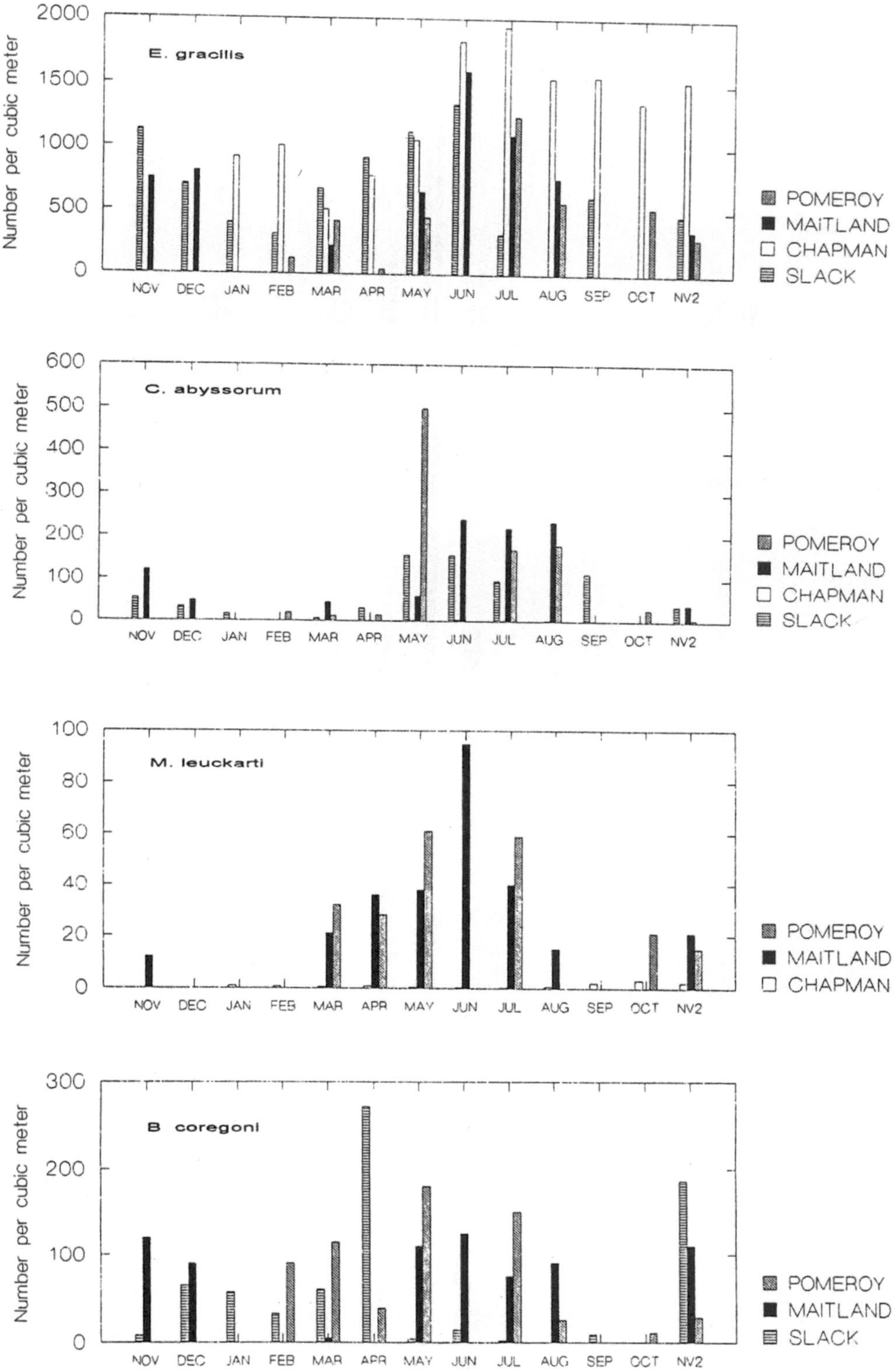

Fig. 2 i. Comparison of zooplankton counts obtained in four studes on Loch Lomond: Slack and Hamilton 1955–1957; Chapman (1965, 1969, 1972); Maitland *et al.* (1981) and Pomeroy (1987). Counts for Maitland *et al.* (1981) and Pomeroy (1987) are the medians of the range of five replicate vertical hauls. Counts for Chapman are for means of up to six replicates of duplicate vertical hauls. Counts for Slack and Hamilton are for a single Hardy sample taken at the surface. Note that samples were not available for every month in each data set.

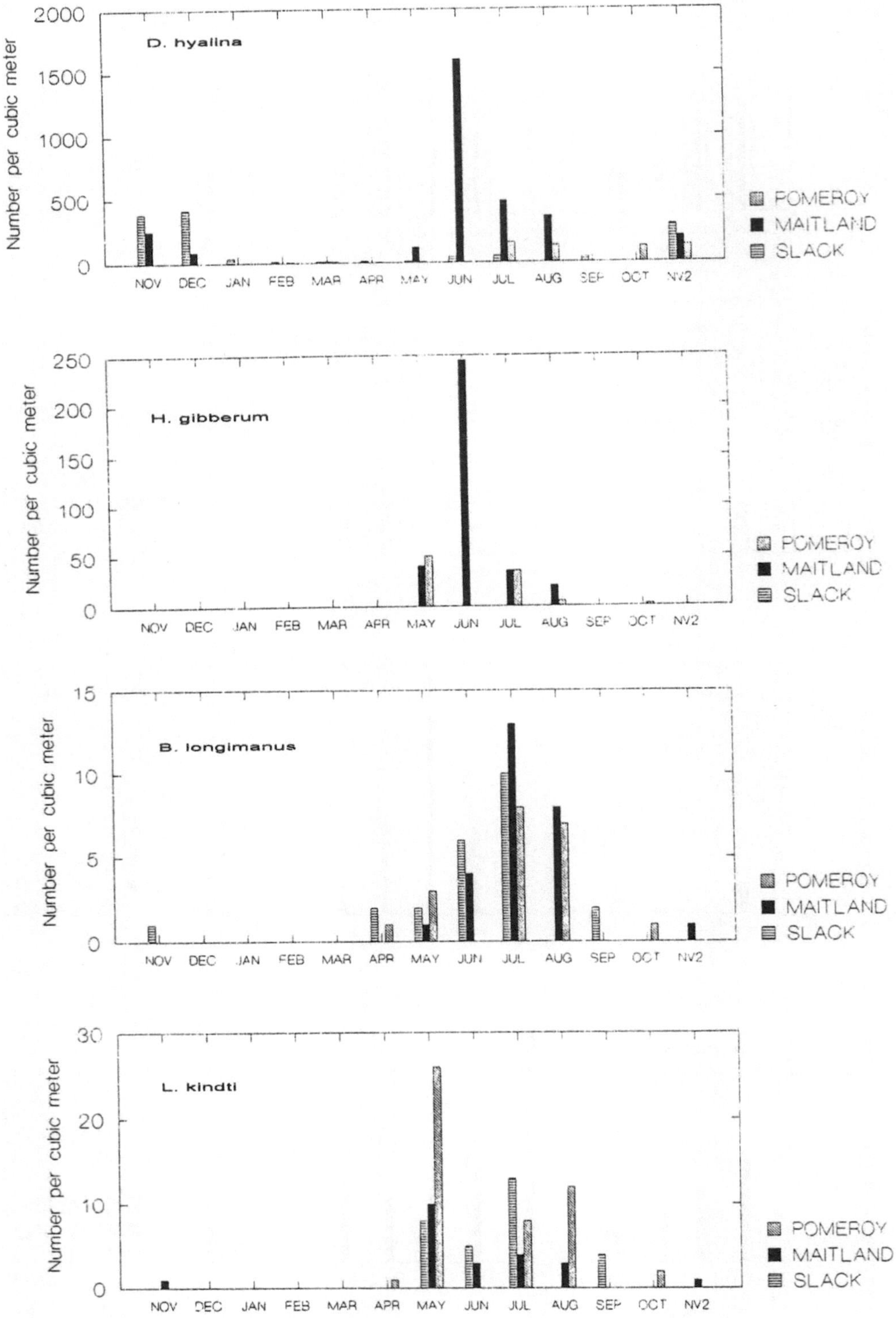
D. hyalina
Number per cubic meter
2000
1500
1000
500
0
NOV
DEC
JAN
FEB
MAR
APR
MAY
JUN
JUL
AUG
SEP
OCT
NV2
POMEROY
MAITLAND
SLACK
H. gibberum
250
200
150
100
50
B. longimanus
15
10
5
L. kindti
30
20
10

Fig. 2 ii.

the 1980's were among the highest recorded in spring (Pomeroy, 1987).

Of the five largest Scottish freshwater lochs studied by Maitland *et al.* (1981), only Loch Lomond contained *M. leuckarti* Chapman (1972) found that the limnetic winter population of *M. leuckarti* consisted of stage IV and V copepodites, with males predominating. The first adults appeared by mid April and egg bearing females were found soon after this. In early June the first three copepodite stages are present in the plankton and by the end of June and into July these groups were the most numerous in the population. By early August the adult population had increased, with more females than males. These adults produced eggs, but young copepodites did not appear until mid September. These copepodites did not develop beyond stage V or V and formed the overwintering population of the next season. After August adults gradually disappeared from the samples and by November all the animals were stage IV or V copepodites. Maitland *et al.* (1981) and Pomeroy (1987) observed similar seasonal trends, but *M. leuckarti* was completely absent from all December 1977, February 1983 and August 1983 samples. The vertical distribution of *M. leuckarti* changed over the season. As autumn progressed stage IV and V copepodites descended from the surface layers to concentrate in the profundal region, with a portion of the population entering a period of diapause in the substrate. Thus the population shows dual behaviour over the winter months, with both phases having suspended development.

Differences in abundances of the other zooplankton species are evident in the data recorded by Slack & Hamilton, Maitland *et al.* (1981) and Pomeroy (1987) in Fig. 2. Abundances of *Daphnia* were variable between surveys, with low numbers recorded by Pomeroy (1987). *H. gibberum* was a relatively common seasonal plankter in the loch in the 1970s and 1980s, although not recorded by Slack & Hamilton, Hamilton (1958) or Chapman (1965, 1969, 1972). The periodicity of the larger Cladocera in the loch is distinct (Fig. 2). Numbers of zooplankton taken in hauls in May 1985 were particularly low compared to those of May in the previous two years (Pomeroy, 1987).

II. Predation of zooplankton by powan in Loch Lomond

Predators are rarely indiscriminate in prey selection. Preferential feeding occurs if the relative frequencies of prey types in a predator's diet differ from the relative frequencies in the environment. Powan *C. lavaretus* in Loch Lomond consume zooplankton as the main part of their diet between May and October (Pomeroy, 1987, 1991). Predation by powan on zooplankton is examined here using samples collected from the mid basin of Loch Lomond from October 1982–May 1984. The role of prey characteristics in determining predator catch was investigated with reference to relative prey abundance, size and visibility.

Methods

The contents of powan stomachs were compared to the prey available in Loch Lomond's plankton. Powan were captured and their stomach contents were examined as detailed in Pomeroy (1991). Plankton samples (vertical hauls) were taken from a small boat between 1100 and 1500 hours on the same day as collection of fish samples, as described above. Additional horizontal tows of standard plankton nets, using a diving plane with nets attached at intervals on the towing line were taken to give vertical distribution data. The duration, distance and location of hauls was standardised. Quantitative samples were obtained using a 5 litre Freidinger-type water sampler lowered to the required depth.

Sub-samples of preserved plankton samples were identified and counted. The body length distributions of *Bosmina* and *Daphnia* were recorded by measuring individuals using a binocular microscope ($\times$ 40) fitted with an ocular micrometer. Animals were grouped into 50 μ m size classes. Measurements of body length included *Daphnia* helmets but excluded caudal spines. Counts were recorded of eggs present. *Daphnia* helmet morphs were recorded separately.

Measuring preferences: preference α and electivity ϵ
Chesson's (1978, 1983) preferences α and electivities ϵ were calculated for Loch Lomond samples between October 1982 and May 1984, using diet data from Pomeroy (1991) to estimate r_i, the relative proportion of prey i in the diet, and relative plankton species abundance n_i, determined from the plankton samples described above. The measures are obtained from a normalized version of Ivlev's (1961) electivity index. The α_i's are relative measures which indicate the preference of a predator for a food type i relative to the other food types present (Chesson, 1983) and the α_i's sum to unity. Using α, the vector with ith component α_i, a

Table 4. Numbers of fish in each sample with preferred zooplankton prey in their diets ($\alpha_i > 1/m$). n = number of fish stomachs examined, m = number of planktonic prey groups available. Eud = *Eudiaptomus gracilis* Bsm = *Bosmina coregoni* Dph = *Daphnia hyalina* Cyc = *Cyclops* spp. Hol = *Holopedium gibberum* Cbr = *Chaoborus flavicans* larvae Byt = *Bythotrephes longimanus* Lpt = *Leptodora kindti*

Date	*n*	m	Eud	Bsm	Dph	Cyc	Hol	Cbr	Byt	Lpt
Oct	30	8	0	0	0	1	0	9	22	4
Nov	29	7	4	12	11	12	0	2	6	2
Dec	17	6	1	3	1	0	0	4	1	0
Feb	43	5	3	4	0	0	0	6	0	0
Mar	24	5	0	0	0	2	0	11	0	0
Apr	33	5	0	0	26	0	0	11	0	0
May	11	8	0	4	0	0	1	5	1	2
Jul	30	8	0	4	17	0	8	3	15	9
Aug	31	8	0	8	2	0	7	0	16	1
Oct	42	6	1	0	1	0	0	25	4	0
Nov	42	6	0	0	21	14	0	8	0	0
Jan	30	5	1	2	2	0	0	9	0	0
May	25	8	0	16	0	0	11	1	5	13

predator's deviation from random prey choice can be estimated. When selection is non-random but different numbers of alternative food types are present in different situations an electivity index can be used to display preferences (Chesson, 1983). The preference α_i for a food type i relative to the average preference for all other food types can be used to calculate electivity ϵ:

$$\varepsilon_i = \frac{m\alpha_i - 1}{(m-2)\alpha_i + 1}, i = 1 \ldots m,$$

where $\alpha_i = 0$, $\epsilon_i = -1$ and where $\alpha_i = 1$, $\epsilon_i = +1$. Electivity for species i, ϵ_i can take any value between -1 and $+1$.

The following prey species were considered here: *E. gracilis*, *B. coregoni*, *D. hyalina*, *C. str. abyssorum*, *H. gibberum*, *B. longimanus*, *L. kindti* and the larvae of *C. flavicans*. Although the latter is not wholly planktonic, it was taken in plankton samples over much of the year and was included in this analysis. Occasionally, a prey species found in fish had been absent from corresponding plankton samples. In such instances the relative abundance of the prey type in plankton was set at 0.01% to calculate preferences.

Individual variation in preference

Standard deviations of the mean preference values calculated can give an estimate of the different preferences exhibited by fish. The 'effective plankton diet width' can be defined as the number of species for which powan show preferences greater than 1/m. For individual fish, the effective plankton diet width will vary:

where $\alpha_i > 1/m$, preference is indicated for species i,

and if $(1-\alpha_i \max) < (1/m)$, 1 prey is preferred ($\alpha_i \max$).

Size selectivity in powan predation

Bosmina and *Daphnia* were not the largest prey taken but were consumed in large numbers by powan. Powan swallow prey whole and in most stomach samples containing these two species, it was possible to measure whole animals and count frequencies of *Daphnia* morphs. Egg counts and relative frequencies of gravid and non-gravid individuals were recorded in samples, although no correction was made for eggs which may have been lost from the shell valves. Measurements of total body length of these two species were recorded from powan stomach contents whenever possible and from each plankton sample. Length frequency distributions were compared using the Kolmogorov-Smirnov 2 sample test, with a null hypothesis of identity for the two sample distributions. It was assumed that no bias was introduced into the measurements by differential damage to either morph or size range of prey in the course of ingestion or subsequent analysis.

Table 5. The number of preferred zooplankton prey groups found in each fish in samples from October 1982 - May 1984, where $\alpha_i > 1/m$. n = number of fish stomachs examined, m = number of planktonic prey groups available

Date	n	m	effective plankton diet width, number of prey groups			
			0	1	2	3
Oct	30	8	0	24	6	0
Nov	29	7	2	11	9	7
Dec	17	6	9	6	2	0
Feb	43	5	32	9	2	0
Mar	24	5	11	13	0	0
Apr	33	5	0	29	4	0
May	11	8	1	8	1	1
Jul	30	8	1	9	12	8
Aug	31	8	2	24	5	0
Oct	42	6	12	29	1	0
Nov	42	6	12	17	13	0
Jan	30	5	16	14	0	0
May	25	8	0	7	14	4

Predator size classes were not differentiated as virtually all powan sampled were very much larger than their biggest prey.

Results

Effect of diel variation in plankton abundance on preference

Preference and electivity values obtained for the stomach and plankton samples of April 1983 were compared. Relative frequencies of the eight plankton species considered in surface plankton samples taken at 1400 and 2100 in April 1983 were significantly different (K-S 2 sample test: Dmax(760,279) = 0.396, $p<0.001$). However, powan preference values for the plankton species at these times were not significantly different ($t_{(5)} = 1.63$, $p>0.1$). Temporal fluctuations in relative numerical composition of plankton species did not significantly alter calculated preference values. This superficially surprising result is largely attributable to the varying contribution to the plankton made by species unimportant to powan as food.

Between sample variation in preference-electivity

Mean electivity values for pooled fish samples are shown in Table 3. The seasonal pattern of prey preference exhibited by powan in Loch Lomond indicates that the large cladoceran species and *Chaoborus* larvae are the preferred prey of the sampled powan population (Table 3).

Individual variation in preference within samples

The numbers of fish in each sample which show preferences for zooplankton prey are given in Table 4. In November 1982, those fish which had *Chaoborus* as the preferred species had eaten many larvae (>50) and contained little else, whilst fish with *Eudiaptomus* as the preferred species had only a few animals in their stomachs. Most powan in the August 1983 sample showed a high preference value for a single prey species but chose different prey types.

Effective plankton diet widths of powan varied seasonally (Table 5). Summer samples tended to have the greatest number of preferred prey species. Where some fish are indicated as having no preferred species, the diet consisted of other (mostly benthic) food, fish in this category were taken frequently in winter.

Interspecific prey size selectivity

Although the maximum body lengths of each prey species considered varied seasonally, the relative sizes of prey species changed little. Subjective prey characteristics suggest that large, slow moving zooplankton are likely to be preferred prey (Table 6).

Intraspecific prey size selectivity

(a) all *Daphnia*

Body length frequency distributions of *Daphnia* eaten by powan and taken in plankton hauls are shown in Fig. 3. In all samples except December 1982 where there was no difference, *Daphnia* in powan stomachs were larger than those in plankton samples of the same sampling period (K-S one-tailed test, $p<0.001$).

(b) *Daphnia* morphs in plankton

In the plankton samples of July 1983, the peaked helmet morph was significantly bigger than the round helmet morph (K-S Dmax(218,78) = 0.339, $0.01<p<0.05$). No significant difference in size distributions of the two morphs was found in the August and November 1983 plankton samples.

(c) *Daphnia* morphs in powan stomachs

Peak helmeted morphs were composed of significantly larger individuals than the round helmet morphs in the July and November 1983 stomach samples (K-S Dmax(333,65) = 0.357, $p<0.01$ and Dmax(216,143) = 0.483, $p<0.001$ respectively) but no significant differences were found in morph size distri-

Table 6. Characteristics of zooplankton in Loch Lomond. Relative visibility compares females which are not bearing eggs.

Species	Body length (mm)	Relative visibility	Relative mobility
Eudiaptomus gracilis	1.2	medium	medium
Bosmina coregoni	1.0	low/medium	low
Daphnia hyalina	2.0	low/medium	low
Cyclops str. abyssorum	1.7	medium	medium
Holopedium gibberum	1.4	medium	low
(including matrix)	(3.0)		
Chaoborus flavicans larvae	12.0	medium/high	high
Bythotrephes longimanus	2.0	medium	medium
Leptodora kindti	10.0	low	medium

butions for the August 1983 stomach samples. *Daphnia* in powan stomachs were significantly larger than those in the plankton samples. This was true of both morphs in all samples examined (K-S one tailed test, $p<0.05$: Fig. 3).

(d) The proportions of peaked and round helmet *Daphnia* morphs counted in the plankton and stomach samples were not independent in the samples of July, August, September, October and November 1983, implying differential predation on the morphs at these times (χ^2 tests, $p<0.05$). In December 1983 there was no indication of differential predation. In July, August and September a greater proportion of round helmets were counted in stomachs than in plankton but in October and November a greater proportion of peaked *Daphnia* were counted in stomachs.

The largest *Bosmina* measured were in the 1.05 mm size class and was therefore the smallest prey species in the powan's diet. Comparison of *Bosmina* body length frequency distributions from powan stomach contents and plankton samples showed that size selection of *Bosmina* occurred in Nov 1982, May and July 1983 and May 1984. *Bosmina* which had been eaten by powan were larger than *Bosmina* in the plankton (Fig. 4). In August and September 1983 no significant differences were found.

Visibility of prey

In all samples examined, the proportion of gravid *Bosmina* in powan stomachs was greater than that in plankton (χ^2 tests, $p<0.001$). In late spring and summer vertical migrations of gravid *Bosmina* occur from deeper water during the day to the surface layers at night (Pomeroy, unpubl.).

Discussion

The zooplankton community of Loch Lomond can be characterised by the features of species which occupy key niches. The loch's most abundant zooplankton species are (i) the relatively small, unpigmented grazing copepod, *E. gracilis*, and (ii) relatively small, unpigmented grazing Cladocera, *B. coregoni* and *D. hyalina*. In the last few decades it has been shown that the zooplankton community of Loch Lomond is unique amongst those of Scotland's largest freshwater lochs. Although each of the other species found in Loch Lomond is present in at least one other large loch, the presence of *M. leuckarti* is the discriminating factor identifying Loch Lomond's plankton (Maitland *et al.*, 1981).

Zooplankton communities have been viewed as being structured either in a 'top-down' manner by predation or a 'bottom-up' manner by food availability (Wells, 1970; Hansson *et al.*, 1990). The theoretical consequences of these alternatives are different. Species which suffer intense predation will tend not to increase in abundance if their food supply increases. However if food availability structures the community, all species should benefit from uniform increases in phytoplankton supply. No study of Loch Lomond to date has collected data simultaneously at each level of the ecosystem to allow a completely integrated analysis of factors controlling cyclical events, but several studies discussed below provide parts of the story.

Differences appear between cyclopoid copepod life cycles in Loch Lomond described by Chapman (1972) and later studies (Maitland *et al.*, 1981, Pomeroy, 1987). *C. str. abyssorum* was recorded by Maitland

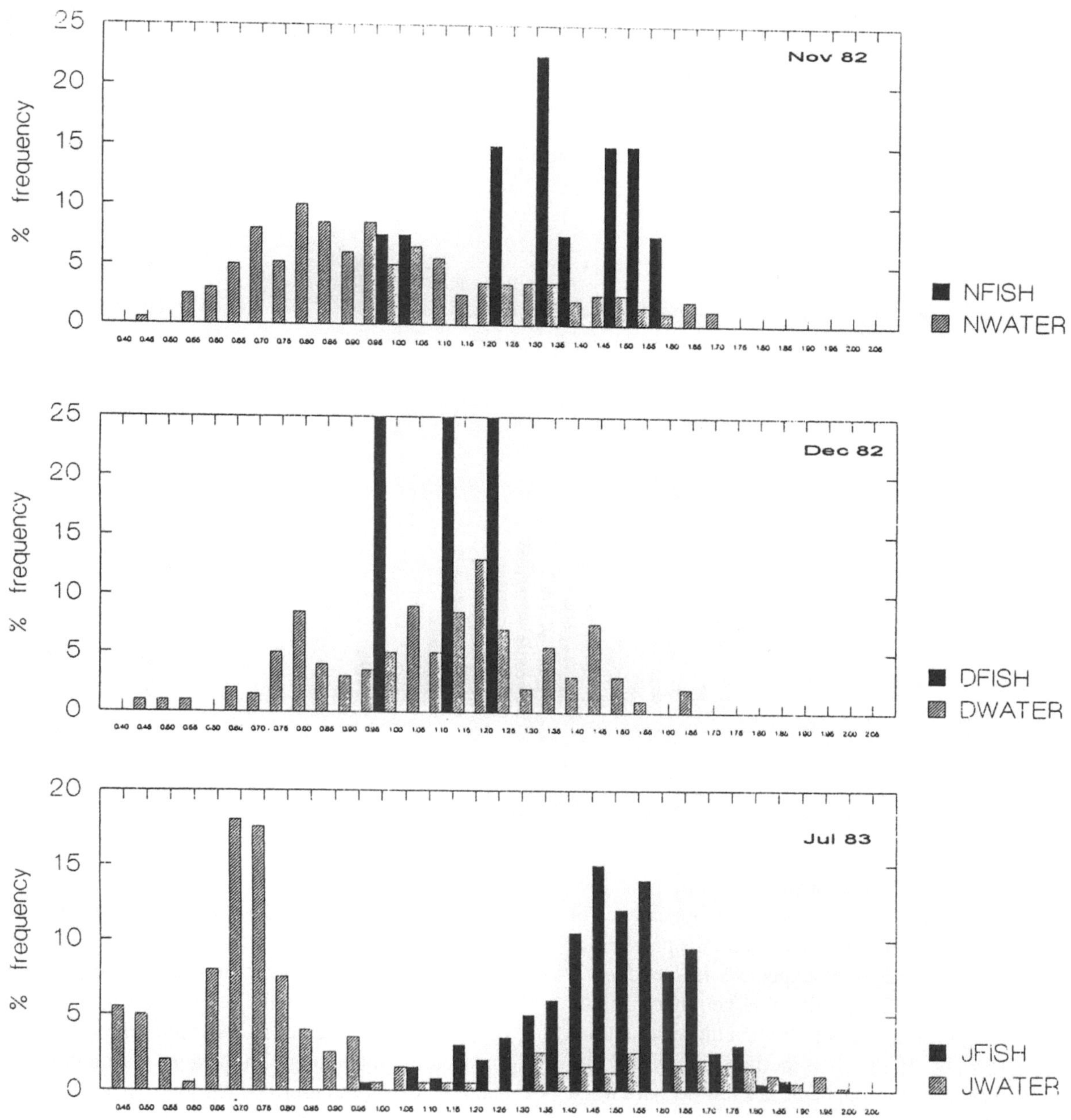

Fig. 3. Daphnia hyalina body length frequencies in plankton samples (hatched bar histogram) and in powan stomachs (solid bar histogram).

et al. (1981) and Pomeroy (1987) throughout the winter in the plankton, although in low numbers. These differences may be real and reflect changed conditions in the loch's environment or they may be due to sampling artefact. Copepod seasonal cycles are subject to intraspecific variation. The population of *M. leuckarti* in Esthwaite Water has no planktonic presence during the winter months (Fryer & Smyly, 1954). In some Russian lakes *M. leuckarti* copepodites enter the bottom deposits during winter, but remain active (Ulomskii, 1953). In more tropical regions *M. leuckarti* is present in the plankton throughout the year (Fryer, 1957). This plasticity of life history characteristics, not only for populations within a region but within populations appears to be a feature common to zooplankton species. However, the mechanisms which control such fundamental characteristics and the factors which influence them remain poorly understood. Changes

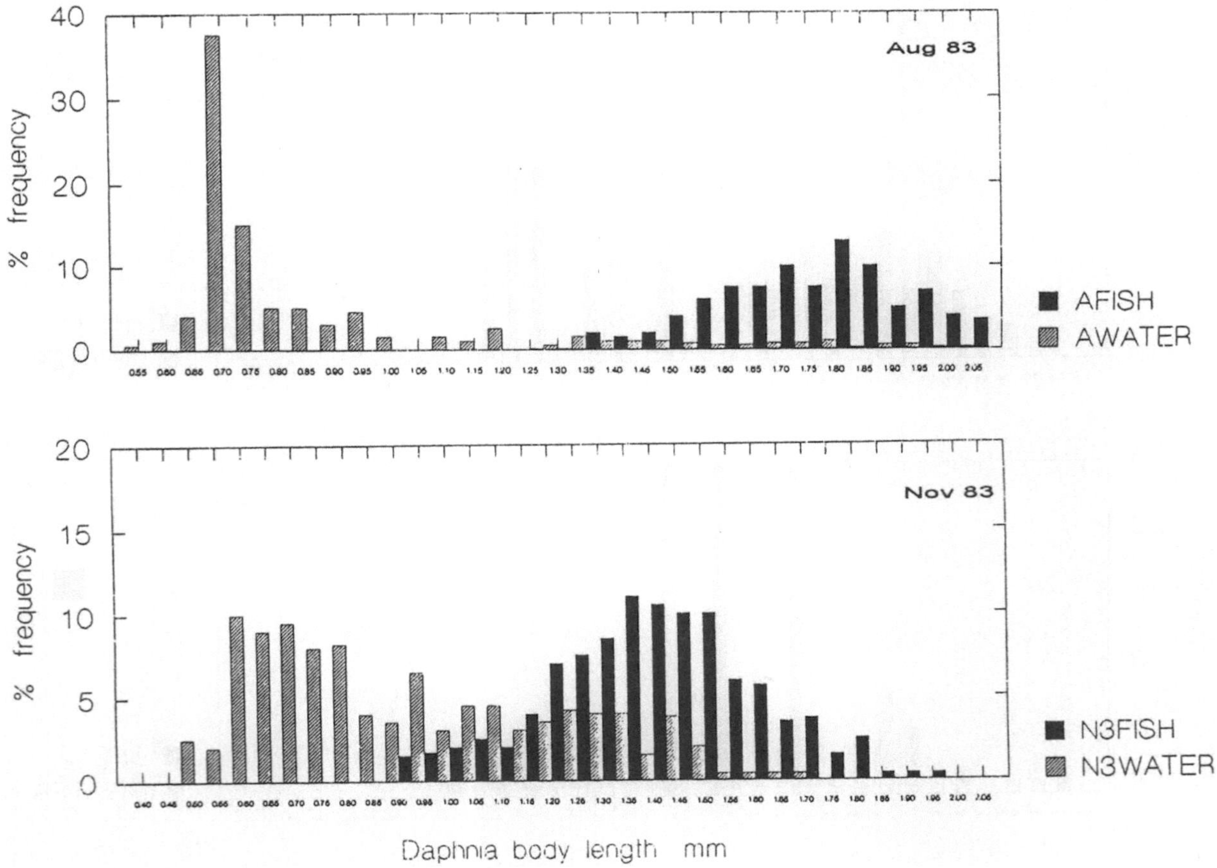

Fig. 3 (continued).

observed in plankton composition may be indicative of abiotic or biotic factors. The recent appearance of *Holopedium gibberum* in Loch Lomond's plankton is contrary to what might be expected. *Holopedium* is known to graze preferentially on large phytoplankton in dilute suspension and is lost when those conditions change (Smyly, 1978). It is perhaps surprising that *Holopedium* has appeared in the last few decades when nutrient levels in the southern loch particularly have increased. An absence of *Holopedium* from samples in the 1950s and 60s may have been due to a long term failure of resting eggs (Smyly, 1977) or substantial changes in the composition of the phytoplankton community, with its subsequent loss from the zooplankton. Examination of cladoceran and vertebrate remains in loch sediments may give a better indication of longer term changes in community structure (Nilssen & Sandoey, 1990).

Predation

In a loch such as Loch Lomond, predation on zooplankton can come from other zooplankton species, other invertebrate predators and vertebrate predators including fish. The extremely clumped distributions of zooplankton make extensive sampling a prerequisite for accurate modelling. The spatial discontinuity of plankton has been described for some freshwater systems (Hutchinson, 1967; George, 1974, 1981). As in this study, Northcote & Clarotto (1975) found that the contents of the stomachs of the salmonid species of six lakes in British Columbia did not always tally with the species taken in contemporaneous plankton samples. This emphasises the small scale temporal and spatial heterogeneity of zooplankton distributions.

Predatory Cladocera have substantial impact on the smaller stages of daphnid populations (De Bernardi & Giussani, 1977). Abundances of predatory Cladocera themselves may be influenced not only by prey availability, but by abundance of other predatory inverte-

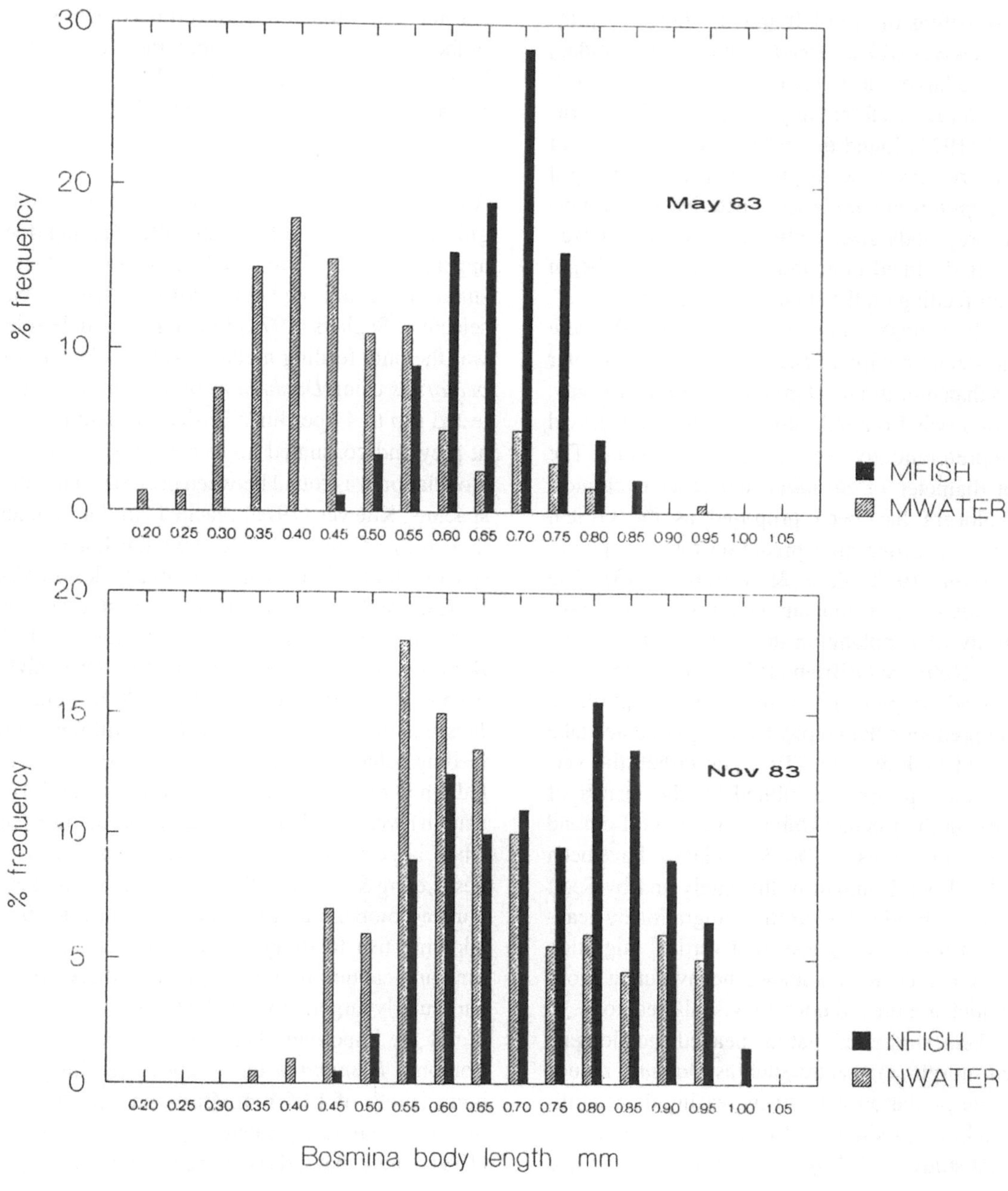

Fig. 4. Bosmina coregoni body length frequencies in plankton samples (hatched bar histogram) and in powan stomachs (solid bar histogram).

brate species as well as vertebrate predators (Lehman & Cáceres, 1993).

Fish predation on zooplankton has been studied widely, with considerable effort devoted to evaluating different aspects of the nature of successional force on the zooplankton community by selective piscine predators, (Brooks & Dodson, 1965; Zaret, 1972) and determination of the mechanisms responsible for observed predation patterns (O'Brien *et al.*, 1976; Confer *et al.*, 1978; Bence & Murdoch, 1986). Some microcarnivorous fish display preferences for large invertebrate prey (Brooks, 1968; Werner & Hall, 1974). While predators may achieve a rapid increase in profit (=energy gain per unit time) by selecting larger prey, detectability of prey also increases with prey body size (Werner & Hall, 1974; Confer *et al.*, 1978). Active selection of the most profitable prey may be confused with the results of simply eating the most frequently detected prey. Captive alewives (*Alosa pseudoharengus*, Clupeidae) removed the largest size classes of the copepod *Eudiaptomus minutus* preferentially (Janssen, 1978). Werner & Hall (1974) found that where equal numbers of *Daphnia* size classes were present, juvenile bluegill sunfish (*Lepomis macrochirus*) electivity was proportional to prey body size. Only when the largest prey size classes declined in abundance did the fish begin significant feeding on the smaller size categories.

Bluegill sunfish (*Lepomis macrochirus*) were shown to detect ephippial *Daphnia* magna at greater distances than non-gravid *Daphnia* (Tucker & Woolpy, 1984). In Loch Lomond, powan consumed gravid *Bosmina* preferentially over non gravid *Bosmina*. The eye-spot diameter of *Bosmina* and other cyclomorphic Cladocera has been proposed as the critical determinant in differential predation rates by planktivores (Zaret, 1972; Zaret & Kerfoot, 1975). The present study suggests that any factor which increases detectability of zooplankton increases predation risk by powan. Kettle & O'Brien (1978) showed that pigmented pond zooplankton were more susceptible to lake trout predation than typically non-pigmented lake forms. Goldspink & Scott (1971) described the vertical migration patterns exhibited by the instars of *Chaoborus* in the Lochan Dubh beside Loch Lomond and similar migrations of *Chaoborus* larvae have been observed in Loch Lomond in this study and by Scott (pers. comm.). Evidence of vertical migration by heavily predated *Bosmina* suggests that vertical migration is a common predator avoidance behaviour in those groups which are preyed upon by visual predators.

It has been suggested that the peaked and elongate helmeted morphs in species such as *Daphnia* reduce invertebrate predation without increasing detectability to vertebrate predators (O'Brien *et al.*, 1978). In the present study, both *Daphnia* morphs suffered size selective predation by powan. In the one plankton sample where the peaked morph was larger than the round helmeted morph (July 1983) the size distribution of *Daphnia* in powan stomachs reflected this. However in November 1983 peaked *Daphnia* in powan stomachs were larger than those in the plankton.

Predator preferences change with time. Murdoch & Oaten (1975) used guppies feeding on fruit flies and worms to demonstrate that fish switched to the most commonly available food, and this was related to experience and learning. However, although the population of predators showed little preference when offered equal proportions of two prey, individual fish showed considerable variation *i.e.* the proportion of specialists in the population changed. Individual powan in Loch Lomond took different preferred prey species when plankton populations were maximal.

Feeding modes

Although most fish planktivores have numerous fine gill rakers on their gill arches, filter feeding does not appear to be a preferred feeding method. In fact, most studies have indicated that freshwater fish are particle selectors. Seghers (1974) found that particle selection was the only feeding method used by *Coregonus clupeaformis* using *Daphnia* as prey at all prey densities tested (up to 44 per litre). Fish took individual 'nips' at prey and consumed up to 600 prey in 15 minutes. No relation was found between prey size and gill raker spacing. Kliewer (1970) reached similar conclusions in a study of the food of lake whitefish and gill raker spacing in populations of from seven lakes in Manitoba, Canada. Powan in Loch Lomond eat large numbers of small prey. One powan stomach contained 49 560 *Bosmina* in August 1983. Although filter feeding has been observed in clupeids (Blaxter & Holliday, 1958; Janssen, 1978), particle selection is still the dominant feeding behaviour. The northern Pacific anchovy filter fed when high densities of small prey items (*Artemia* nauplii) were available, but began particulate feeding when large prey were available, even at low densities (Leong & O'Connell, 1969). This behaviour accumulated biomass at a greater rate than filter feeding, allowing fish to forage over a greater area per unit time and capture prey which could otherwise escape, particularly larger, mobile plankters.

Of the important planktonic prey species in Loch Lomond, *Bosmina* is the smallest, reaching a maximum length of 1.05 mm. Bosmina less than 0.5mm in length were rarely eaten by powan, although Miller (1984) found that gill raker spacing was in the order of 0.7–0.9 mm. The smallest *Bosmina* were eaten when there were few of the species in the plankton (August 1983) and size selectivity was greatest at maximum plankton abundance.

Other workers have reported that the first food of *Coregonus* larvae consisted of copepods and particularly *Eudiaptomus* in the first few weeks after hatching (Bronisz, 1979; Marciak, 1979; Dabrowski, 1981; Teska & Behmer, 1981; Salojarvi, 1982). *Eudiaptomus* may have a superior escape ability to that of many Cladocera, and the reason for the predominance of this species in the diet of newly-hatched *Coregonus* larvae may lie in the relative scarcity of alternative zooplankton prey at hatching time. Indeed, the strength of individual year classes of powan is likely to depend at least in part to the availability of food for fish larvae.

In summary, the zooplankton species and forms which make up Loch Lomond's pelagic community reflect the effects of sustained predation pressure from a number of sources, both invertebrate and fish. Loch Lomond has a particularly diverse fish community, most of which prey on zooplankton at some time. However, the loch's principal zooplanktivore in terms of biomass is the powan. All the pelagic zooplankton species in the loch show some adaptations for predator avoidance, the main ones being lack of pigmentation, small body size and diel vertical migrations. Structuring of zooplankton populations in Loch Lomond is at least partly driven from the top of the ecosystem.

Acknowledgments

I wish to thank all those who helped in this work, particularly the late Dr H. D. Slack and his family. The cooperation and help of the University Field Station and Loch Lomond Angling Improvement Association is acknowledged. J. Hamilton kindly provided access to unpublished data. A. J. Lisle provided invaluable assistance and advice during manuscript preparation. The work was supported by a D.E.N.I. studentship.

References

Bence, J. R. & W. W. Murdoch, 1986. Prey size selection by the mosquitofish in relation to optimal diet theory. Ecology 67: 324–336.

Blaxter, J. H. S. & F. G. T. Holliday, 1958. Herring *Clupea harengus* L. in aquaria. II: Feeding. Mar. Res. 6–22.

Bronisz, D., 1979. Selective exploitation of lake zooplankton by coregonid fry in cage culture. In Cultivation of fish fry and its live food. E. Styczynska-Jurewicz, T. Backiel, E. Jaspers & G. Persoon (eds.), Spec. Publ. Eur. Maricult Soc. 1979 (4), Bredene: Belgium.

Brooks, J. L., 1968. The effects of prey size selection by lake planktivores. Syst. Zool. 17: 273–291.

Brooks, J. L. & S. I. Dodson, 1965. Predation, body size, and composition of plankton. Science 150: 28–35.

Chapman, M. A., 1965. Ecological studies on the zooplankton of Loch Lomond. Unpublished Ph.D. thesis, University of Glasgow.

Chapman, M. A., 1969. The bionomics of *Diaptomus gracilis* Sars (Copepoda, Calanoida) in Loch Lomond, Scotland. J. anim. Ecol. 38: 257–284.

Chapman, M. A., 1972. The annual cycles of the limnetic cyclopoid Copepoda of Loch Lomond, Scotland. Int. Revue ges Hydrobiol. 57: 895–911.

Chesson, J., 1978. Measuring preference in selective predation. Ecology 59: 211–215.

Chesson, J., 1983. The estimation and analysis of preference and its relationship to foraging models. Ecology 64, 1297–1307.

Confer, J. L., G. L. Howick, M. H. Corzette, S. L. Kramer, S. Fitzgibbon & R. Landesberg, 1978. Visual predation by planktivores. Oikos 31: 27–37.

Dabrowski, K. R., 1981. The spawning and early life history of the pollan *Coregonus pollan* Thompson in Lough Neagh, Northern Ireland. Int. Revue ges Hydrobiol. 66: 299–326.

De Bernardi, R. & G. Giussani, 1977. Effects of mass fish mortality on zooplankton structure and dynamics in a small Italian lake (Lago di Annone). Ver. int. ver. Limnol. 20: 1045–1048.

Fryer, G., 1957. Freeliving freshwater Crustacea from Lake Nyasa and adjoining waters. Arch. Hydrobiol. 53: 62–86.

Fryer, G. & W. J. P. Smyly, 1954. Some remarks on the resting stages of some freshwater cyclopoid and harpacticoid copepods. Ann. Mag. nat. Hist. 7: 65–72.

George, D. G., 1974. Dispersion patterns in the zooplankton populations of a eutrophic reservoir. J. mar. Res. 43: 537–551.

George, D. G., 1981. Zooplankton patchiness. 49th Ann. Rep. F.B.A.: 32–44.

Goldspink, C. R. & D. B. C. Scott, 1971. Vertical migration of *Chaoborus flavicans* in a Scottish loch. Freshwat. Biol. 1: 411–421.

Hamilton, J. D., 1958. On the biology of *Holopedium gibberum* Zaddach (Crustacea: Cladocera). Verh. int. Ver. Limnol. 13: 785–788.

Hensson, S., U. Larsson & S. Johansson, 1990. Selective predation by herring and mysids, and zooplankton community structure in a Baltic Sea coastal area. Limnol. Oceanogr. 12: 1099–1116.

Hutchinson, G. E., 1967. A Treatise on Limnology. 3 Vols. Vol. 2: Introduction to lake biology and the limnoplankton. New York: Wile

Ivlev, V. S., 1961. Experimental ecology of the feeding of fishes. New Haven, Conn. Yale Univ. Press.

Janssen, J., 1978. Feeding behaviour repertoire of the alewife *Alosa pseudoharengus* and the ciscoes *Coregonus hoyi* and *Coregonus artedii*. J. Fish. Res. Bd. Can. 35: 249–253.

Kettle, D. & W. J. O'Brien, 1978. Vulnerability of arctic zooplankton species to predation by small lake trout (*Salvelinus namaycush*). J. Fish Res. Bd. Can. 35, 1495–1500.

Kliewer, E. V., 1970. Gillraker variation and diet in lake whitefish (*Coregonus clupeaformis*) in northern Manitoba. In C. C. Lindsey & C. S. Woods, eds., Biology of coregonid fishes. Winnipeg: University of Manitoba Press: 147–165.

Kokova, V. Ye. 1978. The feeding of larvae of the whitefish *Coregonus lavaretus ludoga*, the Baikal omul *Coregonus autumnalis migratolius*, and the 'peled' *Coregonus peled*. J. Ichth. 18: 953–960.

Lehman, J. T. & C. E. Caceres, 1993. Food-web responses to species invasion by a predatory invertebrate: *Bythotrephes* in Lake Michigan. Limnol. Oceanogr. 38: 879–891.

Leong, R. J. H. & C. P. O'Connell, 1969. A laboratory study of particulate and filter feeding of the northern anchovy (*Engraulis mordax*). J. Fish. Res. Bd. Can. 26: 557–582.

Maitland, P. S. (ed.), 1981. The ecology of Scotland's largest lochs: Lomond, Awe, Ness, Morar and Shiel. Monogr. Biol. 44, Dr W. Junk Publishers, The Hague: 290 pp.

Maitland, P. S., B. D. Smith & G. M. Dennis, 1981. The crustacean zooplankton. In Maitland, P. S. The ecology of Scotland's largest lochs: Lomond, Awe, Ness, Morar and Shiel. Monogr. Biol. 44, (ed.), Dr W. Junk Publishers, The Hague: 290 pp.

Marciak, Z., 1979. Food preferences of juveniles of three Coregonid species reared in cages. In Cultivation of fish fry and its live food. E. Styczynska-Jurewicz, T. Backiel, E. Jaspers & G. Persoone (eds). Spec. Publ. Eur. Maricult. Soc. 1979. Belgium: Bredene.

Miller, J., 1984. Gill raker morphology and variation in powan *Coregonus lavaretus* (L.) from Loch Lomond and Loch Eck. Unpubl. thesis, Department of Zoology & Marine Biology, St. Andrews University.

Murdoch, W. W. & A. Oaten, 1975. Predation and population stability. Adv. Ecol. Res. 9: 2–131.

Murray, J. & L. Pullar, 1910. Bathymetrical survey of the freshwater lochs of Scotland. Vols. I–VI Edinburgh.

Nilssen, J. P. & S. Sandoey, 1990. Recent lake acidification and cladoceran dynamics: Surface sediment and core analyses from lakes in Norway, Scotland and Sweden. Philos. Trans. r. Soc. Lond., B, 327 1240: 299–309.

Northcote, T. G. & R. Clarotto, 1975. Limnetic macrozooplankton lakes. Verh. int. Ver. Limnol. 19: 2378–2393.

O'Brien, W. J., N. A. Slade & G. L. Vinyard, 1976. Apparent size as the determinant of prey selection by bluegill sunfish (*Lepomis macrochirus*). Ecology 57: 1304–1310.

Pomeroy, P. P., 1987. The food and feeding of powan, *Coregonus lavaretus* (L.), (Coregoninae: Salmonidae) in two Scottish lochs. Unpubl. Ph.D. thesis, University of St Andrews.

Pomeroy, P. P., 1991. A comparative assessment of temporal variation in diet of powan, *Coregonus lavaretus* (L.), from Loch Lomond and Loch Eck, Scotland, UK. J. Fish. Biol. 38: 457–478.

Salojarvi, K., 1982. Spawning ecology, larval food supplies and causes of larval mortality in the whitefish (*Coregonus lavaretus* L.). Pol. Arch. Hydrobiol. 29: 159–178.

Seghers, B. H., 1974. Role of gill rakers in size-selective predation by lake whitefish, *Coregonus clupeaformis* (Mitchill). Verh. int. Ver. Limnol. 19: 2401–2405.

Scott, T., 1899. The invertebrate fauna of the inland waters of Scotland – report on special investigation. Rep. Fishery Bd. Scot. 17: 133–204.

Scourfield, D. J. & J. P. Harding, 1966. A key to the British species of freshwater Cladocera. Sci. publ. no. 5, Freshw. biol. Ass. UK (3rd edn). Titus Wilson, Kendal.

Slack, H. D., 1954. The bottom deposits of Loch Lomond. Proc. r. Soc. Edinb. B, 65: 213–38.

Slack, H. D., 1957. Studies on Loch Lomond. I. Glasgow: Univ. Publ., 133 pp.

Slack, H. D., F. W. K. Gervers & J. D. Hamilton, 1957. The biology of the powan. In Studies on Loch Lomond. 1, 113–127. Glasgow: Univ. Publ., 133 pp.

Smyly, W. J. P., 1968. Observations on the planktonic and profundal Crustacea of the lakes of the English Lake District. J. anim. Ecol. 37: 693–708.

Smyly, W. J. P., 1977. A note on the resting egg of *Holopedium gibberum* Zaddach. Microscopy. 33: 170–171.

Smyly, W. J. P., 1978. Crustacean zooplankton of Grasmere before and after a change in sewage effluent treatment. Int. Revue ges Hydrobiol. 63: 389–403.

Sokal, R. R. & F. J. Rohlf, 1981. Biometry 2nd edn. San Francisco: W. H. Freeman.

Teska, J. D. & D. J. Behmer, 1981. Zooplankton preference of larval lake whitefish. Trans. am. Fish. Soc. 110: 459–461.

Tucker, R. P. & S. P. Woolpy, 1984. The effect of parthenogenic eggs in *Daphnia magna* on prey location by the bluegill sunfish (*Lepomis macrochirus*). Hydrobiologia 109: 215–217.

Ulomskii, S. N., 1953. New ecological studies of some oar-legged crustaceans (Copepoda), *Mesocyclops leuckarti* CLAUS. C.R. acad. Sci. USSR 90: 295–297.

Wells, L., 1970. Effects of alewife predation on zooplankton populations in Lake Michigan. Limnol. Oceanogr. 15: 556–567.

Werner, E. E. & D. J. Hall, 1974. Optimal foraging and the size selection of prey by the bluegill sunfish (*Lepomis machrochirus*). Ecology 55: 1042–1052.

Zaret, T. M., 1972. Predators, invisible prey, and the nature of polymorphism in the Cladocera (Class Crustacea). Limnol. Oceanogr. 17: 171–184.

Zaret, T. M., 1980. Predation and Freshwater Communities. Yale University Press.

Zaret, T. M. & W. C. Kerfoot, 1975. Fish predation on *Bosmina longirostris*: body-size selection versus visibility selection. Ecology 56: 232–237.

Hydrobiologia **290**: 91–102, 1994.
K. J. Murphy, M. C. M. Beveridge & R. Tippett (eds), The Ecology of Loch Lomond.

The fish community of Loch Lomond, Scotland: its history and rapidly changing status

Colin E. Adams
Fish Behaviour and Ecology Group, University Field Station, Rowardennan, Glasgow G63 0AW, UK

Key words: Loch Lomond, fish community, introductions

Abstract

The fish community of Loch Lomond is of national importance. Its diversity of species and rare populations of powan (*Coregonus lavaretus*) and freshwater feeding river lampreys (*Lampetra fluviatilis*) warrant high conservation status. It is also of value for its sport fisheries for sea-trout (*Salmo trutta*), salmon (*Salmo salar*) and pike (*Esox lucius*). Historical records demonstrate that the species composition of the fish community has remained stable over a very long period until recently when a series of introductions of fish species new to the catchment has resulted in successful colonisation by a number of species. These have resulted in fundamental changes in the ecosystem. Here using historical records the long-term stability of the fish community is examined, recent rapid changes in the fish community are documented and some of the resultant effects of changes in the fish community are demonstrated .

Introduction

The assemblage of species that makes up the fish community of Loch Lomond, Scotland, is unique, of national importance and under threat. In this paper I explore the basis for this claim by reviewing the historical literature, by providing a description of the current status of the fish community and by describing recent changes that have occurred and some of their consequences.

The biological, conservation and recreational importance of the Loch

The importance of the Loch Lomond fish community lies partly in its relative diversity of species. Maitland (1972) lists the 15 species of fish known to occur in the loch and its tribularies at that time. This is more than for any other Scottish loch. This richness of species is partly a result of the dichotomous nature of the loch (see Tippett, (1994) for a full description). The conditions in the narrow, deep, oligotrophic highland north of the catchment generally favouring the adaptable, phenotypically plastic, generalist species, such as the salmonids, and the conditions in the shallower, more nutrient rich, warmer south in general tending to favour more specialised species, such as the cyprinids.

In reality of course, the Loch Lomond catchment is more complex than this and offers a multiplicity of aquatic habitats ranging from invertebrate poor, torrential streams to nutrient rich ox-bow lakes. The result is that this wide range of habitats plus the loch's close link with the sea (the River Leven draining the system into the Clyde estuary, is only 13 km long) provide suitable environments for a wide range of fish species.

Loch Lomond also has nationally important populations of particular species (Lyle & Maitland 1992). The powan (*Coregonus lavaretus*) population is probably the largest of only two populalions in Scotland and only seven in the UK and since 1986 has been afforded protection under Schedule 5 of the Wildlife and Countryside Act 1981 (see Brown & Scott (1994)). Loch Lomond's population of river lampreys (*Lampetra fluviatilis*) is the only known population in the UK that does not migrate to sea but remains in freshwater (see Maitland *et al.* (1994)) probably feeding mainly on powan (Robertson, 1870). The roach (*Rutilus rutilus*) population is also interesting in that a proportion of the population undertakes a spawning migration each year up the River Endrick, the main tributary in the south of the catchment, to breed.

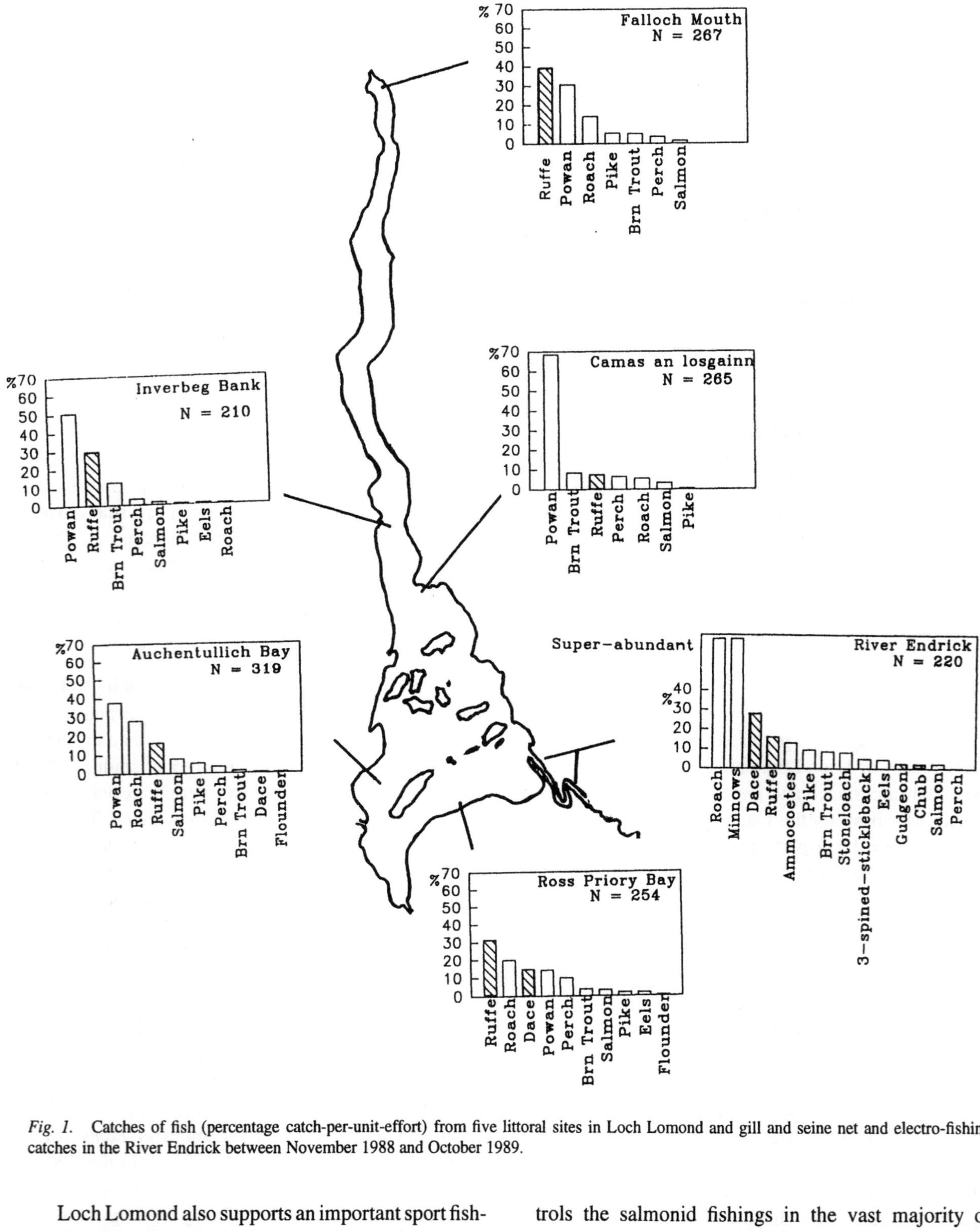

Fig. 1. Catches of fish (percentage catch-per-unit-effort) from five littoral sites in Loch Lomond and gill and seine net and electro-fishing catches in the River Endrick between November 1988 and October 1989.

Loch Lomond also supports an important sport fishery for Atlantic salmon (*Salmo salar*) and sea-trout (*Salmo trutta*) which as been managed since at least 1860. The Loch Lomond Angling Improvement Association is the angling association that currently controls the salmonid fishings in the vast majority of the catchment. The Association was formed in 1895 and is thus one of the oldest fishing clubs in the UK (Lamond, 1931). It currently has around 900 members and employs 5 full-time bailiffs. In recent years Loch

Lomond has also become an important fishery for pike (*Esox lucius*) and in 1947 produced the current Scottish rod-caught record for this species weighing 21.6 kgs (47 lbs 11 oz).

The structure of the fish community

As a result of its importance as a sport fishery and more recently because of scientific and conservation interest, the fish fauna of Loch Lomond has been better documented than any other major inland waterway in Scotland. An examination of the literature shows that the fish community has undergone two distinct and contrasting phases during its recorded history.

Historic accounts of the fish community pre-1970.

The first known full account of the fish fauna of Loch Lomond was in 1795 by Ure (1795), this was followed by accounts by Brown (1891), Lamond (1931) and Hunter *et al.* (1959). These accounts and other records have been comprehensively reviewed by Maitland (1972). Although nomenclature differences accounted for most of the discrepancies between Maitland's list of 15 species of fish occurring in Loch Lomond and previous fish fauna lists (see Table 1) there are some additional records that are worth examining more closely.

Tench (*Tinca tinca*) have been recorded from the loch by several authors (Young, 1870; Brown, 1891; Regan, 1911) and a specimen was exhibited to the Natural History Society of Glasgow on the 19th of December 1870. According to Lumsden and Brown (1895) this species was restricted to the area around the mouth of the River Endrick. There have been no records of this species during the 20th century and it is thought to have died out (Hunter *et al.*, 1959; Maitland, 1972). It is likely that extinction of this species in Loch Lomond was the result of a spawning failure as according to Wheeler (1969), tench only spawn when the water temperature reaches 18 °C, it is unlikely that this temperature would be consistantly achieved in Loch Lomond to enable successful spawning (Slack, 1957).

Arctic charr (*Salvelinus alpinus*) was recorded from Loch Lomond by Bidie (1896) and there is mention of charr in Loch Lomond in the Statistical Account of Scotland (Stewart, 1796). The 1896 record was rejected by naturalists at the time as a mis-identification possibly of a brook charr (*Salvelinus fontinalis*) (Brown, 1896). As there have been no other records of this species either before or since, despite considerable collection effort (Maitland, 1972; Adams & Tippett, 1990), it is most likely that this record did result from a mis-identification.

Brook charr are known to have been introduced to Loch Lomond around 1876 (Brown, 1896) and although the introduced stock maintained its identity, it did not become established and according to Scott & Brown (1901), eventually died out.

Several specimens of the Rainbow trout (*Oncorhynchus mykiss*) have been recorded from the River Endrick pre-1970 (Lamond, 1931; Maitland, 1966a). Maitland (1966a) suggests that these all resulted from cultured escapes from farms or artificially stocked sites and that they are unlikely to become established in the catchment. More recently rainbow trout have been consistently represented in rod-and-line captures by anglers (H. Ward pers. comm.), however the condition factor and fin damage to these fish suggest that they originate from fish farms, there is no evidence to suggest a self-sustaining population.

Thick-lipped mullet (*Crenimugil labrosus*) was recorded in the River Leven by Lumsden & Brown (1895) and by Scott & Brown (1901). The personal recollections of R. McMath (University Field Station, Rowardennan) of mullet (of unknown species) caught in moderate numbers in gill-nets set in Milarrochy Bay in the early 1970's represent the only known record of mullet in the main loch and can not be lightly dismissed. However because of the paucity of records of mullet in the main loch despite relatively intensive collection over many years it seems that this species enters the main loch, at best infrequently.

The record of plaice (*Pleuronectes platessa*) in Loch Lomond by Scott & Brown (1901) is highly dubious as there have been no other records of this species entering freshwater. It is possible that this record resulted from confusion of this species with the relatively common flounder (*Platichthys flesus*).

The fish community post-1970

The end of the 1970's marked a clear watershed for Loch Lomond, as the fish community entered a period of rapid and dramatic change, unprecedented in its history, as five fish species, new to the Loch Lomond catchment were recorded there.

Table 1. Records of fish species from Loch Lomond

Species		Status
Sea lamprey	*Petromyzon marinus*	Native
River lamprey	*Lampetra fluviatilis*	Native
Brook lamprey	*Lampetra planeri*	Native
Salmon	*Salmo salar*	Native
Sea/Brown trout	*Salmo trutta*	Native
Rainbow trout	*Oncorhynchus mykiss*	Occurring regularly in catchment (Lamond 1931; Maitland 1966) presence probably only maintained by farm escapees.
Arctic charr	*Salvelinus alpinus*	2 historic records (Ure 1795; Bidie 1896) neither confirmed
Brook charr	*Salvelinus fontinalis*	Introduced c. 1876 (Brown 1896) but failed to become established and died out
Powan	*Coregonus lavaretus*	Native
Pike	*Esox lucius*	Native
Crucian carp	*Carassius carassius*	Introduced - first recorded 1991 (Adams & Mitchell 1992) establishment success not yet known
Gudgeon	*Gobio gobio*	Introduced - established by early 1980's (Maitland *et al.* 1983)
Tench	*Tinca tinca*	Introduced - (Young 1870; Lumsden & Brown 1895) but failed to become established and died out - no records from the 20th century.
Minnow	*Phoxinus phoxinus*	Native
Roach	*Rutilus rutilus*	Native
Chub	*Leuciscus cephalus*	Introduced - established by mid 1980's (Adams *et al.*, 1990)
Dace	*Leuciscus leuciscus*	Introduced - established by mid 1980's (Adams *et al.*, 1990)
Stone loach	*Noemacheilus barbatulus*	Native
Eel	*Anguilla anguilla*	Native
Three-spined stickleback	*Gasterosteus aculeatus*	Native
Ten-spined stickleback	*Pungitius pungitius*	Native
Eurasian perch	*Perca fluviatilis*	Native
Ruffe	*Gymnocephalus cernuus*	Introduced - established by mid 1980's (Maitland *et al.*, 1983; Maitland & East 1989).
Flounder	*Platichthys flesus*	Native
Thick-lipped mullet	*Crenimugil labrosus*	previously recorded only in River Leven (Lumsden & Brown 1895; Scott & Brown 1901) but personal recollections of R. McMath, University Field Station of mullet of unknown species from gill nets in main loch in early 1970's.
Plaice	*Pleuronectes platessa*	a record of this species in Loch Lomond by Scott & Brown (1901) is highly dubious not normally regarded as a freshwater species

In 1981, a fish species new to the catchment, the gudgeon (*Gobio gobio*) was discovered in the River Endrick. This species apparently made its way into the system from a small loch in the catchment to which it had been deliberately introduced (Maitland *et al.*, 1983).

This record was followed in 1982 by the discovery of another fish species new to Loch Lomond, the ruffe

(*Gymnocephalus cernuus*). This species had not been recorded in Scotland before this and was thus 250 km outwith its previously reported range in the UK (Maitland *et al.*, 1983).

In 1987 yet two more species new to the catchment, dace (*Leuciscus leuciscus*) and chub (*Leuciscus cephalus*) were found in the lower reaches of the River Endrick (Adams *et al.*, 1990). Prior to this, both species had a more southerly UK distribution.

In 1991 one further species not native to the catchment, crucian carp (*Carassius carassius*) was discovered there, (Adams & Mitchell, 1992) as with the other new species, crucian carp is native to more southerly parts of the UK.

Vector of introductions

Studies of isolated communities such as are found in island ecosystems have shown that natural invasions of isolated habitats resulting from dispersal of species is a relatively rare phenomenon (Roughgarden, 1986). For obligate freshwater fish species with poor powers of dispersal, Loch Lomond is an isolated ecosystem. However the rapid changes in the fish community composition have not resulted from an unusual biological event, rather they are almost certainly the result of human activity.

Gudgeon are known to have been deliberately introduced into a pond in the catchment of the River Endrick, from where they made their way into the river system (Maitland *et al.*, 1983). The other four species probably all emanate from the discarded live-bait of pike anglers collected from catchments in England and/or Wales (Carnell, 1987).

The current status of fish community

To assess any change in the fish community structure, following the introduction of these new fish species, a wide ranging survey was undertaken between November 1988 and October 1989. Over this period fish were collected regularly by multipanel, gill net (19–50 mm mesh size) at 5 littoral zone sites from the extreme north to the extreme south of the main loch. Fish were also collected by gill net, seine net and by electro-fishing in the lower reaches of the River Endrick (Fig. 1) (for full details of the survey see Adams & Tippett, 1990).

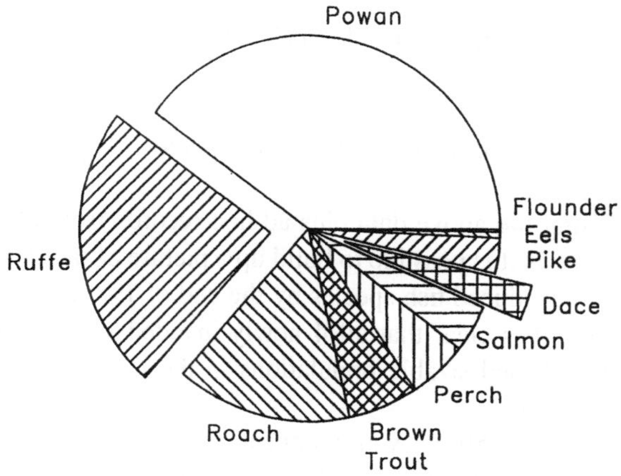

Fig. 2. Catch composition of gill net captures (all sites combined) corrected for catch-perunit-effort. Exploded segments - introduced fish species.

Loch captures

Ten fish species were recorded in catches at the five loch sampling sites. Overall powan dominated catches (Fig. 2), making up 40% of all fish collected, however introduced ruffe were also extremely common making up 24% of catches. Other fish species included in catches were: roach 15%; brown trout (*Salmo trutta*) - 6%; perch (*Perca fluviatilis*) - 5%; Atlantic salmon - 4%; pike - 3% and introduced dace - 3%. Eels (*Angulla angulla*) and flounders made up less than 1% of catches.

Introduced ruffe were the most common species in catches at two of the five loch sites examined, second most common at Inverbeg Bank, and was ranked third most common at the other two sites. Introduced dace were only recorded at the two most southerly sites, (Ross Priory Bay and Auchentllllich Bay) (Fig. 1).

River captures

Figure 1 also shows the catch composition data for sites in the lower reaches of the River Endrick. Because these data result from more than one collection technique they are more comprehensive in their coverage of the species present, however as a result they are also not directly comparable with loch catches.

Fourteen species of fish were recorded in the lower reaches of the River Endrick. Underyearling minnows and roach dominated catches such that on several occasions they could not be counted. All four newly introduced species recorded by 1989 were found in this area. Ruffe made up 16% of catches, dace 28%, chub

2.3% and gudgeon 2.3% of all catches excluding roach and minnows.

The establishment of introduced species

Using the above data plus ad hoc records from other sources it is possible to build up a picture of the status of introduced fish species by the end of 1989. By this time it is clear that all 4 newly invaded species had established sizable populations in the Loch Lomond catchment.

By 1989 it would appear that chub, had not increased their range in the catchment beyond the lower reaches of the River Endrick. A waterfall, the Potts of Gartness 18 km from the mouth of the river presents an impassable barrier to further upstream expansion of the population for this species. There is, however, no barrier preventing distribution downstream, yet to date there have been no records of this species in the main loch. There is some question of the long term viability of this species as according to Wheeler (1969) chub do not spawn until water temperatures reach 15 °C. This is only acheived for a very short period in mid summer (Maitland, 1966b). Despite this, they are commonly caught by coarse anglers in the River Endrick, who regard this species as common there.

Gudgeon have also become well established in the lower reaches of the River Endrick below the Potts-of-Gartness, where they are relatively abundant in electro-fishing catches. This species has recently (1990-91) also moved into the main loch where it is commonly caught in fyke traps set for eels in the Balmaha area. There are no records of this species from other localities.

Dace would appear to have been more successful in establishment than chub and gudgeon. When they were first formally recorded in 1987, it was clear that they were abundant in the lower reaches of the River Endrick below the Potts-of-Gartness. However at this time they were not established in some areas of the Endrick system accessible to them (the River Blane) nor had they been recorded in the main loch. However in the following years, dace expanded their range in the catchment. By 1989 ad hoc electro-fishing in the Blane showed that they were now resident there. In addition, by the end of 1989 dace were commonly collected in gill nets set throughout the south end of the main loch (Fig. 3). Since then dace have continued to extend their range in the main loch and have now been collected as far north as Rowardennan (1991) (pers. obs.).

Of all the introduced species recorded, the one that has shown the most spectacular powers of establishment has been the ruffe. Maitland & East (1989) have demonstrated the remarkable population growth of this species by monitoring the number of dead fish collected from trash screens at the Ross Priory Pumping Station on the south shore of Loch Lomond. The first ruffe were recorded there in 1982, in the years following this, the numbers rose exponentially (Maitland & East 1989). Data presented here (Fig. 1) shows that the dramatic ruffe population explosion occurring at Ross Priory was mirrored throughout the loch. Ruffe are now almost ubiquitous throughout the catchment, only excluded from areas such as the upper reaches of tributaries where they are prevented from invading by impassable barriers and from tributaries with a high water velocity, ruffe are poor swimmers.

There is every reason to suspect that the latest recorded fish introduction, crucian carp, may also now be establishing itself. Electro-fishing at the site of its first discovery showed that more than one year class was present, according to Wheeler (1969) this species is capable of spawning at relatively low temperatures.

Thus, after a long period of stability lasting at least 2 centuries and probably even longer, the fish community, as a result of a series of introductions of new species, is now undergoing a period of instability unprecedented in its recorded history (Fig. 4). In the last comprehensive review of the fish community of Loch Lomond, Maitland (1972) recorded 15 established fish species now, only two decades later, at least 19 (and probably 20) species are established there, 35% of all British freshwater fish species (Maitland, 1977).

Factors affecting colonisation success in Loch Lomond

The colonisation of Loch Lomond by ruffe and to a lesser extent by dace, described here, has clearly been highly successful. Ruffe have extended their range throughout the accessible parts of the catchment and its population size has increased explosively over the decade since its arrival, such that it is now probably the commonest species there. With the benefit of hindsight it is clear that ruffe is a highly successful coloniser species but it is important to consider why has ruffe been such a successful coloniser, when others such as tench and brook charr have not? What are the characteristics of ruffe that have enabled successful colonisation to occur? And are there any generalisations that

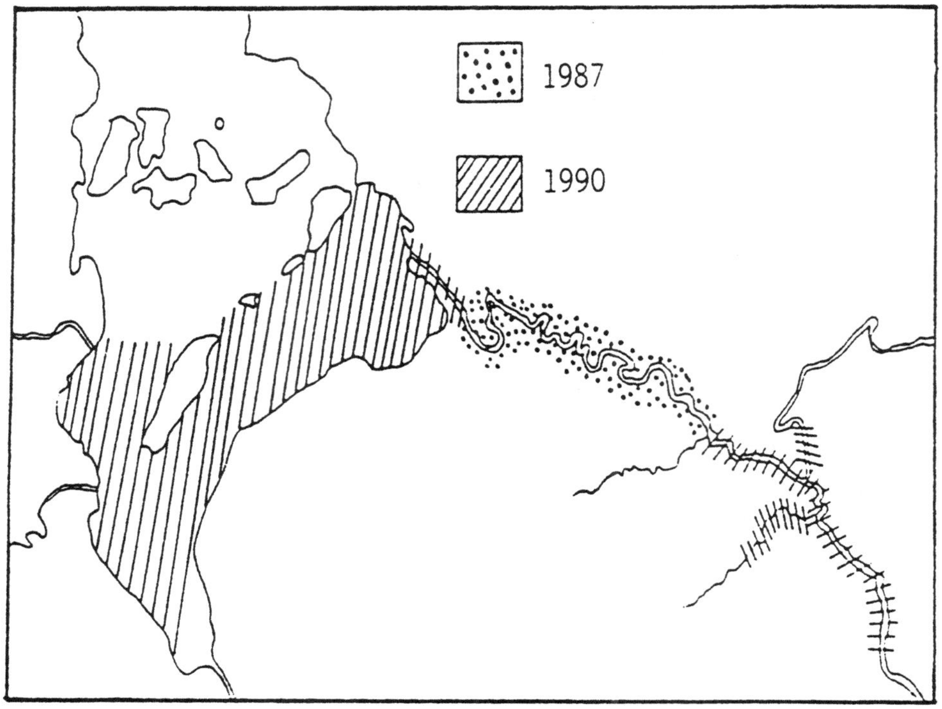

Fig. 3. Range expansion by dace between 1987 and 1989 in the Loch Lomond catchment.

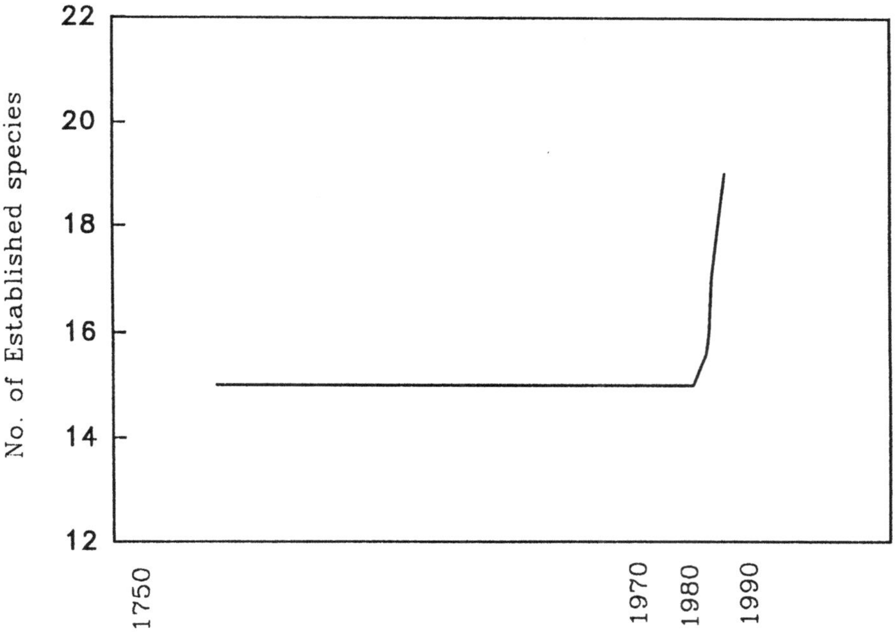

Fig. 4. Historical changes in species composition of the Loch Lomond fish community.

can be made about the characteristics of successfully colonising species?

Predicting which species are likely to be successful colonists following invasion has proved to be highly unsuccessful in the past. It would appear that there is no one single biological characteristic distinguishing those species that are successful colonists from those that are not (Moyle, 1986). However, some general characteristics of successful invaders have been identified, (Ehrlich, 1986; Moyle,1986) and it is interesting to see which of these characters might apply to ruffe.

According to Ehrlich (1986) and Moyle (1986) in general, successful colonisers are likely to be:

1) Omnivorous - ruffe are extremely catholic in their diet, including a wide range of invertebrates, fish ova and fish (Johnson, 1965; Adams & Tippett, 1991).
2) Hardy - able to function in a wide range of environments - ruffe have a relatively high temperature tolerance range with an upper lethal temperature of 31 °C (Varley, 1967) yet they can maintain foraging to temperatures as low as at least 4 °C (Bergman, 1987; Adams & Tippett, 1991).
3) Short generation time/high fecundity - the age at which 50% of individuals spawn in a stable ruffe population has been reported as 1 year old for males and 2 years for females (Varley, 1967) - fecundity is relatively high (Muus & Dahlstrom, 1967).
4) Widely distributed in their original range - (see Wheeler, 1969)
5) Genetically heterogeneous - this is unknown for ruffe.
6) Associated with humans - the translocation of ruffe to Loch Lomond was by man.

It is clear from the above that ruffe fit most of the criteria of a successful colonising species. Thus it is perhaps not surprising that ruffe have recently been accidentally introduced to Lake Superior, North America (Pratt *et al.*, 1992; Selgeby & Ogle, 1991) and to Lake Constance (R. Rosch, pers. comm.) in both lakes they are apparently colonising very successfully.

Successful colonisation however, is not solely a function of the coloniser but also depends on the community it is invading. Therefore it is important to ask what characteristics of the invasion site have enabled colonisation to occur. Several authors have examined the characteristics of communities that make them suseptable to successful invasion. Orians (1986) states that invasion is more likely to be successful if the ecosystem is undergoing a period of disturbance. Evidence would suggest that Loch Lomond was not undergoing a period of disturbance during colonisation by ruffe although one could speculate that community disturbance caused by colonisation by ruffe may have aided other invading species. Moyle (1986) has shown that communities are likely to be more suseptable to invasion by species from within the same country than by exotic (non-endemic) species. All five non-native species recently recorded in Loch Lomond fit into this category. In addition Case (1990) has shown that the probability of successful establishment of an invading species increases with decreasing community size and with increasing average strength of competition between species in the invaded community. Although there is little known about the strength of competition between native species in Loch Lomond, the fish community, as a result of being further from sites of dispersal of freshwater fish following the last ice age, is a relatively impoverished one, compared with communities further south.

Thus it would appear that, *a priori* ruffe would appear to be a potentially invasive species and that the fish community of Loch Lomond, as with most others in Scotland, might be judged to be susceptible to invasion by new species. Thus if the identity of Scottish fish communities is to be maintained, protection from the risk of accidental introductions must be an important consideration.

The ecological impact of the introduction and colonisation by new fish species

Given that now at least four new species have become established in Loch Lomond, it is important to establish what changes in the ecosystem have resulted. The potential effects of the introduction of new fish species on the existing fish species has been explored by Welcomme (1986) who suggests 5 potential effects.

The concomitant introduction of new parasites or disease

Because obligate freshwater fish communities are by their nature isolated from each other, they are also buffered from the effects of the transmission of disease from one freshwater community to another. Thus the introduction of fish obtained from outwith the catchment considerably increases the risk of also transmitting a parasite or disease to which that catchment has not yet been exposed.

This effect has been documented with the introduction of a crustacean parasite *Argulus foliaceous* along with dace into lakes in the English Lake District (Fryer, 1982). This parasite is now also commonly found on fish in Loch Lomond and has been found 'free living' in benthic samples (Adams, unpublished data; S. Rushton-Mellor pers. comm.) where it had not previously been recorded (Cambell, 1971) however it is impossible to determine with certainty if this parasite has simply been overlooked in previous surveys of fish parasites or whether it is an example of concomitant parasite introduction.

Hybridisation with or gene pool changes to native species

Isolated fish populations may frequently show local adaptations to local environmental conditions, much of this adaplation may be genetically based. Introduction of fish of the same species with differing adaptations may result in genetic dilution of locally adaptive traits. In addition many related fish species will relatively readily hybridise. This is particularly prevalent in the cyprinids. On the 11 September 1989 a single specimen of a roach (*Rutilus rutilus*) X bream (*Abramis brama*) hybrid was collected from Loch Lomond. The origin of this fish is uncertain, however one possibility is that it is the first generation hybrid progeny of a previously undelected population of bream and native roach (Adams & Maitland, 1991).

Competition with native fish species

Clearly introductions of competitively superior species with niches overlapping with existing native species may exclude native species from resources that are in short supply such as food, feeding sites, spawning areas, etc.

Comparing the diet of dace with salmon parr shows that the diet of these two species overlaps considerably. Mann (1974) examined the diet of dace in the River Stour, and found 81% of the prey items (by number) were Ephemeroptera, Trichoptera and Diptera, compared with 82% for salmon parr from the River Endrick (Maitland, 1965). In addition dace and salmon parr appear to be occupying similar habitats in the River Endrick (Adams, unpublished data). This clearly points to potential competition for food between these two species.

Environmental change to the ecosystem

This may occur when an introduced fish species changes the environment to the detriment of a native species.

There is no evidence that this effect may be occurring in Loch Lomond.

Changes in the relationship between predator and prey

This effect is certainly the most widely reported effect resulting from the introduction of new fish species to inland waterways. The best known example is that of the deliberate introduction of Nile perch (*Lates niloti-*

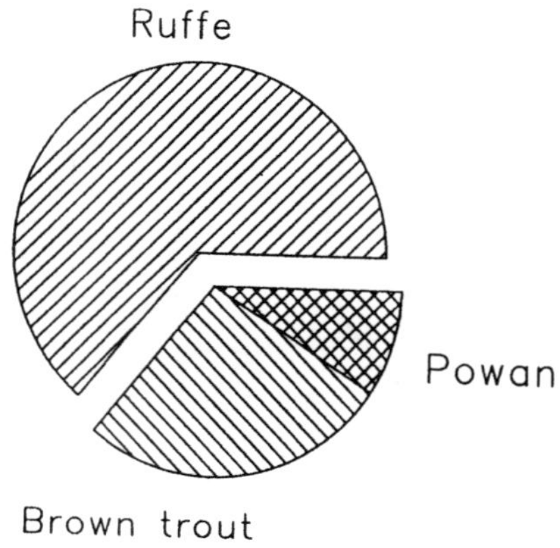

Fig. 5. The proportion of powan ova in stomachs of each of the three main fish predators.

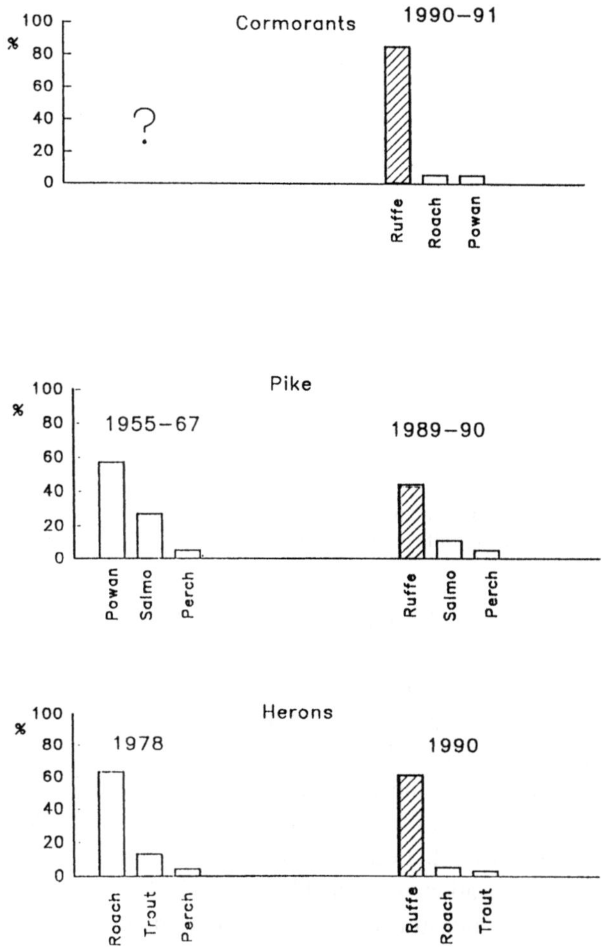

Fig. 6. The occurrence of fish prey in the diets of cormorants, pike and herons, before and after the establishment of new fish species in Loch Lomond.

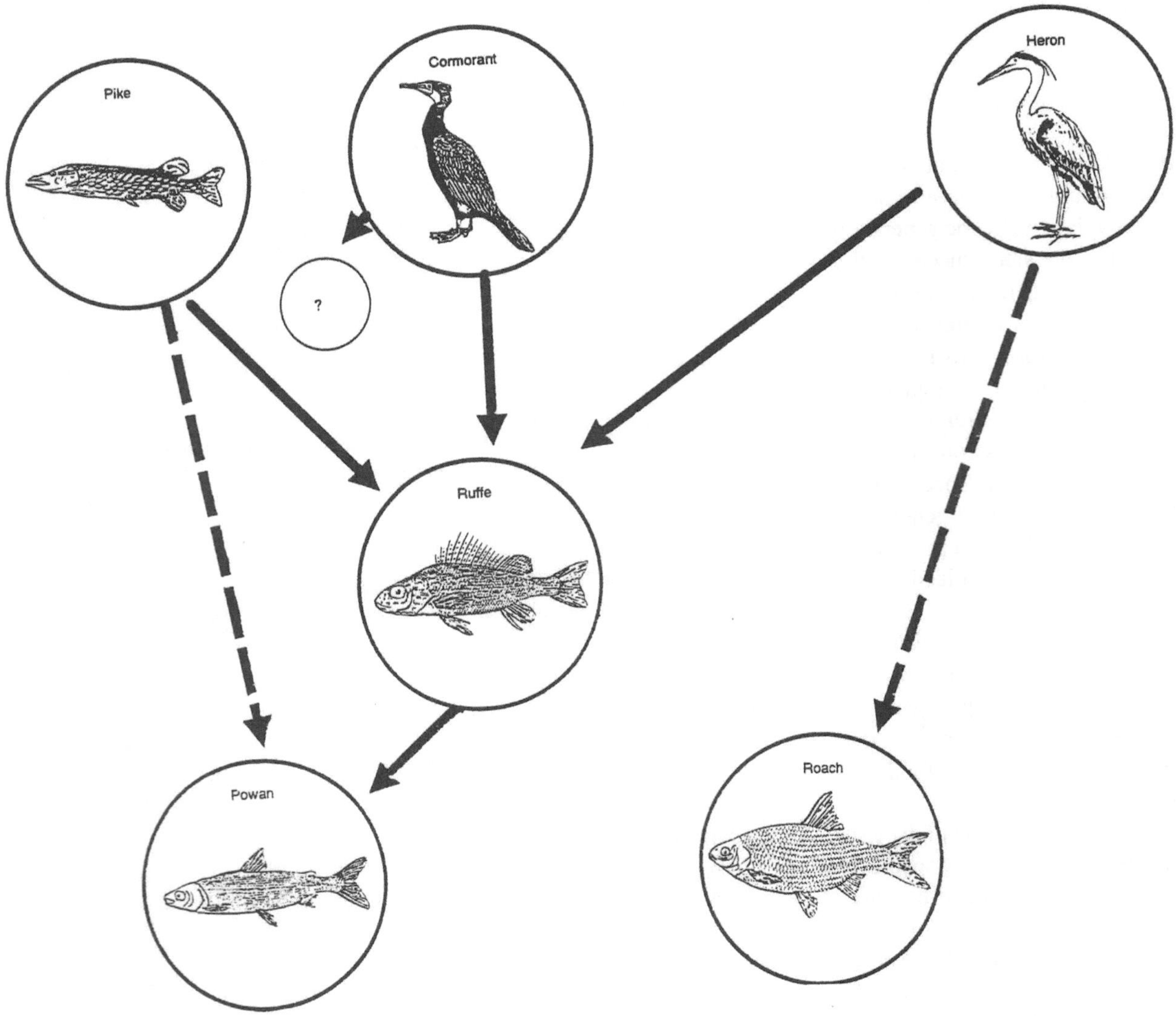

Fig. 7. Interspecific predator/prey interactions that have become newly established (solid lines) or weekened (broken lines) as a result of the introduction of ruffe to Loch Lomond.

cus) to Lake Victoria, East Africa, in 1960. This pisciverous species quickly became established and rapidly devastated commercially important populations of prey species such as the endemic haplochromines and the tilapia, *Oreochromis sp.* driving these species to extinction in some areas (Barel *et al.*, 1985).

When it was discovered that ruffe had been introduced to Loch Lomond, one potential repercussion of colonisation was quickly established. Pokrovskii (1961) demonstrated that ruffe were regulating the population size of *Coregonus albula*, a close relative of the powan, in Russian lakes by affecting spawning success through ova predation. To determine if a similar interaction between ruffe and powan was occurring in Loch Lomond, the diet of fish foraging on the powan spawning grounds at the Ross Isles was examined during the spawning seasons 1988–90.

The results of this study showed that powan themselves as well as native brown trout and introduced ruffe were foraging on powan ova during the incubation period following egg deposition (Adams & Tippett, 1991). However because unlike powan and brown trout, ruffe foraging activity is not curtailed by the low water temperatures found at this time (5– 7 °C) (Bergman, 1987; Adams & Tippett, 1991) ruffe intake of ova was significantly greater than that of the other predators. Ruffe accounting for 64% of all observed ova predation by fish (Fig. 5). Although we do not know the ultimate effects of ruffe predation on ova, it is known that powan ova mortality rate from other sources is very high (see Slack, (1957); Brown & Scott (1994)) and there is a risk that increased mortality from ruffe predation of powan ova may seriously impact on the powan population.

The relationship between powan and its predators is not the only predator-prey relationship to have been affected by new fish introductions. Between 1990 and 1991 the diet of cormorants (*Phalacrocorax carbo*) feeding on Loch Lomond was monitored by regular examination of regurgitated remains recovered at the main roosting site (Fig. 6). This data shows that an estimated 85% of the diet of this species is now ruffe (see Adams *et al.* (1994)). Clearly ruffe represent a large, abundant and available food source for cormorants which did not exist before 1980.

One pisciverous species for which good data exist on diet before recent introductions is pike. Shafi (1969) examined the diet of pike between 1955 and 1967. At this time pike were feeding principally on powan (57%). To look for a change in the feeding strategy following changes in the fish community, I repeated Shafi's study in 1989–90. The results (summarised in Fig. 6) demonstrate a clear change in pike foraging strategy with a shift in their diet from principally powan, a fast moving pelagic species, to ruffe, a poor swimming, benthic species (Adams, 1991).

One other fish eating species that has also changed its diet following new fish introductions, is the heron (*Ardea cinerea*). Giles (1981) examined the diet of herons feeding at the Gartfairn heronry in the south-east corner of Loch Lomond in 1978. At this time, herons were principally feeding on roach (63%). When the diet of herons at this colony was re-examined in 1990 (Fig. 6) it was found that they had shifted there prey choice to primarily ruffe (61%).

It is clear that the introduction of new fish species (principally ruffe) has had major repercussions for the long-term, well established, predator-prey relationships that have been in existence for several centuries. Figure 7 sumarises those predator/prey relationships that are known to have altered, it is clear that there may be many more interspecific interactions between predator/prey; host/parasite; resource competitors, that are also likely to have changed.

I have shown here that Loch Lomond is currently in a phase of rapid flux, resulting from the introduction and successful colonisation of the loch by at least four fish species new to the catchment. It will take many years for all the repercussions for the ecosystem of these introductions to become clear however data sumarized here shows that the effects already quantified are far-reaching, extending throughout the ecosystem, and fundamental, changing the basic long-established relationships between species.

Acknowledgments

The new work described here could not have been achieved without the assistance and support of David Brown. Thanks to Roger Tippett, Rab McMath and numerous other volunteers who gave freely of their time and advice. Much of the data presented here was collected with support from Scottish Natural Heritage.

References

Adams, C. E., 1991. Shift in pike, *Esox lucius* L., predation pressure following the introduction of ruffe, *Gymnocephalus cernuus* (L.) to Loch Lomond. J. Fish Biol. 38: 663–667.

Adams, C. E., D. W. Brown & L. Keay, 1994. Elevated predation risk associated with inshore migrations of fish in a large lake, Loch Lomond, Scotland. Hydrobiologia 290: 135–138.

Adams, C. E., D. W. Brown & R. Tippett, 1990. Dace (*Leuciscus leuciscus* (L.)) and chub (*L. cephalus* (L.)) new introductions to Loch Lomond. Glasg. Nat. 21: 509–513.

Adams, C. E. & P. S. Maitland, 1991. Evidence of further invasions of Loch Lomond by non-native fish species with the discovery of a roach X bream, *Rutilus rutilus* (L.) X *Abramis brama* (L.), hybrid. J. Fish Biol. 38: 961–963.

Adams, C. E. & J. Mitchell, 1992. Introduction of another non-native fish species to Loch Lomond: Crucian carp (*Carassius carassius* (L.)). Glasg. Nat. 22: 165–168.

Adams, C. E. & R. Tippett, 1990. The status of the fish community. Report to the Nature Conservancy Council, Hope Terr., Edinburgh, 29 pp.

Adams, C. E. & R. Tippett, 1991. Powan, *Coregonus lavaretus* (L.) ova predation by newly introduced rufre, *Gymnocephalus cernuus* (L.), in Loch Lomond, Scotland. Aqua. Fish. Mgmnt 22: 239–246.

Barel, C. D. N., R. Dorit, P. H. Greenwood, G. Fryer, N. Hughes, P. B. N. Jackson, H. Kawanabe, R. H. Lowe-McConnel, M. Nayoshi, A. J. Ribbink, E. Trexavas, F. Witte & K. Yamaoka, 1985. Destruction of fisheries in Africas lakes. Nature 315: 19–20.

Bergman, 1987. Temperature dependent differences in foraging ability of 2 percids, *Perca fluviatilis* (L.) and *Gymnocephalus cernuus*. Envirn. Biol. Fish 19: 45–53.

Bidie, G., 1896. Char in Loch Lomond. Ann. Scot. Nat. Hist. p 258.

Brown, A., 1891. The fishes of Loch Lomond and its tributaries. Scot. Nat. 10: 114–124.

Brown, A., 1896. Char in Loch Lomond. Ann. Scot. Nat. Hist. p 192.

Brown, E. A. R. & D. B. Scott, 1994. Life histories of the powan, *Coregonus lavaretus* (L.) (Salmonidae, Coregoninae) of Loch Lomond and Loch Eck. Hydrobiologia 290: 121–133.

Cambell, A. D., 1971. The occurrence of *Argulus* (Crustacea: Branchiura) in Scotland. J. Fish Biol. 3: 145–146.

Carnell, D., 1987. Pike baits have fled the loch. Anglers Mail, 3rd October 1987.

Case, T. J., 1990. Invasion resistance arises in strongly interacting species-rich model competilion communities. Proc. Natl. Acad. Sci. 87: 9610–9614.

Ehrlich, P. R., 1986. Which animals will invade? In The Ecology of Biological Invasions of North America and Hawaii (H. A. Mooney & J. A. Drake, eds) Ecological Studies 58, Springer Verlag, New York.

Fryer, G., 1982. The parasitic copedoda and branchiura of British freshwater fishes. F. B. A. Publication No. 46, Freshwaler Biological Association, Windermere.

Giles, N., 1981. Summer diet of the grey heron. Scott. Birds 11: 153–159.

Hunter, W. R., M. D. Slack, & M. R. Hunter, 1959. The lower vertebrates of the Loch Lomond district. Glas. Nat. 18: 84–90.

Lamond, M., 1931. Loch Lomond. Jackson, Glasgow, pp. 340.

Johnson, P., 1965. Studies on the distribution and food of ruffe (*Acerina cernua*) in Denmark with notes on other aspects. Danmarks Fiskeri-og Huvundersogelser 4: 137–156.

Lumsden, J. & A. Brown, 1895. A guide to the natural history of Loch Lomond and neighbourhood. Bryce, Glasgow, 103 pp.

Lyle, A. & P. S. Maitland, 1992. Conservation of freshwater fish in the British Isles: the status of fish in National Nature Reserves. Aquatic Conservation: Marine and Freshwater Ecosystems 2: 19–34.

Maitland, P. S., 1965. The feeding relationships of salmon, trout, minnows, stone loach and three-spined-stickleback in the River Endrick, Scotland. J. Anim. Ecol. 34: 109–153.

Maitland, P. S., 1966a. Rainbow trout, *Salmo irideus* Gibbons, in the Loch Lomond district. Glasg. Nat. 18: 421–423.

Maitland, P. S., 1966b. Studies on Loch Lomond II - the fauna of the River Endrick. Blackie, Glasgow. 194 pp.

Maitland, P. S., 1972. Loch Lomond: Man's effects on the salmonid community. J. Fish. Res. Bd Can. 29: 849–860.

Maitland, P. S., 1977. A coded checklist of animals living in freshwater in the British Isles. ITE, Edinburgh. 76 pp.

Maitland, P. S. & K. East, 1989. An increase in the numbers of ruffe *Cymnocephalus cernua* (L.) in a Scottish loch from 1982 to 1987. Aqua. Fish. Mgmnt 20: 227–228.

Maitland, P. S., K. East & K. H. Morris, 1983. Ruffe, *Gymnocephalus cernua* (L.) new to Scotland in Loch Lomond. Scott. Nat. 7–9.

Maitland, P. S., K. H. Morris & K. East, 1994. The ecology of lampreys (Petromyronidae) in the Loch Lomond area. Hydrobiologia 290: 105–120.

Mann, R. H. K., 1974. Observations on the age, growth, reproduction and food of dace *Leuciscus leuciscus* (L.) in two rivers in southern England. J. Fish Biol. 6: 237–253.

Moyle, P. B., 1986. Fish introductions into North America: Patterns and ecological impact.In The Ecology of Biological Invasions of North America and Hawaii (Mooney, H. A. & Drake, J. A. eds) Ecological Studies 58, Springer-Verlag, New York.

Muus, B. J. & P. Dahlstrom, 1967. Freshwater fish of Britain and Europe. Collins, London. 148 pp.

Orians, G. H., 1986. Site characteristics favouring invasion. In The Ecology of Biological Invasions of North America and Hawaii (Mooney, H. A. & Drake, J. A. eds) Ecological Studies 58, Springer-Verlag, New York.

Pratt, D. M., W. H. Blust & J. Selgeby, 1992. Ruffe, *Gymnocephalus cernuus*: newly introduced in North America. Can. J. Fish. Aquat. Sci. 49: 1616–1618.

Pokrovski, V. V., 1961. Basic environmental factors determining the abundance of whitefish. Trudy Soveshchanii, 13: 228–234.

Regan, C. T., 1911. British Freshwater Fishes, Methuen, London.

Robertson, D., 1870. On *Petromyzon fluviatilis* and its mode of preying on *Coregonus clupeoides*. Proc. Nat. Hist. Soc. Glasg. 2: 61–63.

Roughgarden, J., 1986. Predicting invasions and rates of spread. p 179–188. In The Ecology of Biological Invasions of North America and Hawaii (Mooney, H. A. & Drake, J. A. eds) Ecological Studies 58, Springer-Verlag, New York.

Scott, T. & A. Brown, 1901. The marine and freshwater fishes. In The fauna, flora and geology of the clyde area (Elliot, G. F., Laurie, M. & Murdoch, J. B., eds) Brit. Assoc. Adv. Sci., Glasgow.

Selgeby, J. H. & D. H. Ogle, 1991. Status of ruffe in Lake Superior. Report to the Great Lakes Fishery Commission, March 1991.

Shafi, M., 1969. Comparative studies of populations of perch (*Perca fluvialilis*) and pike (*Esox lucius*) in two Scottish lochs. Ph.D. thesis, University of Glasgow, 143 pp.

Slack, H. D., 1957. Studies on Loch Lomond -1. Blackie, Glasgow. 133 pp.

Stewart, J., 1796. Parish of Luss. Statistical Account of Scotland 17: 238–271.

Tippett, R., 1994. An introduction to Loch Lomond. Hydrobiologia 290: xi–xv.

Ure, D., 1795. Parish of Killearn. Stast. Acc. Scot. 16: 100–129.

Varley, M. E., 1967. British freshwater fishes - factors affecting their distribution. Fishing News Books, London, 148 pp.

Welcomme, R. L., 1986. International measures for the control of introductions of aquatic organisims. Fisheries 11: 4–9.

Wheeler, A., 1969. The fishes of the British Isles and North-west Europe. Macmillan, London, 613 pp.

Young, J., 1870. Tench from Loch Lomond. Proc. Nat. Hist. Soc. Glasg. 2: 67.

Hydrobiologia **290**: 103–104, 1994.
K. J. Murphy, M. C. M. Beveridge & R. Tippett (eds), The Ecology of Loch Lomond.

The importance of Loch Lomond National Nature Reserve for Fish

A. A. Lyle & P. S. Maitland
Institute of Freshwater Ecology, Penicuik EH26 OQB, UK; Fish Conservation Centre, Stirling FK7 9QX, UK

Key words: Loch Lomond, nature reserves, fish conservation

Abstract

Loch Lomond NNR with 19 fish species is ranked top in fish conservation value of 235 NNRs in Great Britain due to species diversity, presence of rare species and unusual forms and combinations.

A recent survey of freshwater fish in 235 National Nature Reserves (NNRs) throughout Great Britain identified Loch Lomond NNR as being of outstanding importance for fish (Lyle & Maitland, 1992).

From the survey results, a fish conservation value was assessed for each NNR based on species diversity, rare species, unusual forms and unusual combinations (see Maitland, 1985). Very few NNRs are valuable for fish conservation, but outstanding amongst them is Loch Lomond NNR, which alone has all these four requirements.

The reserve contains various freshwater types including lowland river reaches (River Endrick), large and small streams and ponds, marsh areas and a small part of Loch Lomond itself. The total of 19 species of fish present there (Table 1) is not only very high for that geographic area, it is also the greatest of any NNR in Great Britain. Loch Lomond is well known for the rare species, Powan, however, the NNR covers only a very small part of the loch, excluding the main Powan habitat and spawning grounds. The River Endrick and Loch Lomond have a unique, exclusively freshwater race of River Lampreys, which form an unusual association with the Powan, upon which they prey (Maitland, 1980).

The above factors make Loch Lomond NNR one of the most valuable and important sites for freshwater fish in Great Britain. However, although the high diversity of fish species present contributes significantly to the sites' conservation status, four species (Table 1) are recent introductions and are considered to degrade the integrity of the fish community. Indeed, one such introduction, that of Ruffe (Maitland *et al.*, 1983), is now so abundant and potentially harmful (by egg predation) to the sites' only rare species, Powan (Adams & Tippett, 1991), that this conservation factor, plus the Powan/River Lamprey interaction factor, may become seriously threatened.

The management action needed to conserve the scientific value of this important site must aim primarily at protecting the habitat, both in terms of water quality and the physical habitat. In addition, the status of the biological community should be protected as far as possible by preventing any further fish introductions.

As well as habitat conservation the needs of the rarer fish must be examined and direct action taken if necessary (Maitland & Lyle, 1991). Thus, since there are only two sites for Powan in Scotland (*i.e.* also Loch Eck) attempts have been made by the authors to establish safeguard populations for the Loch Lomond stock at two sites within the Lomond catchment (Maitland & Lyle, 1990). Consideration is also being given to the possibility of establishing another stock of River Lampreys, either in combination with one of these new populations of Powan, or at a separate new site.

This study was funded by the Nature Conservancy Council.

References

Adams, C. E. & R. Tippett, 1991. Powan *Coregonus lavaretus* (L.) ova predation by newly introduced ruffe *Gymnocephalus cernua* (L.) in Loch Lomond, Scotland. Aquat. Fish. Mgmnt. 22: 261–267.

Lyle, A. A. & P. S. Maitland, 1992. Conservation of freshwater fish in the British Isles: the status of fish in National Nature Reserves. Aquat. Conserv. 2: 19–34.

Table 1. Fish species present in Loch Lomond NNR (*indicates recently introduced species)

Sea lamprey	*Petromyzon marinus*	Chub*	*Leuciscus cephalus*
River lamprey	*Lampetra fluviatilis*	Dace*	*Leuciscus leuciscus*
Brook lamprey	*Lampetra planeri*	Stone loach	*Noemacheilus barbatulus*
Atlantic salmon	*Salmo salar*	Eel	*Anguilla anguilla*
Brown trout	*Salmo trutta*	Three-spined stickleback	*Gasterosteus aculeatus*
Powan	*Coregonus lavaretus*	Ten-spined stickleback	*Pungitius pungitius*
Pike	*Esox lucius*	Perch	*Perca fluviatilis*
Gudgeon*	*Gobio gobio*	Ruffe*	*Gymnocephalus cernua*
Minnow	*Phoxinus phoxinus*	Flounder	*Platichthys flesus*
Roach	*Rutilus rutilus*		

Maitland, P. S., 1980. Scarring of whitefish (*Coregonus lavaretus*) by European River Lamprey (*Lampetra fluviatilis*) in Loch Lomond, Scotland. Can. J. Fish. Aquat. Sci. 37: 1981–1988.

Maitland, P. S., 1985. Criteria for the selection of important sites for freshwater fish in the British Isles. Biol. Conserv. 31: 335–353.

Maitland, P. S., K. East & K. H. Morris, 1983. Ruffe *Gymnocephalus cernua* (L.), new to Scotland, in Loch Lomond. Scott. Nat. 7–9.

Maitland, P. S. & A. A. Lyle, 1990. Practical conservation of British fishes: current actions on six declining species. J. Fish Biol. 37: 255–256.

Maitland, P. S. & A. A. Lyle, 1991. Conservation of freshwater fish in the British Isles: the current status and biology of threatened species. Aquat. Conserv. 1: 25–54.

Hydrobiologia **290**: 105–120, 1994.
K. J. Murphy, M. C. M. Beveridge & R. Tippett (eds), The Ecology of Loch Lomond.

The ecology of lampreys (Petromyzonidae) in the Loch Lomond area

Peter S. Maitland[1], Kenneth H. Morris[2] & Kenneth East[3]
[1] *Fish Conservation Centre, Easter Cringate, Stirling FK7 9QX, UK*
[2] *Bankton Park East, Murieston, Livingston EH54 9BN, UK*
[3] *27a Greenhill Gardens, Morningside, Edinburgh EH10 4BL, UK*

Abstract

All three lamprey species occurring in western Europe are found in the Loch Lomond area: Sea Lamprey *Petromyzon marinus* L., River Lamprey *Lampetra fluviatilis* (L.) and Brook Lamprey *Lampetra planeri* (Bloch). There are major differences in ecology among the three species and in each some features seem to be unique to the Lomond populations. Available data on the ecology of the three species in the area are presented and existing threats reviewed. Future research and conservation requirements are outlined.

Introduction

Three species of lamprey occur in western Europe (Maitland, 1980a) and all of them are found in the Loch Lomond area: Sea Lamprey *Petromyzon marinus* L., River Lamprey *Lampetra fluviatilis* (L.) and Brook Lamprey *Lampetra planeri* (Bloch). Each species differs from the others in major aspects of its ecology and, at Loch Lomond, some features seem to be extremely unusual compared to other populations in the British Isles.

Most lampreys have a similar life cycle, which involves the migration of adults upstream to the spawning areas – normally stony or gravelly stretches in running water. There they spawn in pairs or groups, laying eggs in shallow depressions created by lifting away stones with their suckers. After hatching, the larvae, known as ammocoetes, swim or are washed downstream by the current to areas of sandy silt in slow-flowing water where they burrow and spend the next few years in tunnels. They are blind, the sucker is incomplete and the teeth are undeveloped. They feed by using cilia inside the mouth to create a current which draws organic particles and minute plants into the pharynx where they become entwined in a mucus string which is swallowed by the larva.

The metamorphosis from larva to adult takes place in a relatively short time – usually a few weeks – after several years of larval life. The rim of the mouth develops into a full sucker inside which are the rasping teeth; the skin becomes much more silvery and opaque except over the eyes where it clears to give proper vision for the first time. The transformed lampreys then migrate, usually downstream, away from the nursery areas.

Some species of lamprey, such as the Brook Lamprey, never feed as adults – after metamorphosing they spawn and then die – but most are parasitic on various other fish. Both Sea Lamprey and River Lamprey normally migrate to estuaries and the sea where they feed on various species of fish. However, at Loch Lomond, virtually all River Lampreys and even some Sea Lampreys, apparently uniquely in the British Isles, remain in Loch Lomond and feed there, mainly on coregonids and salmonids. They attach to the sides of these fish and rasp away the skin, ingesting it together with body fluids and underlying muscle. The prey may never recover from such an attack (especially if the body cavity is penetrated) and in some waters lampreys are serious pests of commercial fish stocks.

Relatively little work has been done on lampreys in Scotland, though all three native species occur in various rivers south of the Great Glen. Most of the Scottish references relate to short observations of natural history and there has been no serious study, prior to the present project, comparable to the work of Hardisty (1944, 1961, 1969, etc.) and others in England and Wales. Unfortunately, the work carried out previously in the Loch Lomond area by MacDonald (1959a, 1959b) is, for a variety of reasons, suspect and better ignored (Hardisty & Potter, 1971; Maitland, 1980a).

Methods

Most of the work for this study was carried out between 1980 and 1985, as part of a long-term programme of research on lampreys in Scotland (Maitland, 1980a, 1980b; Schoonoord & Maitland, 1983; Maitland *et al.*, 1984; Morris & Maitland, 1987; Morris, 1989). The population of River Lampreys in Loch Lomond is unique in the British Isles and so emphasis has been given to work on this species.

Studies of distribution in the Lomond area (Fig. 1) were carried out by electrofishing all main tributaries of the loch and its outflow, the River Leven. Some of this work was done in 1979–80 (Rosie & Maitland, 1980) and the rest during 1983–85. In addition, observations of key spawning areas were made from 1983–85 and questionnaires were issued to anglers and others in the area who were likely to come across lampreys. The questionnaires were especially directed at observations of lampreys feeding on fish.

In Loch Lomond, studies of the feeding behaviour of lampreys have been in progress for many years (Maitland, 1980b). Detailed studies of fish attacked by lampreys were carried out by the authors from 1980–83. Sampling of fish in Loch Lomond was by means of mixed-mesh gill nets (Maitland, 1981) and all fish taken were examined for fresh and healed lamprey wounds.

Trapping for downstream transformers was based on a trap design supplied by the U.S. Fish & Wildlife Service. It consisted of a large fine mesh (3 mm) net whose mouth was 2 m across and 1 m deep; the net was 3 m long and ended in a funnel device leading into a jar which could be emptied with ease. This net was fished continuously for 24 hours once each month during 1983 off the bridge at Woodend (site a in Fig. 1). Trapping, marking and population studies were carried out mainly during 1983–85 and were limited to the Rivers Endrick and Leven which are believed to be the most important migration routes, spawning sites and larval nursery areas for all three species of lamprey. A variety of traps were tested initially and eventually one standard design was selected and used in subsequent work (Morris & Maitland, 1987). Most of the trapping was carried out at the key site below Drymen Bridge (site b in Fig. 1), but traps were also used in the main river at Woodend, at Gartness and in the River Blane at Blane Bridge and at Dumgoyach.

Parallel work was also carried out at various sites to the east in the neighbouring catchments of the Rivers Forth and Teith.

Marking studies of larval lampreys were carried out at Alexandria on the River Leven and just below Drymen Bridge on the River Endrick, using a latex injection technique (Schoonoord & Maitland, 1983). Larvae were collected by electrofishing each site monthly from February 1984 to February 1985 using a backpack electrofisher; they were then transferred to the laboratory where they were anaesthetised, measured for length, marked with a code specific to the site and date and then returned to the site of capture.

All the animals caught at monthly intervals were then weighed, measured for length and various other body and fin measurements were taken. The weights of gonads of lampreys trapped at Drymen Bridge were recorded during the 1984–85 trapping period.

Adults trapped at Drymen Bridge were anaesthetised and then fin clipped by removing small semicircles from the edges of the dorsal and tail fins in positions which were unique to the date of capture. These lampreys were then released back into the river some 500 m below the trapping site and all adults captured subsequently were examined for such fin clips.

Some experimental work in the laboratory was carried out during 1983–85 in order to study the effect of fin clipping and to measure changes in body and fin size and shape over the period before spawning. For this study, 12 early upstream migrants (four females and eight males) were trapped and kept in the laboratory from 10 October 1984 to early April 1985 when they spawned and then died. Six animals were kept in running water in an indoor aquarium and six were kept in an outside shaded tank. Three animals in each tank were marked individually by removing semicircles from the edges of the fins of anaesthetised animals.

Sea Lamprey

Distribution

The Sea Lamprey is the largest but the least common of the three lampreys in the Loch Lomond area. Although adults of this species have been recorded from the River Endrick by Hunter, Slack & Hunter (1959) and Maitland (1966) none was recorded here during the present study. Adults do still ascend the River Leven, however, and there have been several independent reports of the spawning in the upper River Leven, between the barrage and the footbridge at Balloch. This is the only known spawning site in the Lomond catchment.

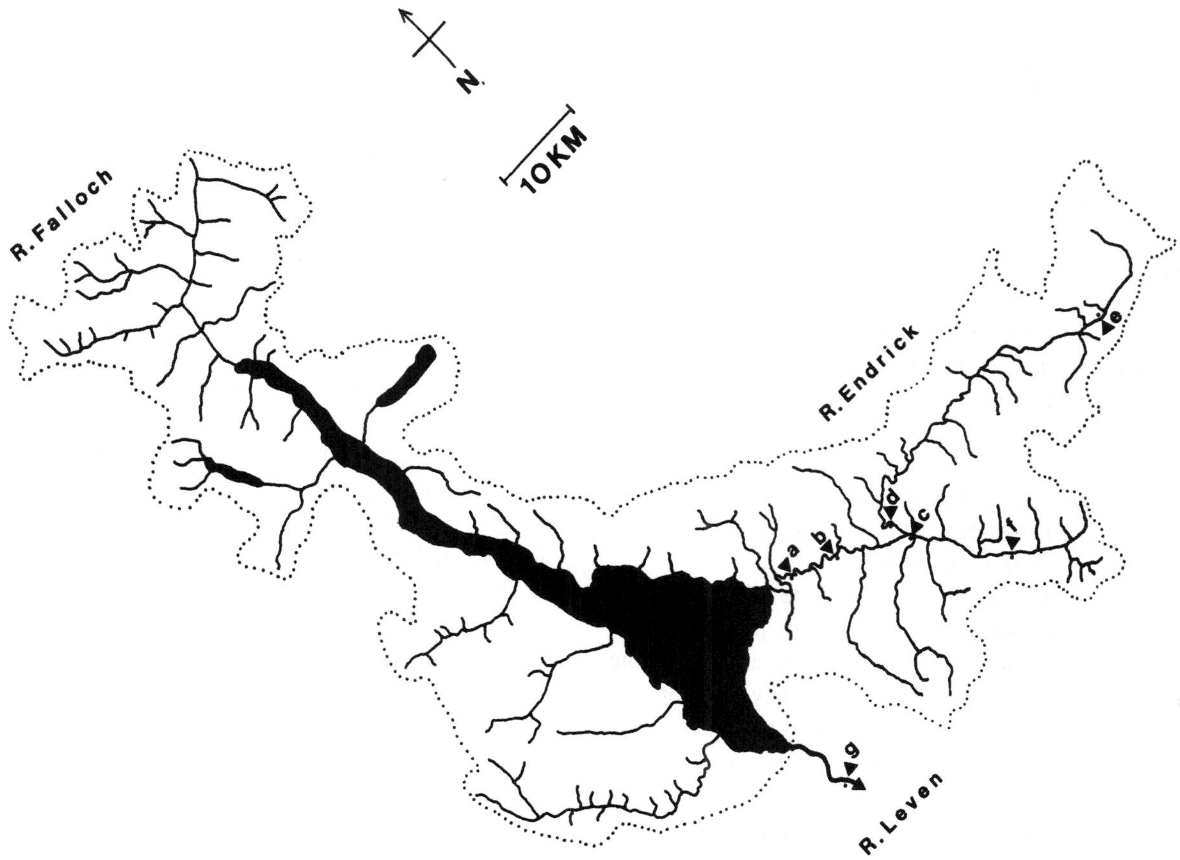

Fig. 1. Loch Lomond and its catchment, showing the major sites mentioned in the text. On the River Endrick: (a) Woodend, (b) Drymen Bridge, (c) Blane Bridge, (d) Pot of Gartness, (e) Loup of Fintry, (f) Dumgoyach. On the River Leven: (g) Alexandria.

Larval Sea Lampreys were found infrequently during the study, always in the River Leven. In all samples they were, however, only a small proportion of the catch of ammocoetes at any site.

Several anglers reported lampreys attached to Salmon *Salmo salar*, Sea Trout *Salmo trutta* and Powan *Coregonus lavaretus* in Loch Lomond. In addition, experienced net fishermen who have seine netted Loch Lomond over several decades have regularly seen lampreys attached to fish coming in with their nets both on Powan and on Salmon and Trout. Occasionally, several have been seen on one fish. Those on Powan are assumed to be River Lampreys, but this may not be the case with Salmon and Sea Trout for a specimen retrieved by an angler from a Salmon proved to be a feeding Sea Lamprey of 200 mm. In addition, the majority of lamprey wounds reported from salmonids have been below the lateral line, as is the case where Sea Lampreys attack fish in the Great Lakes (Lennon, 1954). This is in contrast to River Lampreys which wound mainly on the back and sides of fish (Maitland, 1980b) and thus the feeding biology of Sea Lampreys in Loch Lomond requires further investigation.

River Lamprey

Distribution

Adult River Lampreys are common in the lower River Endrick between October and March. It appears that they stop feeding in Loch Lomond at the end of the summer and then start to migrate into the river in the autumn and during the winter. They have been found in the River Endrick as far upstream as Gartness and in the River Blane (a tributary of the River Endrick) as far upstream as Dumgoyach. There are no records from any other rivers in the Lomond area.

Almost all the adult River Lampreys taken in the River Endrick were the dwarf form and ranged from 170–250 mm but on three occasions in 1984 single animals of the normal type were captured by hand net:

a male of 333 mm (52.5 g) on 9 February, a male of 289 mm (51.2 g) on 21 April and a female of 301 mm (78.0 g) on 28 April. These large lampreys (comparable to those caught on the River Teith at the same time and ranging from 280–360 mm) are assumed to have matured in the Clyde estuary and then to have migrated up the River Leven and through Loch Lomond to the River Endrick.

Larvae must obviously be present in the River Endrick in some numbers but, because they cannot be distinguished from Brook Lampreys, no data are available. Attempts to trap transformers at Woodend on their downstream migration proved unsuccessful, but on two occasions transformed River Lampreys occurred in traps at Drymen Bridge – one of 141 mm (3.0 g) on 6 December, 1983 and a second of 127 mm (2.6 g) on 21 February, 1984.

Feeding River Lampreys must be relatively common in most parts of Loch Lomond, judging by the numbers of fresh wounds on Powan in all parts of the loch (Maitland, 1980b). However, they have proved very difficult to catch in the loch and specimens have been taken on a few occasions only. Data are available for only two of these – one of 231 mm (17.8 g) taken on 19 July 1983 and another of 203 mm (10.4 g) taken on 12 October, 1983.

Adult feeding

The first clear account of attacks by the River Lamprey on Powan was by Robertson (1875) who found dead Powan with lamprey wounds, some with living lampreys still attached to them. He described both the location and the nature of the bites on a few fish. Subsequently Lamond (1922) diagnosed a supposed 'disease' of Powan as being the scars left on the fish after lamprey attacks. Slack (1955) reported that about 17% of Powan examined showed fresh, or more often, healed, wounds made by lampreys. Maitland (1980b) found that the incidence of lamprey scarring on Powan ranged from 26–50% in different years; scars occurred throughout the year but fresh wounds are apparent only in the summer. Very few fish other than Powan carried wounds or scars.

Virtually all fresh lamprey wounds observed on Powan caught regularly during 1981 and 1982 occurred between May and November, with a definite peak around July (Fig. 2).

Spawning migration

Trapping at Drymen Bridge provided the main evidence of activity by adult River Lampreys (Fig. 3). They are apparently absent here during the summer months (none was trapped nor electrofished during this period) but start to appear in late September and October. They occur sporadically during the winter months until April when they are active at spawning time; the numbers rise again at this time and then none are seen until the following autumn.

During the migration period, there is a substantial decrease in the mean length of animals trapped in the field (Fig. 4). In 1983–84, this changed from over 203 mm in November 1983 to less than 189 mm in April 1984. In 1984–85, the decrease was from 221 mm in October 1984 to less than 193 mm in April 1985.

Pickering (1970) found a marked weight difference in River Lampreys from the River Trent and from the River Severn in England, and in both rivers early migrants weighed more than late migrants.

Captive animals

Over the period of the experiment there was a length reduction of 9.2% in the eight males (Fig. 5a) and 13.5% in the four females (Fig. 5b). In the males (Fig. 6a), the immediate prebreeding weight loss was 5.4%; after breeding it reached 12.5%. In the females, the greatest prespawning reduction of 10.8% occurred after 14 weeks (Fig. 6b). They remained like this for four weeks and then increased in weight by 0.9%, just prior to spawning, 23 weeks after capture. After spawning their weight had fallen by 25%.

Substantial changes in the size and shape of the two dorsal fins occurred during the period of the experiment. In the males especially, as previously noted by Larsen (1980) and other workers, both fins grew substantially in height and in length so that the gap between the two fins gradually closed near spawning time. In spite of these changes, the identifying fin clips could still be clearly seen, not only in the upper and lower parts of the caudal fin but also in both the first and second dorsal fins. Little or no regrowth of fin tissue had occurred at the edge of the fins over a period of 25 weeks, and the healed edges of the clip showed up clearly as orange bordered semicircles.

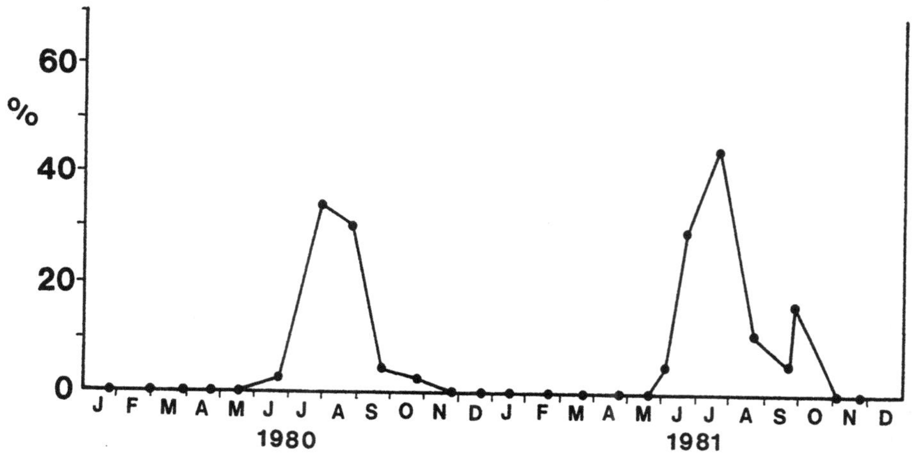

Fig. 2. Percentage of lamprey wounds on Powan caught in Loch Lomond in 1980 and 1981 which were fresh at capture.

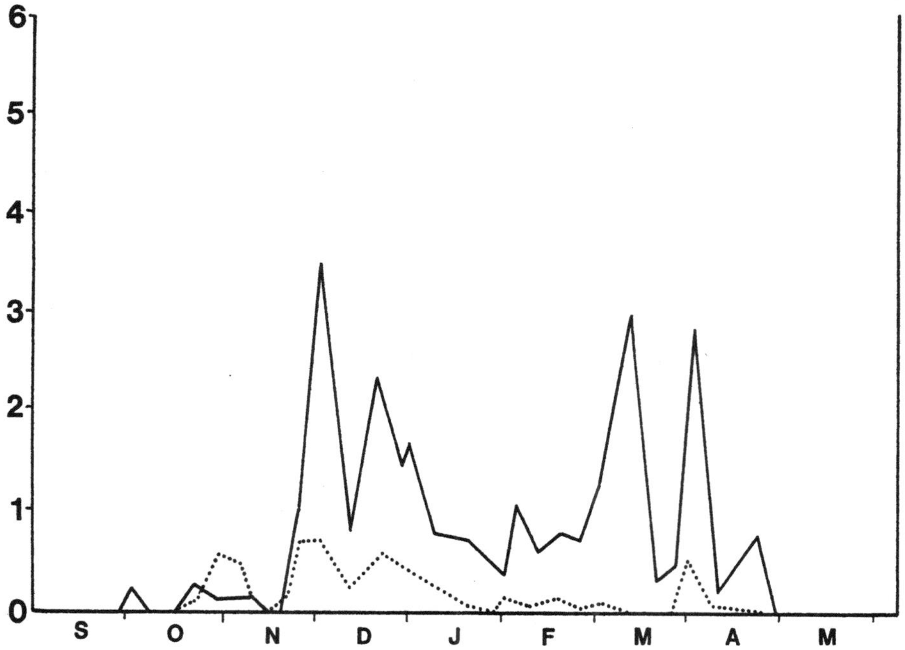

Fig. 3 a. Numbers (means per trap night) of adult River (dotted line) and Brook (solid line) Lampreys trapped at Drymen Bridge in 1983–84.

Gonad development

The gonadosomic index of males showed a steady rise from a mean of 4.3% in early November to 12.6% in late January (Fig. 7a). Spent males taken in April had very low indices (3.2%).

A similar pattern was observed in females (Fig. 7b). Indices were low in November (ranging from 4.7–9.6%), but rose steadily towards the spring to reach a maximum mean value of 18.2% in early April.

Spawning

The adults kept captive in tanks spawned successfully both indoors and outdoors (Fig. 8a). Breeding took place in the spring and was first observed on 3 April when lampreys were seen swimming around the outdoor tank and clinging to the sides; previously they had always remained hidden under stones. Eggs were then seen in some numbers among the gravel in the tank, though no obvious nest was built.

Spawning in the indoor tank commenced on 4 April and typical nest building and spawning behaviour

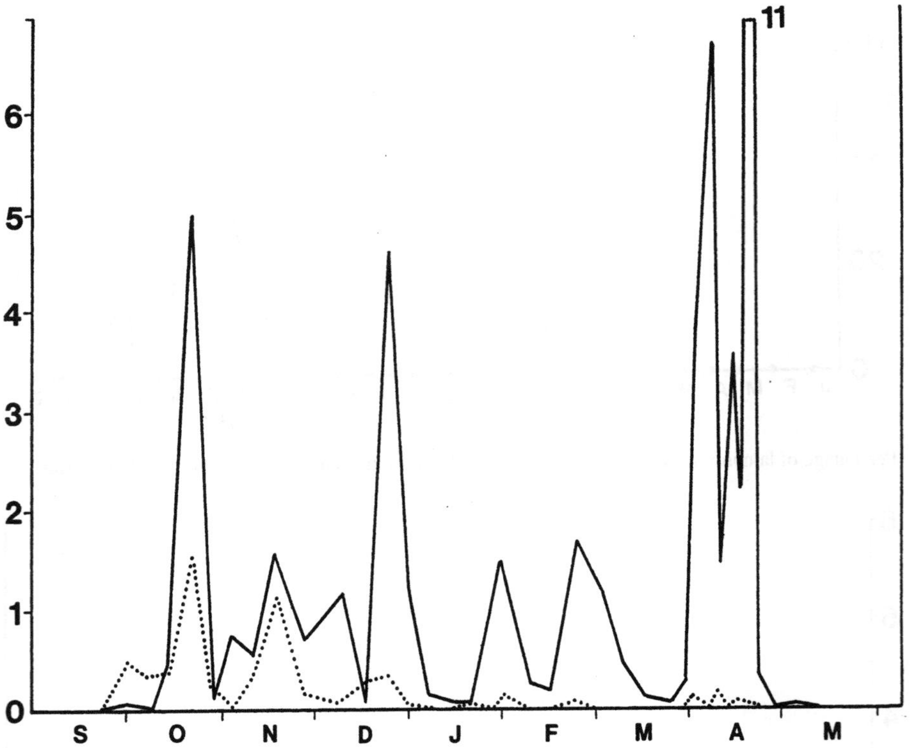

Fig. 3 b. Numbers (means per trap night) of adult River (dotted line) and Brook (solid line) Lampreys trapped at Drymen Bridge in 1984–85.

(Hagelin & Steffner, 1958; Hagelin, 1959) was observed.

There was substantial correlation among the dates of spawning of animals kept under different conditions in captivity and (from observations of redds and ripe and spawned females) those in the field. The duration of spawning appears to be very short. Eggs were first seen in the outside tank on 3 April when the temperature was 9.0 °C, having risen from 6.7 °C 48 hours earlier. Indoors, the lampreys started nest building on 3 April (6.5 °C) and spawned on 4 April (7.0 °C).

In the field (Fig. 8b), the first redds attributed to River Lampreys were seen in the River Endrick on 27 March when the temperature was 5.1 °C, having risen from 3.0 °C on 20 March. A spent female was collected on 15 April, the maximum temperature reached by the river before then being 9.2 °C on 7 April.

Thus it appears that this population of River Lamprey spawns at a much lower temperature than the typical larger form elsewhere in Great Britain, whose spawning begins at 10–11 °C (Hardisty & Potter, 1971). Observations on the normal form of spawning River Lampreys in the River Forth supported this conclusion; spawning here took place on 25 April 1984 when the water temperature was 11.9 °C.

Brook Lamprey

Distribution

Adult Brook Lampreys are common during winter and spring in the lower parts of virtually all the tributaries feeding the south and middle basins of Loch Lomond. They are most obvious during spawning in April and May when hundreds may congregate on suitable gravel beds. They are uncommon, but do occur in tributaries feeding the northern part of the loch, having been recorded from the Cailness Burn and Douglas Water. Larvae were not found during electrofishing of the Cailness, Ardess and Bhdlain Burns in 1980. However, they are common in the Auchentullich, Milarrochy and Mar Burns. In the River Leven they are found from the loch outflow downstream to the tidal limit. They are also common in small streams running into the River Leven (e.g. the Murroch Burn). In the River Endrick they occur from the mouth upstream as far as Fintry (Maitland, 1966). This is well above the waterfalls at Gartness which are probably not negotiable by either Sea or River Lampreys.

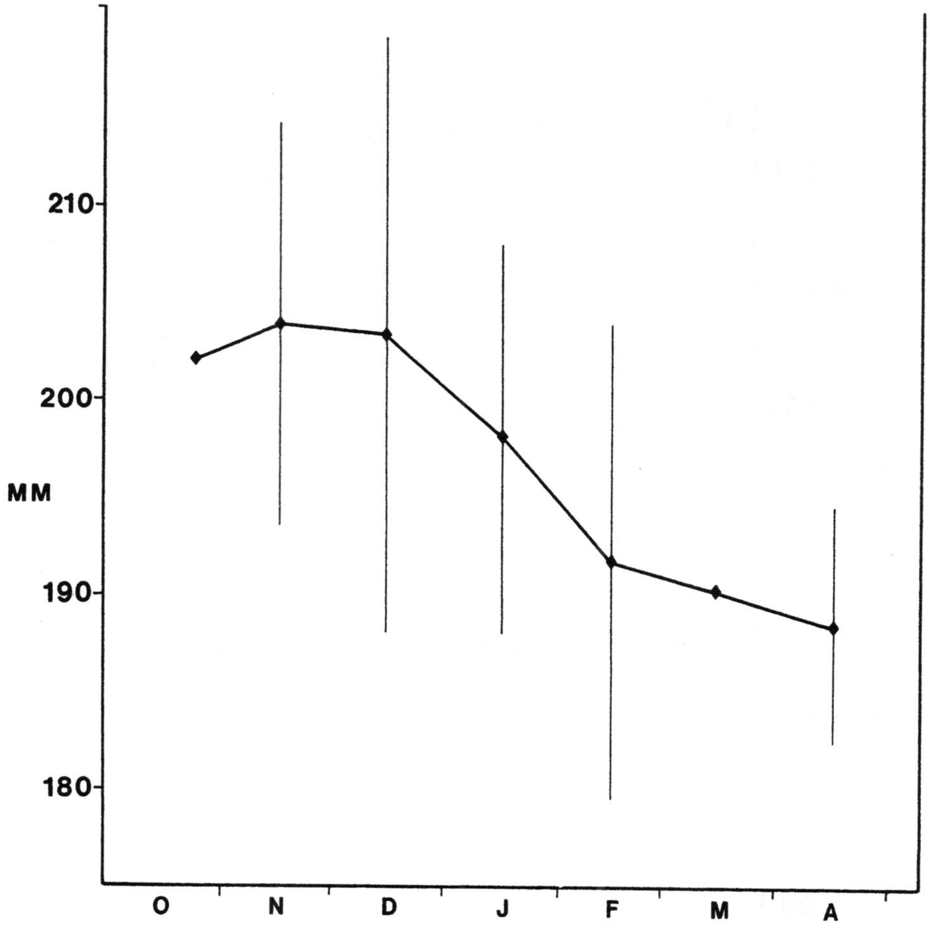

Fig. 4. Seasonal variation in mean length (±SD) of adult River Lampreys collected at Drymen Bridge in 1983–84.

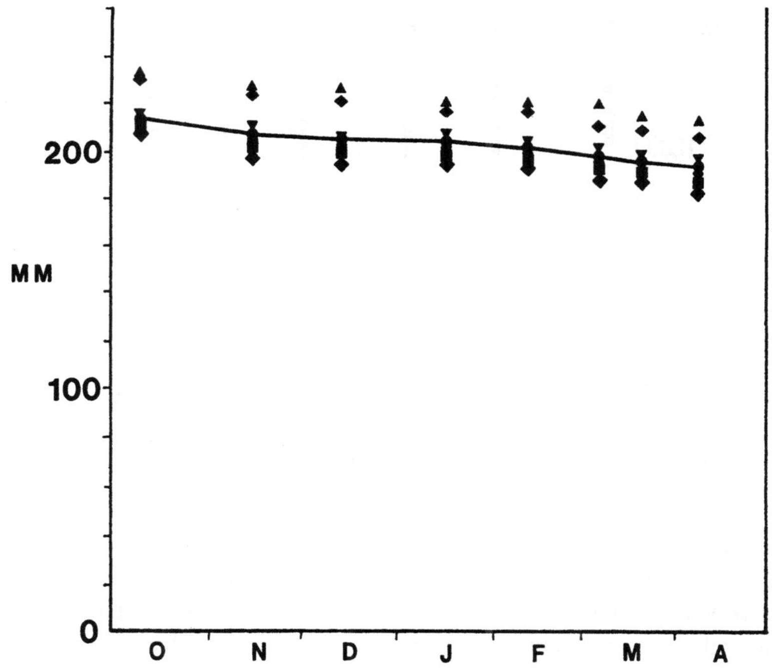

Fig. 5 a. Changes in length of six male River Lampreys kept in aquaria from October 1984 to April 1985. The line joins the mean lengths.

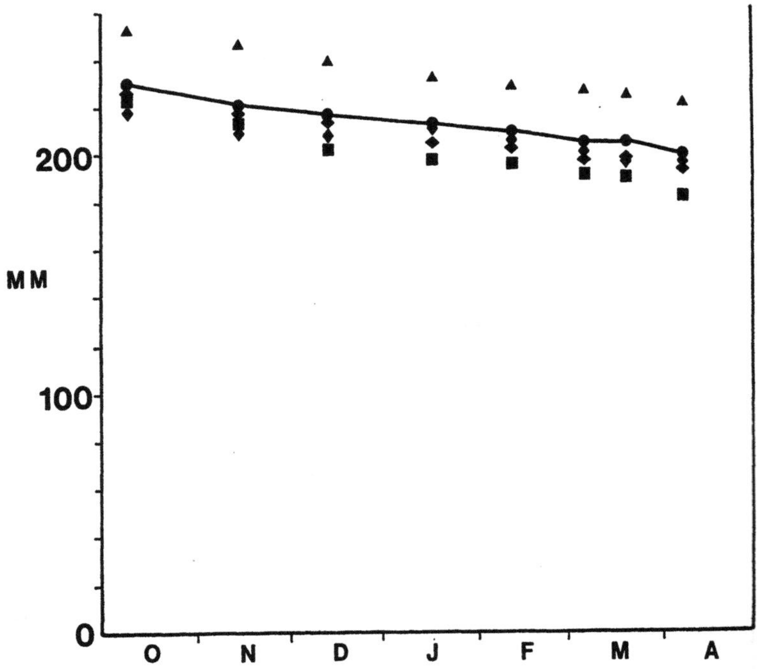

Fig. 5 b. Changes in length of four female River Lampreys kept in aquaria from October 1984 to April 1985. The line joins the mean lengths.

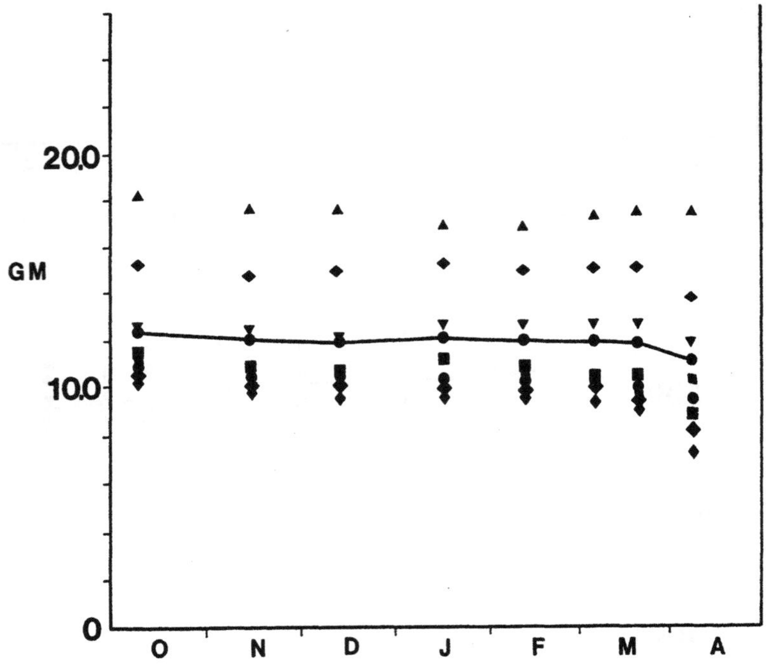

Fig. 6 a. Changes in weight of six male River Lampreys kept in aquaria from October 1984 to April 1985. The line joins the mean weights.

On several occasions large larvae without the normal pigmentation were collected. This 'golden' form is not albino but is very distinctive.

Transformed animals are common in these streams after October. Larvae occur in suitable stretches of water in the same areas and it is evident that they are also found in some numbers in silt beds in Loch Lomond, for ammocoete larvae are not infrequently found in grab samples of silt taken from littoral areas in the south basin. Since River Lampreys transform at a relatively small size (Bird & Potter, 1979), all

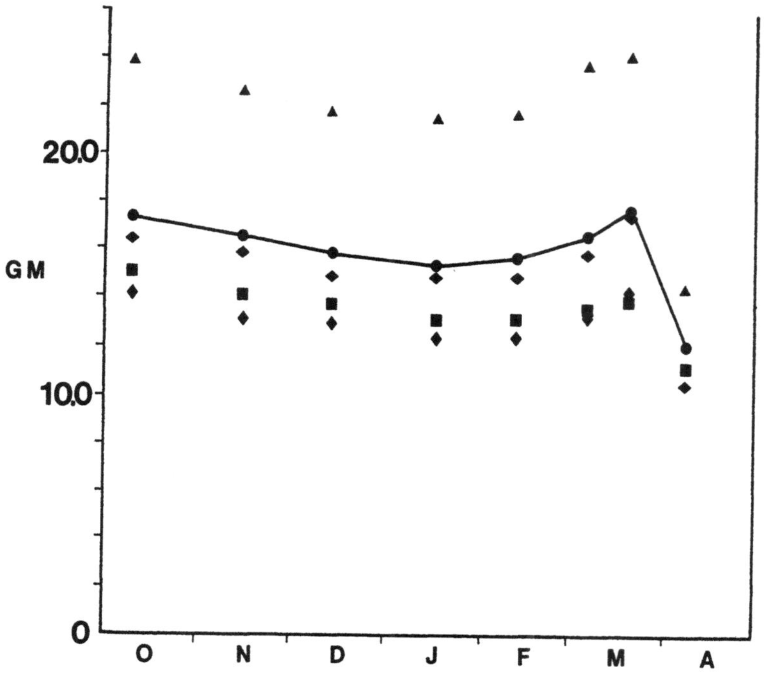

Fig. 6 b. Changes in weight of four female River Lampreys kept in aquaria from October 1984 to April 1985. The line joins the mean weights.

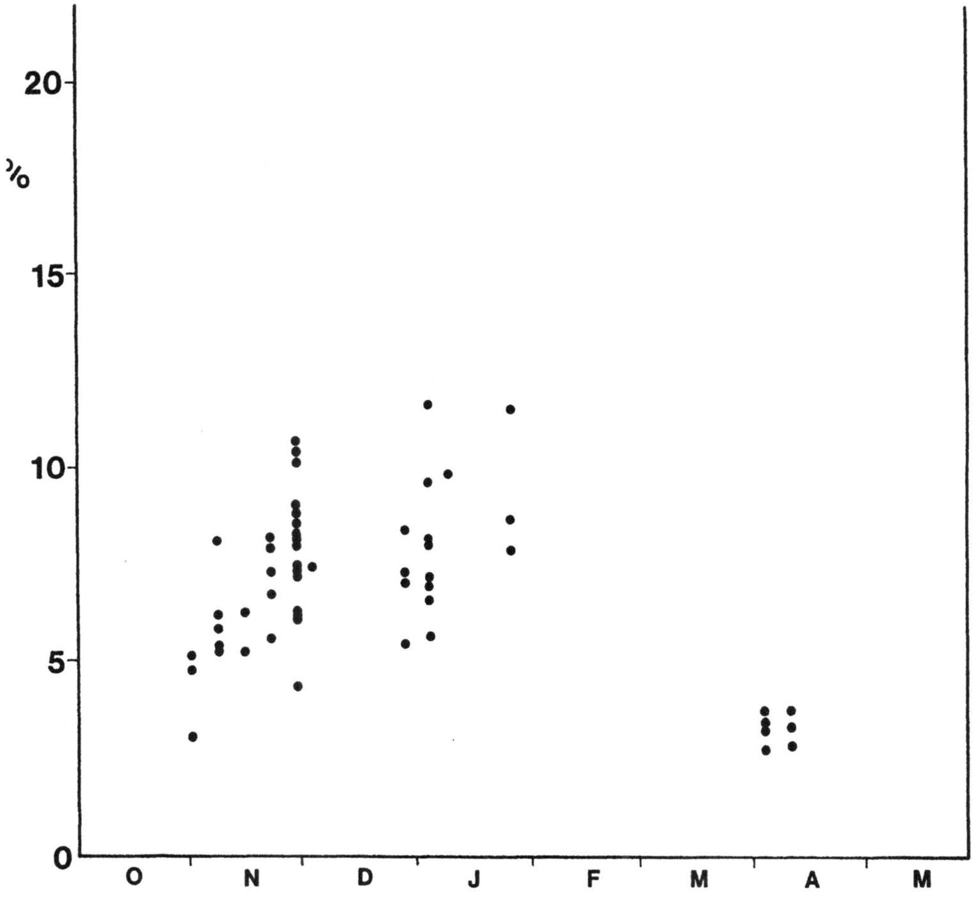

Fig. 7 a. Gonadosomic ratios of male River Lampreys trapped at Drymen Bridge 1983–84.

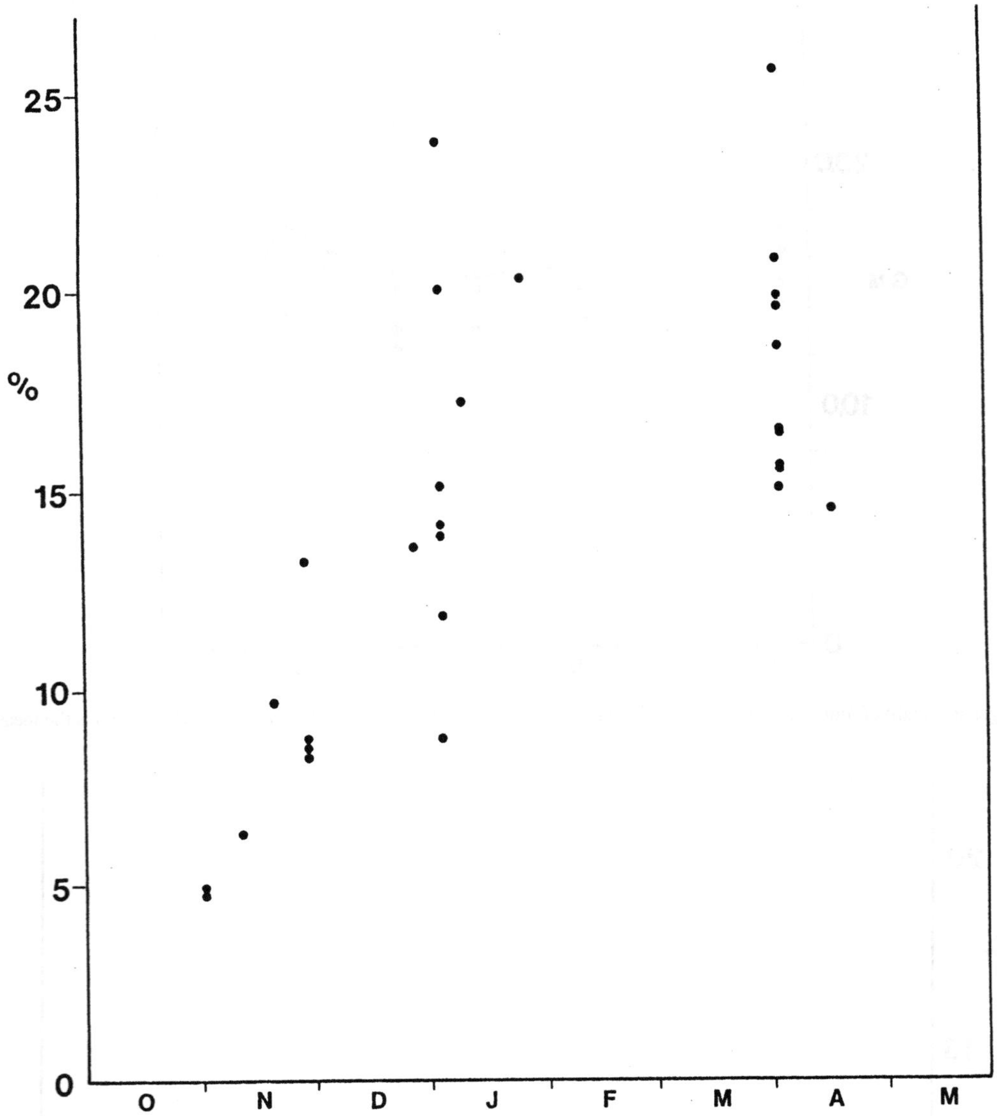

Fig. 7 b. Gonadosomic ratios of female River Lampreys trapped at Drymen Bridge 1983–84.

larvae longer than 150 mm were assumed to be Brook Lampreys.

Size

Adult Brook Lampreys from the River Endrick ranged from 115- –170 mm. Thus the length of the largest thus is only just shorter than that of the smallest of the River Lampreys found – though the great majority were much smaller than this. Malmqvist (1980) found no male Brook Lampreys longer than 170 mm, but in one stream in Sweden (Lansmansbacken) he found females up to 180 mm in length.

Spawning migration

As with River lampreys, trapping at Drymen Bridge provided the main evidence of activity by adult Brook lampreys (Fig. 3). These are apparently absent here during the summer months (none were trapped nor electrofished during this period) but start to appear in late September and October. They are much more abundant than River Lampreys and are trapped during the winter months in very variable numbers until April after which very few appear to be present until the following autumn.

During the migration period, there is a substantial decrease in the mean length of animals trapped in the field (Fig. 9). This changed from over 160 mm in Jan-

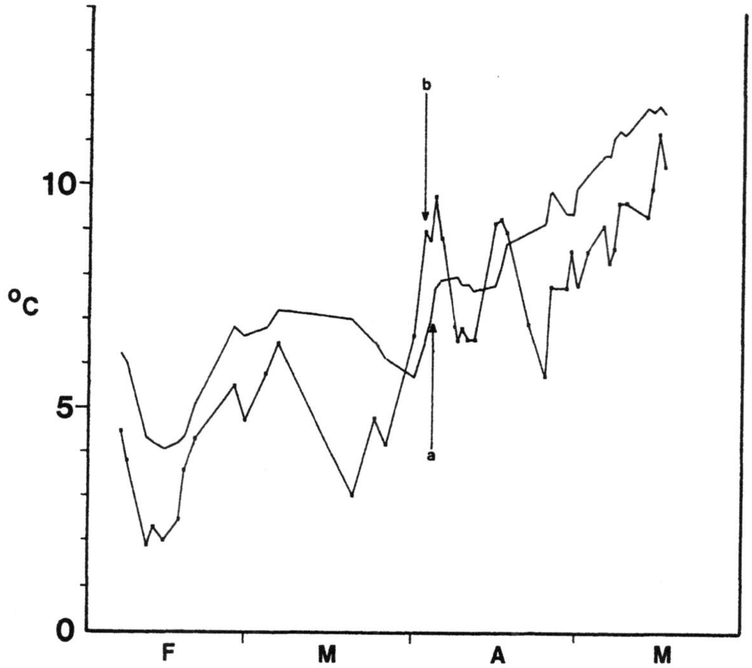

Fig. 8 a. Temperatures in indoor (a) and outdoor (b) aquaria in 1985, indicating first spawning dates of River Lampreys.

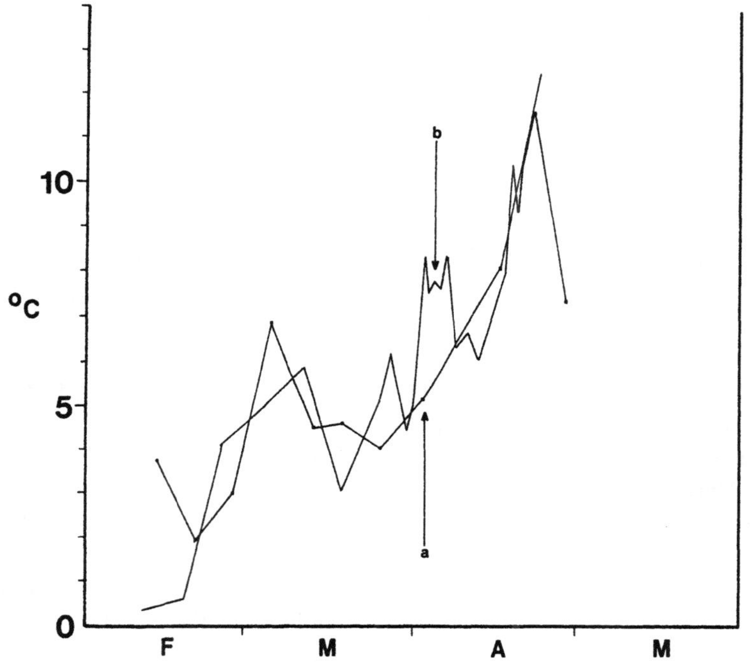

Fig. 8 b. Temperatures at Drymen Bridge in 1984 (a) and 1985 (b) indicating first dates of capture of ripe River Lampreys in traps.

uary 1984 to less than 119 mm in May. Malmqvist (1980) found a similar situation in Sweden where in both the years of his study, the length of adult Brook Lampreys decreased with time.

Gonad development

The pattern of gonad development in Brook Lampreys trapped at Drymen Bridge appeared to be similar, though a little later than that in River Lampreys (Fig. 10). The gonadosomic index of males rose steadi-

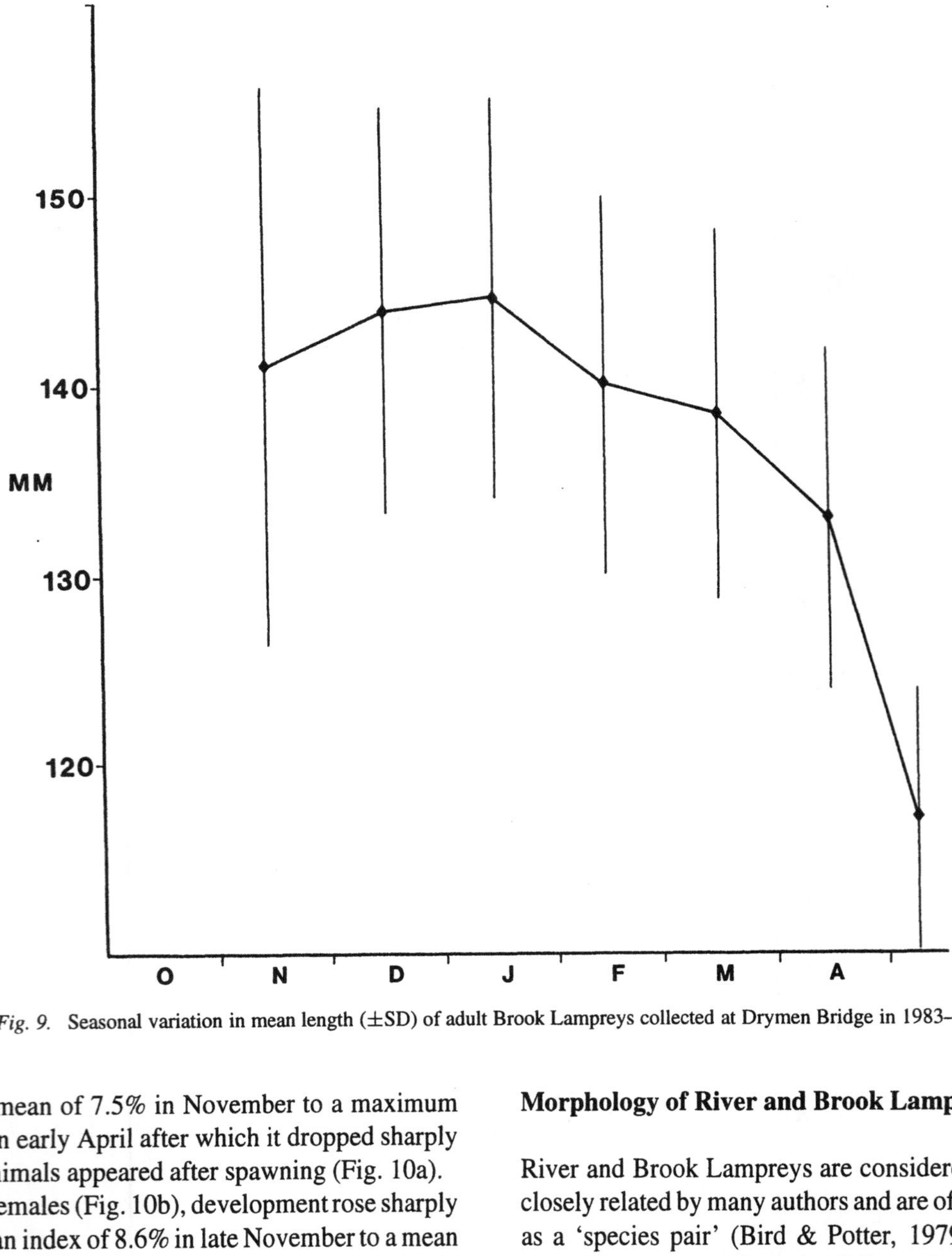

Fig. 9. Seasonal variation in mean length (±SD) of adult Brook Lampreys collected at Drymen Bridge in 1983–84.

ly from a mean of 7.5% in November to a maximum of 11.9% in early April after which it dropped sharply as spent animals appeared after spawning (Fig. 10a).

In the females (Fig. 10b), development rose sharply from a mean index of 8.6% in late November to a mean of 21.5% in mid-April. The few animals remaining in May were fully spent with a low mean index of 1.2%.

Spawning

Lampreys are at their most obvious at spawning time and Brook Lampreys were observed nest building and spawning at many sites in the Lomond catchment. Below Drymen Bridge, spawning was always later than that of River Lampreys and at higher temperatures. At the bottom end of the large pools there, dozens of occupied nests can often be seen in April and May.

Morphology of River and Brook Lampreys

River and Brook Lampreys are considered to be very closely related by many authors and are often described as a 'species pair' (Bird & Potter, 1979). However, the adults are normally quite different in size and this has sometimes been assumed to prevent cross breeding between the two species because of difficulties in copulating. In the Loch Lomond system, however, the River Lampreys are exceptionally small and superficially much closer to Brook Lampreys (Fig. 12).

Morris (1989), using univariate and multivariate statistics, has shown that the Lomond River Lamprey is in some ways intermediate between the normal forms of River and Brook Lampreys and suggested that it represents an intermediate stage between these two species. Table 1 shows morphological data obtained

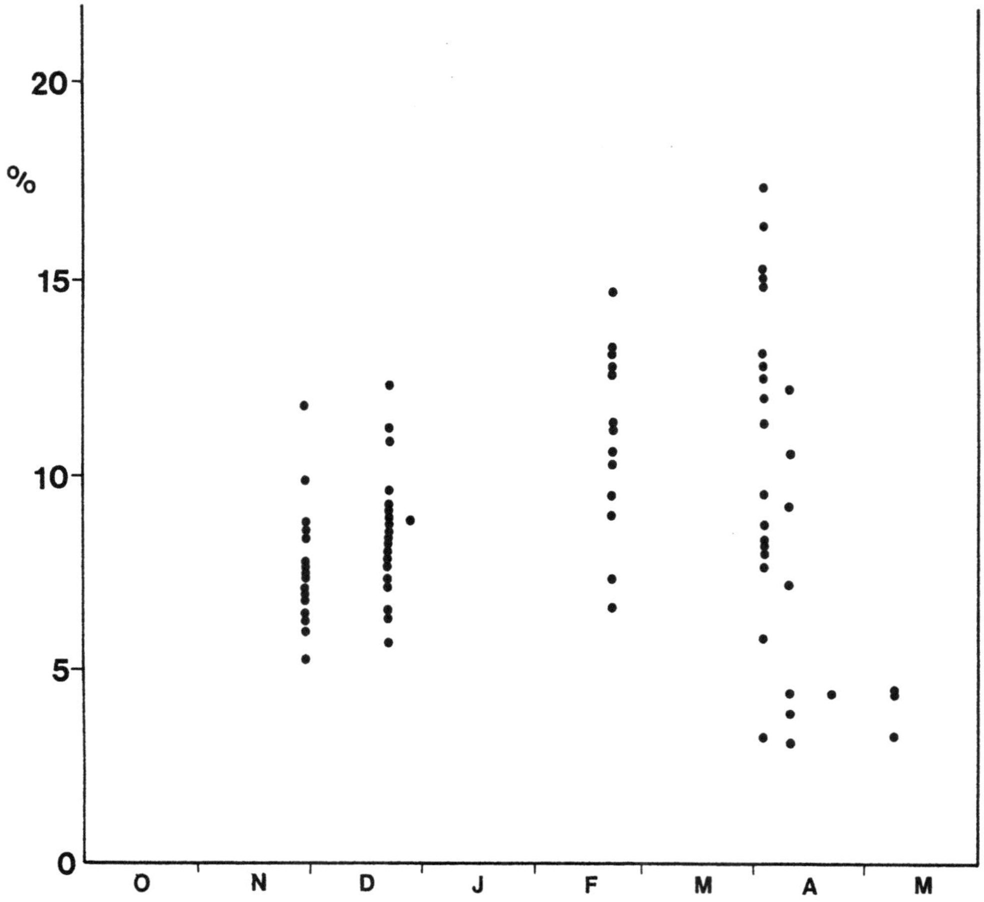

Fig. 10 a. Gonadosomic ratios of male Brook Lampreys trapped at Drymen Bridge 1983–84.

from freshly anaesthetised lampreys from three stocks: Brook and River Lampreys from the River Endrick and normal River Lampreys from the River Teith (whose catchment neighbours that of Loch Lomond).

The intermediate size of the Lomond River Lampreys is clear from Table 1 and Fig. 11. However, this form differs substantially from the Brook and normal River forms in other ways, having a very large sucking disc, prebranchial area and eye. In addition, the Lomond River Lamprey is markedly darker than either of the other two forms being almost black in colour.

There are also ecological features which characterise the Lomond River Lampreys from the normal form. Feeding entirely in fresh water is one of these and the period of feeding appears to be restricted to a few months instead of 15–18 months. In addition, however, as already noted, the Lomond race spawns earlier and at a much lower temperature than the normal form.

Mark/recapture experiments

Between December 1983 and February 1984 38 River Lampreys and 388 Brook Lampreys were trapped at Drymen Bridge, marked by fin clipping and released 500 m below the trapping point. One River Lamprey and eight Brook Lampreys were subsequently recovered, some within 7 days, others after a period of 56 days. Most of the recoveries were made at Drymen Bridge, but one marked Brook Lamprey was recaptured at Blane Bridge on the River Blane, some 5 km further upstream.

In a mark/recapture experiment on Brook Lampreys in a Swedish stream, Malmqvist (1980) marked a total of 50 lampreys and subsequently recovered two of them further upstream. The maximum distance travelled was 2 km.

In May 1969, a mark-recapture experiment was carried out by D.B.C. Scott with Brook Lampreys to establish numbers in the Auchentullich Burn, a trib-

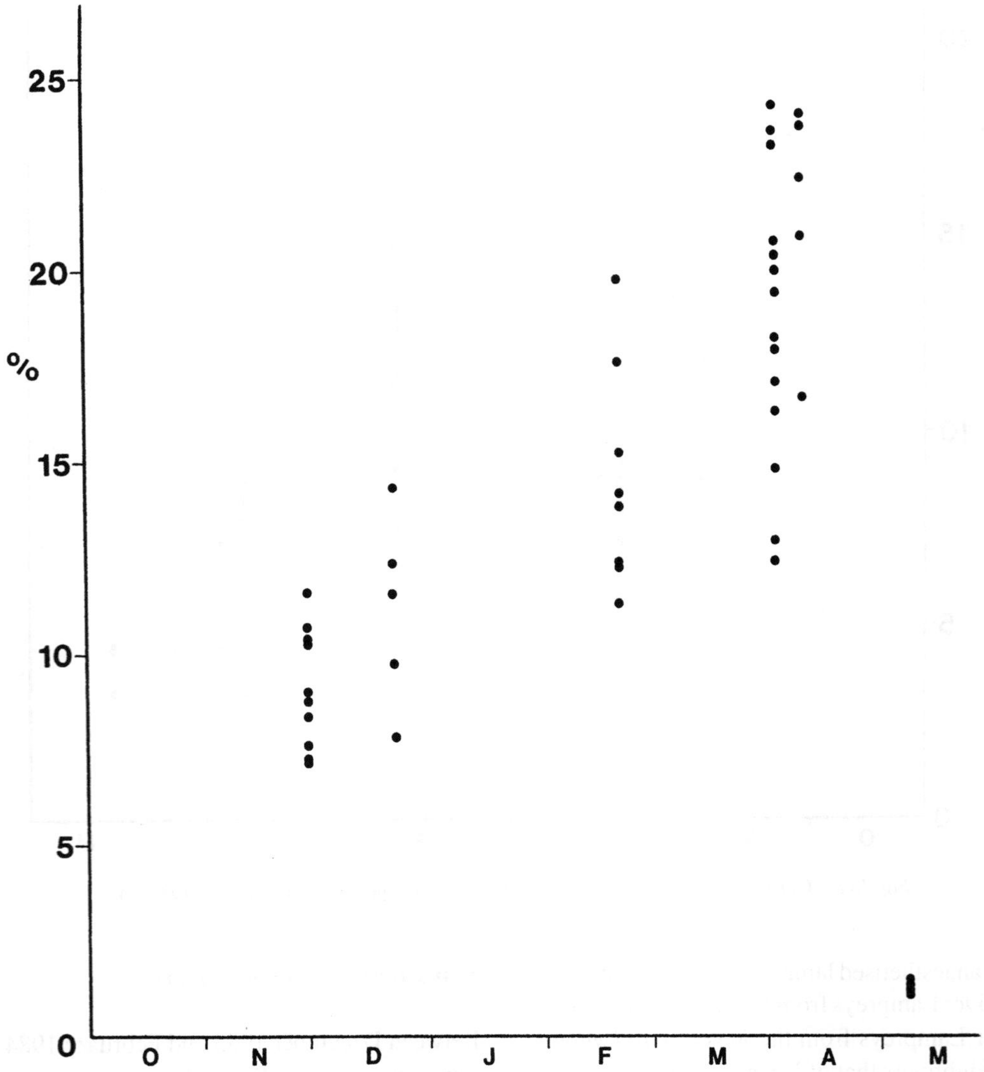

Fig. 10 b. Gonadosomic ratios of female Brook Lampreys trapped at Drymen Bridge 1983–84.

Table 1. Morphological comparison of Brook Lampreys from the River Endrick and River Lampreys from the River Endrick and from the River Teith. Note: the disc length, prebranchial length and eye diameter are expressed as a percentage of total length.

	Endrick *planeri*				Endrick *fluviatilis*				Teith *fluviatilis*			
	Mean	SD	Min	Max	Mean	SD	Min	Max	Mean	SD	Min	Max
Length (mm)	143.3	16.8	118	170	201.5	16.4	170	243	312.3	26.5	250	362
Weight (g)	3.9	1.5	1.9	6.7	10.4	3.0	6.3	19.1	51.5	12.3	26.8	76.7
Disc length	4.8	0.4	4.1	5.5	6.2	0.3	5.5	7.0	4.9	0.3	4.5	5.6
Prebranchial length	10.3	0.5	9.3	11.5	13.6	0.4	13.1	14.3	10.2	0.4	9.9	10.8
Eye diameter	2.1	0.1	1.8	2.3	2.5	0.2	2.2	3.0	1.6	0.2	1.3	1.8

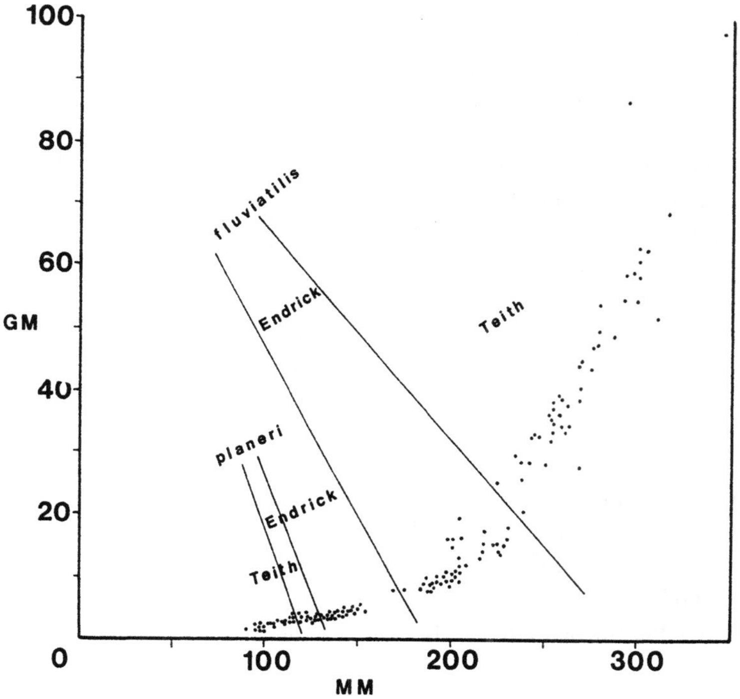

Fig. 11. Length/weight relationships of adult River and Brook Lampreys from the River Endrick and the River Teith; specimens captured in 1983–1984.

utary entering the west bank of Loch Lomond. The results indicated a population of about 200 ammocoetes and 1000 adults present in a 150 m stretch of burn.

Conservation and future research needs

Although it is well known that Sea Lampreys attack fish in fresh water in the North American Great Lakes and some other waters there, this habit appears to be virtually unknown in Europe. Apart from the situation in Loch Lomond, the only other record known to the authors is at Llandegfedd Reservoir in Wales (F. H. Jones, personal communication) where two feeding Sea Lampreys were recovered from Trout – one in July 1972 (35 cm; 84 g) and the other in July 1981 (18.5 cm; 7.6 g). These were collected some 9 and 18 years respectively after the reservoir was constructed after which it was reckoned very difficult for lampreys to reach it from the sea.

The Sea Lamprey is not common in the Loch Lomond area and if the spawning site below the barrage on the River Leven is the only one then the species must be very vulnerable. The proposed dredging of this stretch of the river in the near future is therefore a cause for concern.

The population of River Lampreys appears to be unique in the British Isles and certainly warrants conservation status because of this (Maitland & Lyle, 1991). There appear to be no immediate threats to this population but the fact that it depends for feeding on a very rare species (the Powan, which itself is in need of conservation measures) and apparently uses just one short stretch of one river system in the whole of the Lomond catchment for spawning is a cause for concern. Also, the fact that this stretch of the river is subject to pollution from Drymen sewage works and is also liable to intense public pressure during good weather does make it vulnerable. Similar populations to the Lomond one are known from a few other large lakes in western Europe – in Finland (Lake Saimaa) and Russia (Lake Ladoga).

Conservation of these important stocks of lampreys must be given some consideration. Obviously water quality in important stretches of river must be kept as high as possible. Also, the actual spawning sites should be given some protection – especially during the vital spawning and egg incubation period each year – June and July for the Sea Lamprey and March and April for the River Lamprey.

The creation of new 'safeguard' populations of the stock of River Lamprey may be possible in parallel with the new stocks of Powan which are at present being established (Maitland & Lyle, 1990) and this possibility is under consideration at the moment.

Further research on these important populations of lampreys should address the status and biology of feeding Sea lampreys in Loch Lomond, and the biology of River Lampreys in the River Endrick and Loch Lomond from transformation to adult.

Acknowledgments

We are grateful to many Loch Lomond anglers and others who have supplied observations on lampreys in the area. Mr F. H. Jones passed on details of 'landlocked' Sea Lampreys in Wales and Dr D. B. C. Scott provided data on Brook Lampreys in the Auchentullich Burn.

References

Bird, D. J. & I. C. Potter, 1979. Metamorphosis in the paired species of lampreys, *Lampetra fluviatilis* (L.) and *Lampetra Planeri* (Bloch). 1. A description of the timing and stages. Zool. J. Linn. Soc. Lond. 65: 127–143.

Hagelin, L. O., 1959. Further aquarium observations on the spawning habits of the River Lamprey (*Petromyzon fluviatilis*). Oikos. 10: 50–64.

Hagelin, L. O. & N. Steffner. 1958. Notes on the spawning habits of the River Lamprey (*Petromyzon fluviatilis*). Oikos. 9: 221–238.

Hardisty, M. W., 1944. The life history and growth of the Brook Lamprey (*Lampetra planeri*). J. anim. Ecol. 13: 110–122.

Hardisty, M. W., 1961. The growth of larval lampreys. J. anim. Ecol. 30: 357–371.

Hardisty, M. W., 1969. Information on the growth of the ammocoete larvae of the anadromous Sea Lamprey *Petromyzon marinus* in British rivers. J. Zool. Lond. 159: 139–144.

Hardisty, M. W. & I. C. Potter, 1971. The biology of lampreys. London, Academic Press.

Hunter, W. R., H. D. Slack & M. R. Hunter, 1959. The lower vertebrates of the Loch Lomond district. Glasg. Nat. 18: 84–90.

Lamond, H., 1922. Some notes on two of the fishes of Loch Lomond: the Powan and the lamprey. Fish. Scotl. Salmon Fish. 1923: 3–10.

Larsen, L. O., 1980. Physiology of adult lampreys, with special regard to natural starvation, reproduction and death after spawning. Can. J. Fish. aquat. Sci. 37: 1762–1769.

Lennon, R. E., 1954. Feeding mechanism of the Sea Lamprey and its effect on host fishes. Fish. Bull. Fish. Wildl. Serv. 56: 247–293.

MacDonald, T. H., 1959a. Estimates of length of larval life in three species of lamprey found in Britain. J. anim. Ecol. 28: 293–298.

MacDonald, T. H., 1959b. Identification of ammocoetes of British lampreys. Glasg. Nat. 18: 91–95.

Maitland, P. S., 1966. The fauna of the River Endrick. Glasgow, Blackie.

Maitland, P. S., 1980a. Review of the ecology of lampreys in northern Europe. Can. J. Fish. aquat Sci. 37: 1944–1952.

Maitland, P. S., 1980b. Scarring of whitefish (*Coregonus lavaretus*) by European River Lamprey (*Lampetra fluviatilis*) in Loch Lomond, Scotland. Can. J. Fish. aquat. Sci. 37: 1981–1988.

Maitland, P. S., 1981. The ecology of Scotland's largest lochs: Lomond, Awe, Ness, Morar and Shiel. The Hague, Junk.

Maitland, P. S. & A. A. Lyle, 1990. Practical conservation of British fishes: current action on six declining species. J. Fish Biol. 37: 255–256.

Maitland. P. S. & A. A. Lyle, 1991. Conservation of freshwater fish in the British Isles: the current status and biology of threatened species. Aquatic Conservation. 1: 25–54.

Maitland, P. S., K. H. Morris, K. East, M. P. Schoonoord, B. van der Wal & I. C. Potter, 1984. The estuarine biology of the River Lamprey, *Lampera fluviatilis*, in the Firth of Forth, Scotland, with particular reference to size composition and feeding. J. Zool. Lond. 203: 211–225.

Malmqvist, B., 1980. The spawning migration of the Brook Lamprey, *Lampetra planeri* Bloch, in a south Swedish stream. J. Fish Biol. 16: 105–114.

Morris, K. H., 1989. A multivariate morphometric and meristic description of a population of freshwater-feeding River Lampreys, *Lampetra fluviatilis* (L.), from Loch Lomond, Scotland. Zool. J. Linn. Soc. 96: 357–371.

Morris, K. H. & P. S. Maitland, 1987. A trap for catching adult lampreys (Petromyzonidae) in running water. J. Fish Biol. 31: 513–516.

Pickering, A. D., 1970. Osmoregulation and endocrinology of lampreys. PhD Thesis, University of Nottingham.

Robertson, D., 1875. On *Petromyzon fluviatilis* and its mode of preying on *Coregonus clupeoides*. Proc. Nat. Hist. Soc. Glasg. 2: 61–62.

Rosie, A. J. & P. S. Maitland, 1980. Salmonid nursery streams in the Loch Lomond area: an assessment. Institute of Terrestrial Ecology Report to the North of Scotland Hydro-Electric Board.

Schoonoord, M. P. & P. S. Maitland, 1983. Some methods of marking larval lampreys (Petromyzonidae). Fish. Mgt. 14: 33–38.

Slack, H. D., 1955. Factors affecting the productivity of *Coregonus clupeoides* Lacepede in Loch Lomond. Ver. int. Verh. Limnol. 12: 183–186.

Hydrobiologia **290**: 121–133, 1994.
K. J. Murphy, M. C. M. Beveridge & R. Tippett (eds), The Ecology of Loch Lomond.

Life histories of the powan, *Coregonus lavaretus* (L.) (Salmonidae, Coregoninae) of Loch Lomond and Loch Eck

E. A. R. Brown & D. B. C. Scott
School of Biological and Medical Sciences, The University, St. Andrews, Fife KY16 9TS, UK
Address for correspondence: Beley Brig, Dunino, St. Andrews, Fife KY 16 8LX, UK

Key words: powan, *Coregonus lavaretus*, Loch Lomond, Loch Eck, life history

Abstract

Four aspects of the life histories of the two populations of powan *Coregonus lavaretus* (L.) in Scotland are described: growth (Eck powan are shorter and with greater year to year variance than Lomond); sexual maturation (Eck powan mature younger, but at similar weight to Lomond); spawning (timing in Eck varies, but is consistent in Lomond); and recruitment/mortality (fecundity, sex ratios, and mortality also vary in the short term). Short term differences between the physiological ecology of the populations can be ascribed to the size and topography of the lochs. Long term differences are more difficult to account for, and are more important in that they may signal changes in sustainability. Conservation of powan must be considered in terms of their synecological relationships, not in isolation.

Introduction

Of the four species of coregonine whitefish which survive in the British Isles, three are glacial relict populations isolated in freshwater lakes. *Coregonus pollan* Thompson (pollan) occurs in Loughs Neagh, Erne, Ree and Derg in Ireland. It is more closely related to the ciscoes of North America than to other British coregonines (Ferguson, 1974; Ferguson *et al.*, 1978). *Coregonus albula* L. (vendace) is now restricted to Bassenthwaite and Derwentwater in Cumbria, having recently become extinct in Scotland. *Coregonus lavaretus* (L.) occurs in Red Tarn, Ullswater and Haweswater in Cumbria (schelly); in Llyn Tegid in Wales (gwyniad); and in two lochs in Scotland, Lomond and Eck (powan or freshwater herring (Plate I)). The earliest reference to powan in Loch Lomond is by Buchanan (1582) who in describing "Lominius lacus" wrote: "Praeter multitudinem aliorum piscium habet & sui generis esu non insuaves Pollacas vocant". Buchanan was a native of Strathblane, and probably described powan (he called it pollac) from personal experience. Thirty years later Monipenny (1612), in what reads like a translation of Buchanan, wrote: 'This loch, besides an abundance of other fishes, hath a fish of the owne, called Powan, very pleasant to eate'. The earliest reference to powan in Loch Eck is in The Old Statistical Account of Scotland: 'The most remarkable thing with regard to Loch Eck is its abounding with freshwater herring. They are a dry, insipid fish; and are not discovered in any other lake on the West coast, excepting this and Lochlomond' (Stewart, 1799). The first scientific studies of the ecology of powan are due to Lamond (1922). From the 1950's Slack included the biology of powan as a part of his studies of Loch Lomond, publishing the seminal paper on 'The Biology of the Powan' (Slack *et al.*, 1957). Maitland extended these studies with particular emphasis on reproduction and fecundity, predation and conservation. Recently Adams (1994) has investigated the effect of the introduction of alien species to Loch Lomond on the powan population. Scott and his colleagues at St. Andrews University have made a comparative study of the growth processes of powan in Lochs Lomond and Eck (see bibliography in Treasurer and Mills, 1993). The powan population of Loch Lomond was fished irregularly for several centuries; the last commercial fishery being organised by the Loch Lomond Angling Association and taking 256 200 powan from 1917 to 1926. They were sold in the English midlands as 'pollen' or 'grayling' (Lamond, 1931). Loch Lomond powan were again fished briefly during the 1939–46 war, at the same time as a fishery on Loch Eck. Since then both populations have remained unexploited, and thus ideal subjects for fundamental physiological and ecological research. Although there is now no fishery, an even more serious risk to the powan populations has arisen through human exploitation of their habitats (Hamilton, 1988).

The primary aim of this paper is to bring up to date our knowledge of the biology of the powan populations of Loch Lomond and Loch Eck, with particular reference to recent advances in four major aspects of their life history: somatic growth, achievement of sexual maturation, spawning, and some other factors which affect recruitment and mortality. A corollary is to consider how knowledge of their life histories can be utilised in their conservation.

Materials and methods

Powan were caught in multifilament benthic gill nets 200 to 400 m long, with mesh sizes ranging from 6.25 to 40 mm knot to knot. These nets were selective, taking virtually no 0^+ powan, and very few 1^+ (Plate I(a)). They were set in depths from 1 to 25 m. In Loch Lomond from 1979 to 1988 nets were set in the Sallochy region at approximately monthly intervals throughout the year, and also between the nearby Ross Islands at spawning time. Both these locations are on the eastern shore of the middle basin of Loch Lomond (Fuller *et al.*, 1976). In Loch Eck from 1984 to 1987 nets were likewise set between Rudha na Seamraig on the western, and Coire Ealt on the eastern shore throughout the year, and also from Coire Ealt southwards parallel to the shore in 1 to 4 m depth at spawning time; these sites are near the northern end of the loch. In addition between October 1991 and February 1992, as part of a survey of spawning sites in Loch Eck supported by Scottish Natural Heritage, nets were set throughout the loch. All nets were set during the afternoon, lifted 18 to 24 hr later, and the fish to be used anaesthetized in MS222 (Sandoz Pharmaceuticals) and killed immediately. They were transferred to the laboratory and processed within 24 h. Total length and fork length (nearest mm); total weight (nearest 2 g); and gonad and liver (gall bladder emptied) weights were recorded. Other organs were weighed and tissues preserved in accordance with the requirements of the study in progress. Fecundity was assessed shortly before spawning by counting the oocytes in weighed aliquots of ovary (approximately 10%), preserved in 4% formaldehyde. Spawning status was assessed by inspection of the gonads.

Age was determined by counting checks, and growth was assessed by back calculating length at age from scale reading using the Fraser-Lee equation (Brown, 1989). Ten scales were removed from a standard position above the lateral line and level with the posterior edge of the dorsal fin (Plate I(b)). At least three scales from each fish were read, usually four. As there was no significant difference between the sexes, they were combined. Year-class strengths were calculated using the method of Mills & Mann (1985), taking the beginning of the year-class as 1 April. Scales and data of powan collected from Loch Lomond by F. W. K. Gervers in July 1952 and by P. S. Maitland in December and January 1965–67 were generously provided by P. S. Maitland, and their length at age calculated.

Results and conclusions

Growth

The calculated length at age of powan in Loch Lomond and Loch Eck are shown in Fig. 1. There were significant year to year variations in length at age of each year-class from 1972 to 1986 at ages up to 6+; above this age powan were so rare that statistical analysis was not feasible. There is least year-to-year variation at one year old, increasing with age. Variation was usually higher in Eck than in Lomond, but the growth of 0 to 1 year olds in Loch Lomond in 1984 was exceptionally high. Prior to this unusually fast growing year-class, the length range in the first year was only 13 mm, lower than at any later age (coefficient of variation 2.62). Lomond fish were consistently longer at a given age than Eck fish, but the weight of Loch Eck fish was consistently greater than Loch Lomond fish (Plate I(b), (c)) (Brown & Scott, 1990). There was no significant difference between the length at age of the samples of fish taken in 1952 and 1965–67, but the growth rate of the latter were significantly lower than those of 1980's powan in their first 3 years of life. Because the scales for these samples were taken from sub-optimal places on the fish, and because only 39 fish were available for 1952, the results are suspect; but powan before the population crash of 1968 were 'small, and of poor quality' (Roberts *et al.*, 1970).

Plate 1. (a) Immature powan, Loch Eck, 30 January 1987. Total length 98 mm, weight 8 g, age 0^+. (b) Adult male powan, Loch Eck, 30 January 1987. Total length 357 mm, weight 494 g, somatic condition factor 1.06, age 5^+. Scales removed from standard position near posterior end of dorsal fin. *Saprolegnia* infection in region of cloaca. (c) Adult male powan, Loch Lomond, 1 February 1987. Total length 332 mm, weight 258 g, somatic condition factor 0.7, age 5^+. Scattered *Saprolegnia* infection above lateral line.

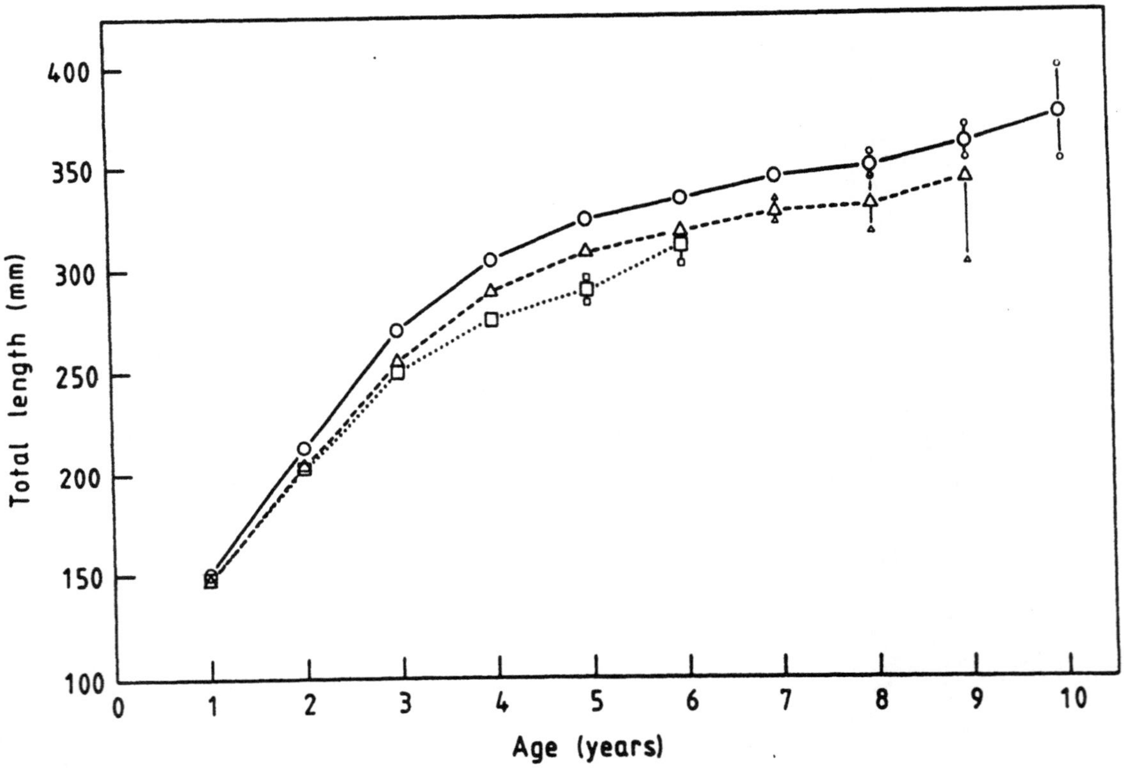

Plate 1. Mean calculated length (mm) at age in Loch Lomond (circles); Loch Eck (triangles); and the sample of 1965–67 (squares) (±standard deviation).

Sexual maturation

Recognition of sexual maturity in powan is complicated by the fact that there are two distinct categories of juvenile powan: besides immature juveniles with minute gonads, there are adolescent juveniles in which gametogenesis begins but is not completed and the process ends in ovarian regression, not spawning (Brown & Scott, 1991). Fig. 2 shows the percentage of powan which are at the immature, adolescent, and adult stage at each age from 1^+ to 4^+, with their mean total lengths and somatic weights at each stage.

Age, length, or weight might induce progress from one stage to the next. In Loch Lomond, a few exceptional females become adolescent at 1^+ but none become adult until 2^+, usually 3^+. Virtually all females are adult by 4^+. In males, exceptionally some fish become fully adult at 1^+, though 2^+ is more usual. Virtually all males are adult at 3^+. In Loch Eck, the percentage of adolescent and adult 1^+ females is similar to Loch Lomond, but a higher percentage is adult at 2^+, and virtually all are adult at 3^+, though a few still adolescent at 5^+ were caught. No Eck males are adult at 1^+, but a higher percentage is adult at 2^+ than in Loch Lomond, and none remains immature at 3^+. Thus in both lochs, males tend to mature one year earlier than females and, with the exception of the adult 1^+ males, both sexes mature at a younger age in Loch Eck than in Loch Lomond. The apparently earlier maturation of Loch Lomond males is an anomaly due to year-to-year variations in growth rate. Both the numbers and the first year growth rate of the 1984 year-class in Loch Lomond were exceptionally high, biassing the results. There was no such effect in Loch Eck (Brown, 1989). Age therefore, cannot be the factor which determines when maturation occurs.

In females in Loch Lomond, either length or weight of the fish may be correlated with the attainment of maturity. The mean total length (mm) of immature females in Loch Lomond ranges from 205 at 1^+ to 219 at 3^+; of adolescents 227 at 1^+ to 250 at 3^+, and of adults 264 at 2^+ (none adult at 1^+) to 285 at 3^+. Corresponding values for Loch Eck females are: immature 188 at 1^+ to 204 at 3^+; adolescent 177 at 1^+ to 241 at 3^+, adult 230 at 2^+ (none adult at 1^+) to 258 at 3^+. The mean weight (g) of immature females in Loch Lomond ranges from 72 at 1^+ to 73 at 3^+; of adolescents 89 at 1^+ to 117 at 3^+, and of adults 136

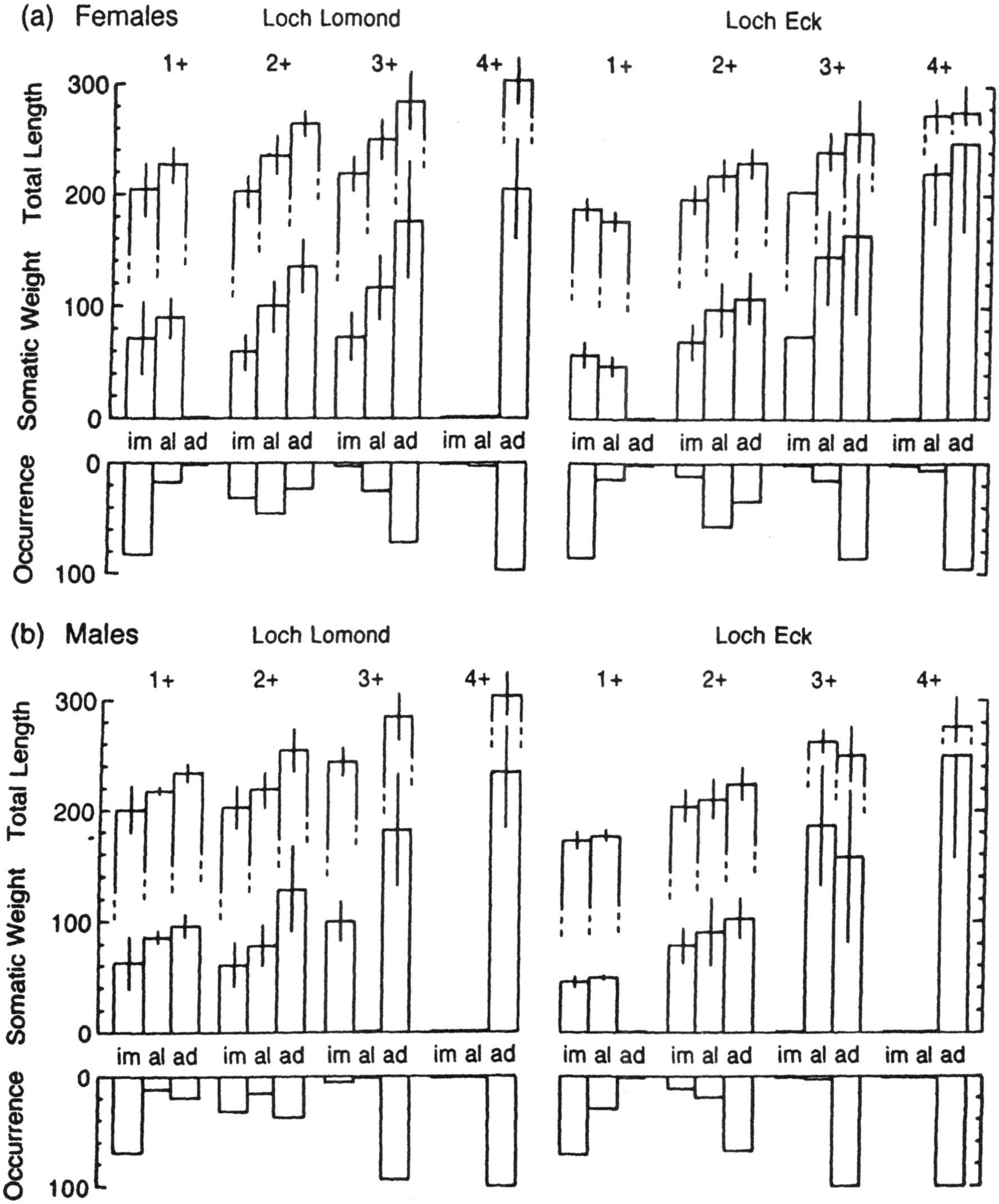

Plate 2. Percentage occurrence of immature, adolescent and adult powan at ages 1^+ to 4^+, and corresponding mean total lengths (mm) and somatic weights (g) (±standard deviations); (a) females; (b) males.

at 2^+ (none adult at 1^+) to 177 at 3^+. Corresponding values for Loch Eck females are: immature 58 at 1^+ to 76 at 3^+; adolescent 48 at 1^+ to 146 at 3^+; adult 109 at 2^+ (none immature at 1^+) to 165 at 3^+.

So Loch Eck females progress to each maturity stage at much the same weight as Loch Lomond females, but at a significantly shorter length. Weight, therefore, rather than length, appears to be the main determinant of sexual maturity. Because it is difficult to identify adolescent male powan macroscopically, corresponding data for this sex are suspect, but the pattern follows that of females. The mean length of immature males in Loch Lomond ranges from 201 mm at 1^+ to 244 mm at 3^+; of adolescents 218 mm at 1^+ to 219 mm at 2^+ (no 3^+ adolescents caught), and of adults 234 mm at 1^+ to 285 mm at 3^+. Corresponding values for Loch Eck males are: immature 173 mm at 1^+ to 205 mm at 2^+ (none immature at 3^+); adolescent 177 mm at 1^+ to 263 mm at 3^+, adult 225 mm at 2^+ (none adult at 1^+) to 250 mm at 3^+. The mean weight of immature males in Loch Lomond ranges from 63 g at 1^+ to 100 g at 3^+; of adolescents 86 g at 1^+ to 79 g at 2^+ (no 3^+ adolescents caught), and of adults 96 g at 1^+ to 182 g at 3^+. Corresponding values for Loch Eck males are: immature 45 g at 1^+ to 79 g at 2^+ (none immature at 3^+); adolescent 50 g at 1^+ to 187 g at 3^+; adult 103 g at 2^+ (none adult at 1^+) to 158 g at 3^+.

Recruitment and mortality

Sex ratios
Table 1 shows the sex ratios of individual year-classes during the period 1974 to 1984. In Loch Lomond only two years deviated significantly from a 1:1 ratio; 1975 when males predominated, and 1984 when females predominated. In Loch Eck ratios were more variable, with a period of male dominance from 1976 to 1982. The overall effect was that routine samples in Loch Lomond consisted of equal numbers of each sex, whereas samples before 1984 in Loch Eck were as high as 3:1 male:female, approaching closer to 1:1 in samples after 1986 (Fig. 3).

Fecundity
Studies of fecundity were made in Loch Lomond in 1982–83 and 1992–93, and in Loch Eck in 1991–92. Fig. 4 shows the results, with the sample of 1966–67 (Maitland, 1969) recalculated in terms of total length. Table 2 shows the corresponding regression data. There is little difference between samples, but analysis is complicated by variations in fish size in different year classes. In the Loch Eck sample of 1990–91 there is an apparent bifurcation in the upper range of total length. In both cases a likely explanation is that there are year-to-year variations in fecundity.

Year-class strengths
Figure 5 shows year-class strengths during the period 1976 to 1984. In Loch Lomond 1984 was exceptionally strong (1.6), and accounted for 51% of the total catch in samples in 1986–87. High indices also occurred in 1976, 1980 and 1981, but even the worst year, 1983, still shows a reasonably high index of 0.6. In Loch Eck, the strongest year class was 1982 (1.4), followed by 1976 and 1981. The weakest year classes were 1977 and 1983 (<0.1), the latter being an almost complete failure.

Spawning

Spawning in Loch Lomond takes place on offshore gravel banks such as the MacDougall Bank in the south basin (Slack *et al.*, 1957). During the present study, the shallows around the Ross Islands were identified as a major spawning ground, as also (though less important) was a shallow bank across the mouth of Sallochy Bay, westwards from Camas an Losgainn. Spawning grounds can be identified by the high numbers of males which congregate there. For example, in Loch Lomond a catch between the Ross Islands on 10 January 1985 showed a characteristically high ratio of 4.7m:1f, as did a catch on 9 January 1986 on the Sallochy Bay ridge: 2.6m:1f (Fig. 3). There are no offshore banks in Loch Eck. To identify spawning grounds there, nets were set all round the loch before and during spawning time in 1991–92 in 0 to 20 m of water with various substrates. Net catches at 1 to 2 m depth often with some growth of *Isoetes*, contained a high percentage of males, almost all with empty stomachs. The few females on these spawning grounds were ovulating or close to it. Growths of *Myriophyllum* occurred at the lower limit of suitable spawning depth, about 2 m. Nearby nets in 2 m or deeper water contained a high percentage of preovulating females; 2 m seems to be the critical lower limit. Suitable sites, however small, all round the loch were used opportunistically; for example, a spawning ground at Coire Ealt in 1983–86 was not used in 1991–92, perhaps because of constructional changes along the shoreline. However overall, the highest concentration of spawning activity was on the western shore from Stuck southwards to Allt na Gaibhre, in particular the promontories of Bernice and Coirantee; and on the east shore the shoreline north and south of Coylet. In the extreme south end of the loch there was little spawning activity.

Males migrate to the spawning grounds several weeks before the females. This movement of males is reflected in deviations from the normal sex ratios in open water catches. The sex ratios in catches from 1984 to 1987 are shown in Fig. 3. In Loch Eck the m:f ratio offshore was already low by September 1984, and fell further until the end of January 1985 (spawning started at the beginning of February). There was a similar pat-

Table 1. Sex ratios (excluding spawning period) of powan in individual year-classes from 1974 to 1984 in Loch Lomond and Loch Eck. Total number of each sex in year-class, and p-value from χ^2 test; n.s. = $p>0.05$.

	Loch Lomond			Loch Eck		
Year-class	Females	Males	*p*	Females	Males	*p*
1984	134	102	<0.05	31	19	n.s.
1983	44	45	n.s.	7	19	n.s.
1982	51	64	n.s.	143	146	n.s.
1981	61	81	n.s.	99	129	<0.05
1980	59	72	n.s.	74	115	<0.01
1979	45	59	n.s.	30	85	<0.01
1978	22	23	n.s.	19	47	<0.01
1977	9	9	n.s.	7	28	<0.001
1976	19	14	n.s.	7	32	<0.001
1975	10	21	<0.05	7	12	n.s.
1974	5	2	n.s.	1	4	n.s.

Table 2. Simple regression analysis of fecundity of: 1992–93 powan; 1991–92 powan; 1982–83 powan; 1967–68 powan. *n* = number of fish; no. of ova = extrapolated number of ova in fish of 300 mm total length.

	Spawning	*n*	*y*	*r*	No. ova
(a)	1991–92	83	−4766 + 31.253x	0.44	4610
(b)	1990–91	176	−3051 + 26.282x	0.55	4833
(c)	1966–67	100	−6936 + 41.731x	0.47	5582
(d)	1982–83	74	−2918 + 28.151x	0.554	5527

tern from September 1985 to February 1986 (spawning occurred between the last week of December and the last week of January).

Females move on to the spawning grounds as they ovulate, leave immediately after oviposition, and all have spawned within a period of four weeks. Powan spawning dates in Loch Lomond since 1979 and in Loch Eck since 1984 are shown in Fig. 6. Spawning in Loch Lomond varied by ±2 weeks, beginning at the earliest in mid-December (1978/79/, 1981/82), and at the latest in the first week of January (1979/80, 1984/85). In most years, spawning began towards the end of December. In Loch Eck spawning time was more variable, beginning near the end of December in 1985/86 and 1986/87, which coincided with spawning in Loch Lomond; but mid- to late January in 1983/84, and the end of January in 1984/85, being in these years almost exactly one month later than in Loch Lomond.

The timing of actual spawning is determined by the females, with a high degree of precision. Though Scott (1990) holds that salmonids are essentially 'long day' animals, their reproductive cycles being initiated in spring, with 'short days' serving only to accelerate the final stages of the cycle, it seems more probable that an environmental timing cue is involved. To account for the many variations in the spawning times of different races of salmonids on a purely photoperiodic basis requires arguments of labyrinthine complexity. Slack *et al.* (1957) postulated that powan spawn when 'the falling temperature of the water approaches close to 6 °C'. Maitland (1969) (who believed that spawning did not begin until January) postulated 'high water levels, low temperatures (normally about 5 °C or lower) and short, but increasing, daylength'. It is difficult to conceive that fish would react to an absolute temperature, rather than to a change, especially as the rate of change of temperature in Loch Lomond prior to spawn-

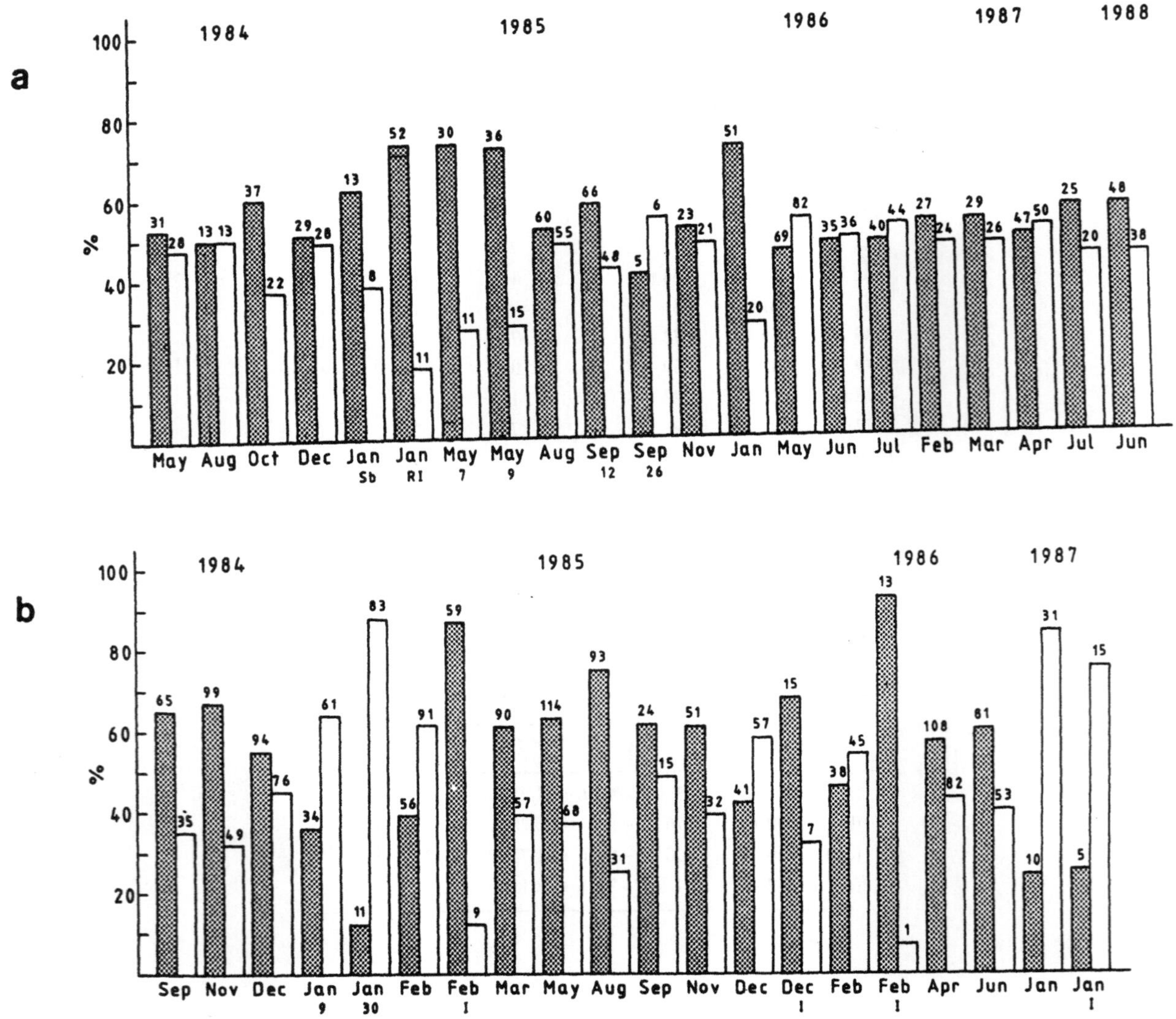

Plate 3. Sex ratios of powan catches in Loch Lomond (a) and Loch Eck (b) from 1984 to 1988. Sb = south basin of Loch Lomond; RI = Ross Islands spawning ground; I = inshore spawning ground. Males stippled. Numbers of powan are shown above each column. (In Loch Eck the overall sex ratio is variable, from 3 m:1 f prior to 1984 to nearer equality recently; in Loch Lomond it is consistently 1 m:1 f).

ing is negligible. Increase in daylength is minimal at the solstice and spawning in some years may precede it or not occur until two months later. We postulate that there is a cue for spawning, but that it is not a 'long-range' one based on the solar cycle, but a more precise one, based on the shorter term lunar cycle. Females, entrained to lunar cycles throughout the year, would be cued to ovulate by the first appropriate moon phase after they attain some critical state of 'readiness'. In Loch Lomond, where the critical summer feeding does not vary much from year-to-year, 'readiness' is likely to be achieved at the same time each year, and spawning would occur on the nearest appropriate moon phase, which is (nowadays) close to the solstice. In the smaller and more variable Loch Eck, readiness might not have been reached by the same moon phase, in which case spawning would be postponed until the next. In Loch Lomond, there is a good degree of correlation between the beginning of spawning and moon phase over this period. Because the lunar year and the solar year do not coincide, the date of full moon closest to the winter solstice fluctuates by a maximum of ±2 weeks over a period of years, and so does the spawning time of powan. The hypothesis seems to be corroborated in Loch Eck where spawning time was the same as in Loch Lomond in 1985/86 and 1986/87, but almost exactly one month later in 1983/84 and 1984/85.

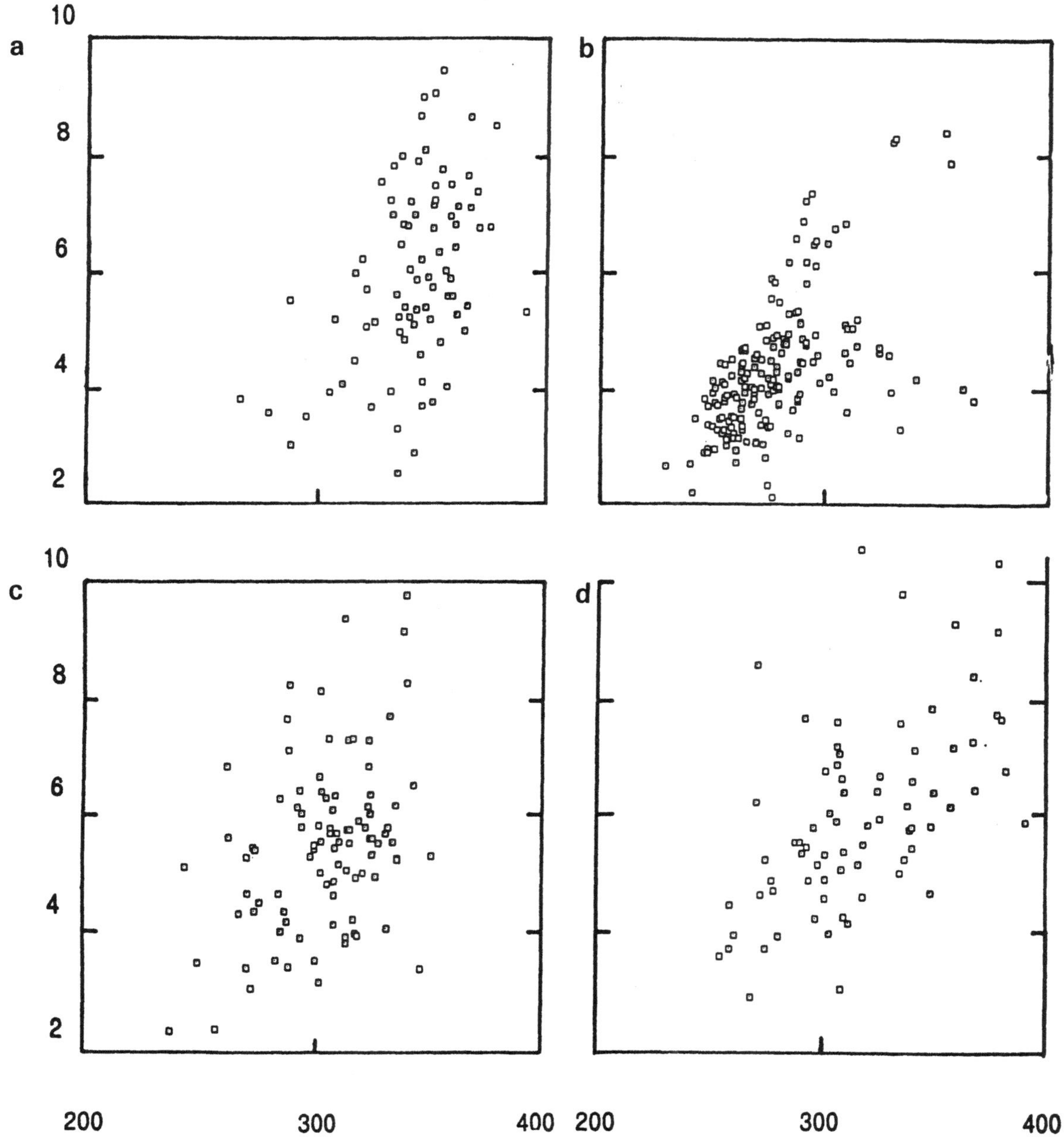

Plate 4. Regression of total length v number of oocytes in: (a) Loch Lomond 1991–92; (b) Loch Eck, 1990–91; (c) Loch Lomond 1966–67; (d) Loch Lomond 1982–83.

Lunar cues have been described in other species, and are of maximum efficacy at northerly latitudes in winter. In winter the full moon is in the sky during much of the night, whereas in summer it is mostly obscured by daylight. Moreover at 56°N, the meridian altitude of the sun at the winter solstice is at its lowest in the year (<10 °), whereas that of the moon is at its highest, 53°. Winter moonlight is consequently bright, and reflectance from the water surface low (<3%). Dabrowski (1981) relegates to a footnote an observation that commercial fishermen in Lough Neagh believe that spawning of pollan (*C. pollan* Thompson) is related to moon phase.

General discussion

A comparative approach to the biology of both powan populations is proving to be more revealing than separate studies of each, because although the fish are

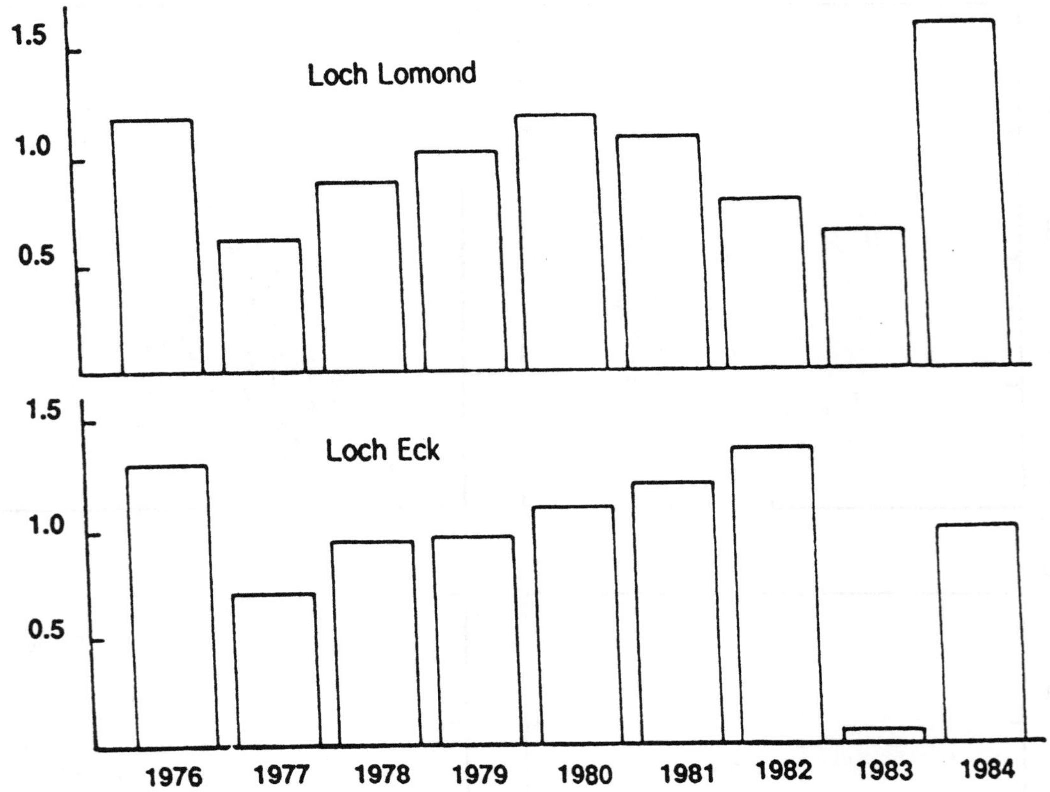

Plate 5. Year-class strengths in Loch Lomond and Loch Eck, 1976–86.

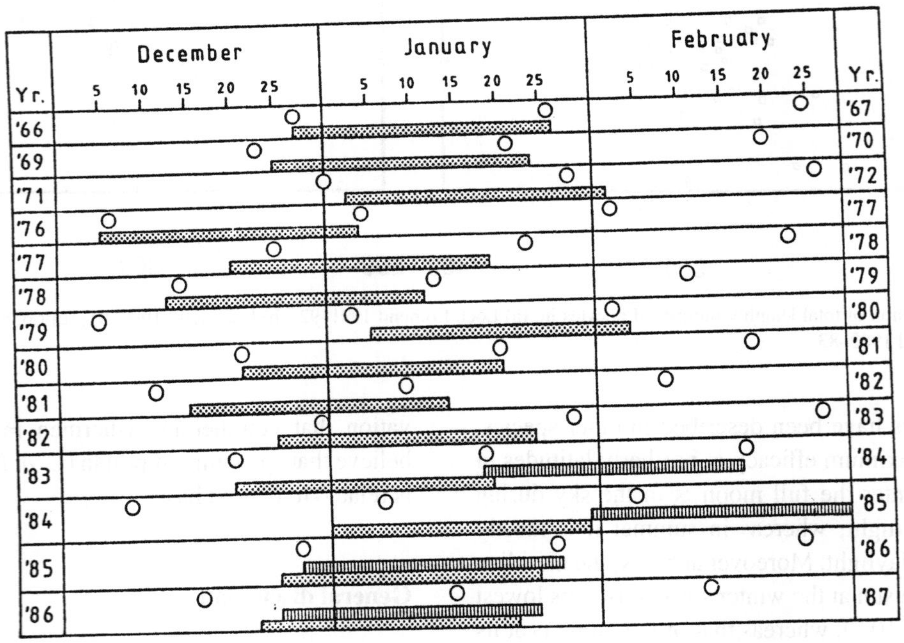

Plate 6. Spawning periods of powan in Loch Lomond (stippled) and Loch Eck (vertical bars). Circles represent date of full moon. Data for 1969 to 1979 from O'Connell, 1984. Data for 1966–67 from Maitland, 1969.

Table 3. Dimensions of Loch Lomond and Loch Eck (from various sources).

	Loch Lomond	Loch Eck
Length (km)	36	10.5
Surface area (km^2)	71	4.5
Volume (10^6 m^3)	2628	68
Mean depth (m)	37	15
Maximum depth (m)	190	42
Catchment (km^2)	696	99
Height above sea level (m)	8.5	9

genetically alike (McEwen, 1985), and the lochs are similar in major limnological respects, the life history strategies of the populations differ markedly. Loch Lomond and Eck are only 22 km apart, though in separate catchments. Both enjoy much the same climate (Birse & Dry, 1970; Birse & Robertson, 1970), and fauna (Slack, 1957; 1976). We speculate that the basic reason for the differences in the life histories of the two populations is that Loch Eck is smaller than Loch Lomond (Table 3) and topographically simpler; it lacks the many islands and sheltered bays of the bigger loch, and lies in a narrow, steep-sided channel bounded by precipitous mountains, akin to the north basin of Loch Lomond. Powan in Loch Eck are probably more exposed to year-to-year environmental variations than those in Loch Lomond. This is reflected in greater variance in all the growth processes described in this paper.

The total number of powan in Loch Lomond is not known, but they are very numerous. Lamond (1922) remarked that culling a quarter of a million from Loch Lomond had no appreciable effect on the total population. The only documented change in abundance occurred in 1968, when the population suffered massive mortality associated with fungal infection and an epidemic of 'bald spot disease' (Roberts *et al.*, 1970). Prior to this crash, powan were 'small and of poor quality', improving rapidly afterwards. Since growth is related to abundance, a change would be expected at such a time. The only recorded large catch of juvenile powan (>200) was made on 14 July 1969; 62 of these fish were measured, but not aged. Length-frequency analysis suggests that they comprised two main age groups, the mean total lengths of each being 98±6.9 mm and 132±4.6 mm (Fig. 7). In July half the growing season remains, so that these fish must have been either very fast growing 0^+ and 1^+ fish, or very slow growing 1^+ and 2^+ fish. In either case, a change in growth rate is indicated. Apart from short term fluctuations such as this, the length of powan seems to have been consistent for a long time. MacGibbon (1799) described them as being 'the size of a large herring'. Parnell (1938) described a specimen 14 inches (356 mm) long, and stated that they occasionally grew to 16 inches (406 mm). Lamond (1911) recorded the biggest specimen seen by him as 17 inches (432 mm), with 'hearsay evidence' of bigger. The biggest powan caught during the present study was 455 mm in total length, which suggests that the maximum size has not changed much in the long term. Although their absolute numbers in the lochs are not known, Brown *et al.* (1992) have estimated the proportions in each age group, and their age-specific fecundity and mortality data. Samples of 0^+ powan are badly needed. So far there is no indication of any long term change in the growth patterns of Loch Lomond powan such as might be expected if ruffe predation were altering numbers significantly. However, the fact that 62% of reproduction takes place in four years (and almost 90% in two of them) makes any two or more consecutive years of poor year classes a matter of concern in sustainability of the population.

Only one long term difference between the life-histories has been established (Pomeroy (1991; 1994). In Loch Lomond, powan feed intensively on zooplankton from May to September, and for the rest of the year take only small amounts of benthos (mainly *Pisidium* and chironomid larvae). From October to April many individuals have empty alimentary canals. Loch Eck powan feed exclusively on benthos throughout the year. They stop feeding only briefly at spawning time. The reason for the differences in feeding is not clear; they are probably complex (Huntingford, 1994). Loch Lomond powan, transferred to a much smaller loch, Sloy, were also found to be feeding solely on benthos in June (P. S. Maitland, pers. comm.). On the other hand, in a study of 17 Loch Eck powan in September, Slack *et al.* (1957) found all their stomachs to be full of zooplankton. Changes in one life history characteristic might well lead to the establishment of others. Lamond (1931) believed that Loch Lomond powan spawned from October to December, and Slack *et al.* (1957) considered that he was misled by the fact that males gather on the spawning grounds long before the females. With lunar cues, it is conceivable that changing longterm conditions in the lochs might lead to actual changes in spawning time.

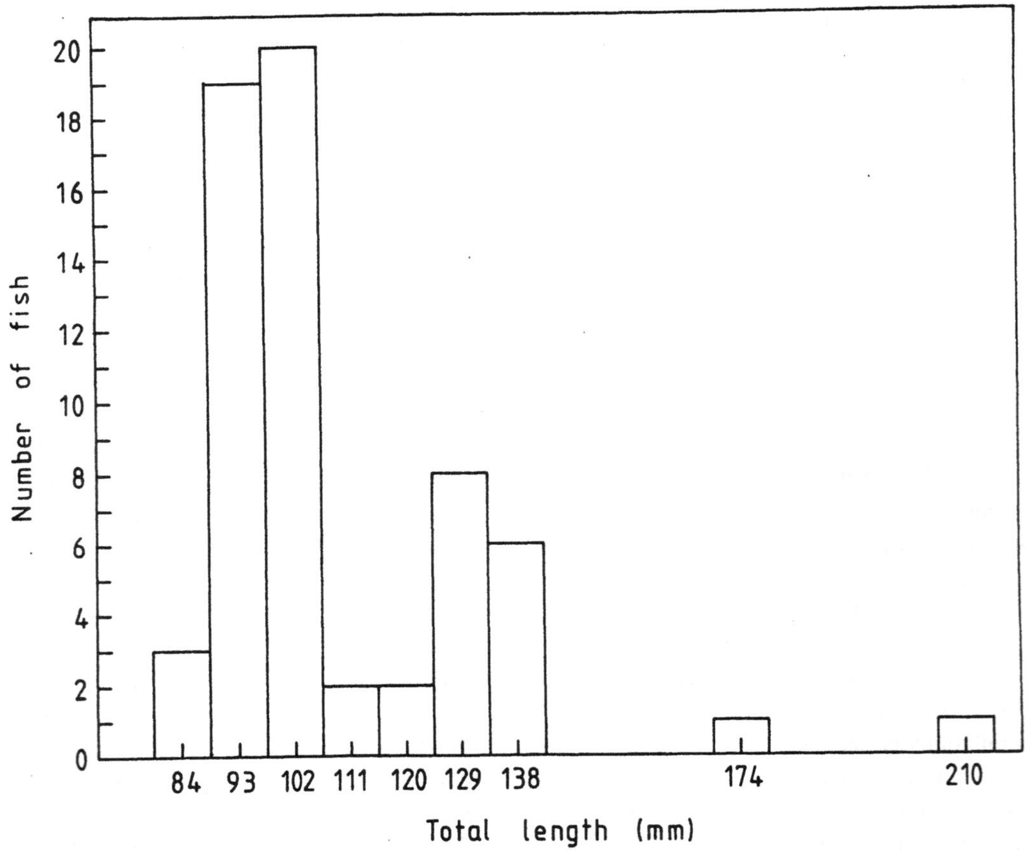

Plate 7. Length-frequency distribution of 62 juvenile powan caught in Sallochy Bay, Loch Lomond, on 14 July 1969 by A. P. Scott and DBCS.

Conservation of the powan populations is irrevocably linked with conservation of their ecosystems as a whole, rather than by 'plantings' in other waters, since their life history strategies are clearly environmentally determined. The relationship between lampreys and powan in Loch Lomond, is not applicable to the Loch Eck population. Similarly, Loch Eck is one of the only two British lakes in which sympatric populations of *Salvelinus* and coregonines occur; it is generally accepted that the taxa are mutually exclusive, with coregonines tending to dominate (Svärdson, 1976). There is some evidence that a charr-powan sympatry may have existed in Loch Lomond in the recent past. Stuart (1799) records that charr occurred then in the parish of Luss, and Lamond (1931) translates the name of an islet in Loch Lomond as 'Charr Island'.

Acknowledgements

Our thanks are due to the University Field Station at Rowardennan for their unstinting provision of facilities during 25 years; to Scottish Natural Heritage (formerly the Nature Conservancy Council) for permission to catch powan in Lochs Lomond and Eck; to Messrs Salar Properties Management for permission to work on Loch Eck. The study of spawning sites in Loch Eck in 1991–92 was commissioned by the Nature Conservancy Council (now Scottish National Heritage) as part of its programme of research into nature conservation. The studies of fecundity in Loch Eck and in Loch Lomond in 1992 were carried out as part of the requirement for Honours B. Sc. in St. Andrews University. In particular we owe a great deal to the late Mrs Flora Slack for allowing us to use the Yett of Ptarmigan, as a home from home and laboratory during many happy years.

References

Adams, C. E., 1994. The fish community of Loch Lomond, Scotland: its history and rapidly changing status. Hydrobiologia 290: 91–102.

Birse, E. L. & F. I. Dry, 1970. Assessment of climatic conditions in Scotland, 1. Soil Survey of Scotland; the Macaulay Institute for Soil Research, Aberdeen.

Birse, E. L. & L. Robertson, 1970. Assessment of climatic conditions in Scotland, 2. Soil Survey of Scotland; the Macaulay Institute for Soil Research, Aberdeen.

Brown, E. A. R., 1989. Growth processes in the two Scottish populations of powan, *Coregonus lavaretus* (L.) (Euteleostei, Salmonidae). Ph. D. thesis, School of Biological and Medical Sciences, University of St. Andrews, Scotland.

Brown, E. A. R. & D. B. C. Scott, 1990. Anabolic adaptiveness in the two Scottish populations of *Coregonus lavaretus* (L.) (Salmonidae, Coregoninae). J. Fish. Biol. 37 (suppl. A): 251–253.

Brown, E. A. R. & D. B. C. Scott, 1991. Adolescence in female powan. In Proceedings of the 4th International Symposium on the Reproductive Physiology of Fish, Norwich 1991. (Scott, A. P., J. P. Sumpter, D. E. Kime & M. S. Rolfe (eds): 163. Sheffield: FishSymp 91.

Brown, E. A. R., N. Finnigan & D. B. C. Scott, 1992. A life table for powan, *Coregonus lavaretus* (L.) in Loch Lomond, Scotland: a basis for conservation strategy. Aquatic Conservation: marine and freshwater ecosystems 1: 183–187.

Buchanan, G. 1582. Rerum Scoticarum historia. Fol. Edinburgi apud Alexandrum Arbuthnetum. (In Latin)

Dabrowski, K. R. 1981. The spawning and early life history of the pollan (*Coregonus pollan* Thompson) in Lough Neagh, Northern Ireland. Internationale Revue der gesamten Hydrobiologie und Hydrographie 66: 299–326.

Ferguson, A. 1974. The genetic relationships of the coregonid fishes of Britain and Ireland indicated by electrophoretic comparison with other holarctic Coregoninae. J. Fish Biol. 6: 311–315.

Ferguson, A., K.-J. M. Himberg & G. Svärdson 1978. Systematics of the Irish pollan (*Coregonus pollan* Thompson): an electrophoretic comparison with other holarctic Coregoninae. J. Fish Biol. 12: 221–233.

Fuller, J. D., D. B. C. Scott & R. Fraser 1976. The reproductive cycle, *Coregonus lavaretus* (L.) in Loch Lomond, in relation to seasonal changes in plasma cortisol concentration. J. Fish Biol. 9: 105–117.

Hamilton, J. D. 1988. Recent human influences on the ecology of Loch Lomond, Scotland. Verh. int. Ver. Limnol. 23: 404–413.

Huntingford, F. A. 1994. Can cost-benefit analysis explain fish distribution patterns? J. Fish Biol. 43: 289–308.

Lamond, H. 1911. 'The Gentle Art'. Murray, London.

Lamond, H. 1922. Some notes on two of the fishes of Loch Lomond: the powan and the lamprey. Fisheries, Scotland. Salmon Fish. No. 2.

Lamond, H. 1931. 'Loch Lomond'. Jackson, Wylie & Co., Glasgow.

MacGibbon, D. 1799. Parish of Buchanan. In The Statistical Account of Scotland (Sinclair, J. ed.) 8; Reissue 1978 (Withrington, D. J. & I. R. Grant eds.): 397–420. EP Publishing Ltd., Wakefield, England.

McEwen, I. 1985. Isozyme variation in Scottish populations of the powan, *Coregonus lavaretus* (L.). B. Sc. thesis, Department of Zoology and Marine Biology, University of St. Andrews.

Maitland, P. S. 1969. The reproduction and fecundity of the powan, *Coregonus clupeoides* Lacépède, in Loch Lomond, Scotland. Proc. Roy. Soc. Edinburgh B 70: 233–264.

Mills, C. A. & R. H. K. Mann 1985. Environmentally-induced fluctuations in year-class strengths and their implications for management. J. Fish Biol. 27 (suppl. A): 209–226.

Monipenny, I. 1612. The abridgement or summarie of the Scots Chronicles. Simon Stafford, London.

O'Connell, W. D. 1984. Environmental timing and control of reproduction in the powan of Loch Lomond, *Coregonus lavaretus* (L.) (Teleostei) in relation to its pineal organ. Ph. D. thesis, Department of Zoology and Marine Biology, University of St. Andrews, Scotland.

Parnell, R. 1938. Observations on the Coregoni of Loch Lomond. Ann. Nat. Hist. 1: 161–165.

Pomeroy, P. P. 1991. A comparative assessment of temporal variation in diet of powan, *Coregonus lavaretus* (L.) from Loch Lomond and Loch Eck, Scotland, U.K. J. Fish Biol. 38: 457–478.

Pomeroy, P. P., 1994. Zooplankton in Loch Lomond: perspectives, predation and powan. Hydrobiologia 290: 75–90.

Roberts, R. J., J. Leckie & H. D. Slack 1970. 'Bald spot' disease in powan *Coregonus lavaretus* (L.). J. Fish Biol. 2: 103–105.

Scott, A. P. 1990. Salmonids. In Reproductive seasonality in teleosts. (Munro, A. D., A. P. Scott & T. J. Lam, eds.) Florida: CRC Press, Inc.: 33–51.

Slack, H. D. 1957. The fauna of the lake. In Studies on Loch Lomond I (Slack, H. D. ed.): 33–48. Glasgow Univ. Publ. Blackie & Son Ltd.

Slack, H. D. 1976. Marine and freshwater life. In Argyll Forest Park Forestry Commission Guide. (Edlin, H. L. ed.). HMSO, Edinburgh.

Slack, H. D., F. W. K. Gervers & J. D. Hamilton 1957. The biology of the powan. In Studies on Loch Lomond I (Slack, H. D. ed.), 113–127. Glasgow Univ. Publ. Blackie & Son Ltd.

Stewart, C. 1799. United parishes of Strachur and Strathlachlan. In The Statistical Account of Scotland (Sinclair, J. ed.) 8; Reissue 1978 (Withrington, D. J. & Grant, I. R. eds.): 397–420. EP Publishing Ltd., Wakefield, England.

Stuart, J. 1799. Parish of Luss. In The Statistical Account of Scotland (J. Sinclair, ed.) 9, Reissue 1978 (Withrington, D. J. & Grant, I. R. eds.): 83–116. EP Publishing Ltd., Wakefield, England.

Svärdson, G. 1976. Interspecific population dominance in fish communities of Scandinavian lakes. Inst. Freshwater Research, Drottningholm Rep. 55: 144–171.

Treasurer, J. W. & D. H. Mills 1992. An annotated bibliography of research on coarse and salmonid fish (excluding salmon and trout) found in fresh water in Scotland. Freshwater Forum 3: 202–236.

Hydrobiologia **290**: 135–138, 1994.
K. J. Murphy, M. C. M. Beveridge & R. Tippett (eds), The Ecology of Loch Lomond.

Elevated predation risk associated with inshore migrations of fish in a large lake, Loch Lomond, Scotland

Colin E. Adams, David W. Brown & Lorraine Keay
Fish Behaviour and Ecology Group, University Field Station, Rowardennan, Glasgow G63 OAW, U.K.

Key words: migration, predation, Loch Lomond, ruffe, roach

Introduction

For wild fish populations, predation is an ever present threat. In consequence fish have developed a range of morphological devices and behavioural strategies to reduce their risk of becoming prey. This risk however, may not be constant and may change significantly for individuals of many species over short time spans, depending on the activities that the individual is engaged in. Investigations into the conflict that occurs between the need to perform some high risk activity and the need to avoid being preyed upon, have focused almost exclusively on laboratory studies of the relationship between foraging and the risk of predation in individuals (Milinski & Heller, 1978; Milinski 1984, Metcalfe *et al.* 1987a, 1987b) and there is a paucity of data relating to natural populations and on the predation risk of activities other than feeding.

Here, using data from a field study we examine the influence of the activity patterns of two species of freshwater fish, the ruffe *Gymnocephalus cernuus*, and the roach, *Rutilus rutilus*, on their risk of predation from freshwater feeding cormorants, *Phalacrocorax carbo*, in Loch Lomond, Scotland.

Methods

To estimate the predation rate of cormorants on each fish species, the diet of cormorants feeding on Loch Lomond was estimated from an examination of the otolith sagittae of fish, recovered from regurgitates collected from the principal roosting site of cormorants on Loch Lomond, 'The Stott', a small rock islet in the south basin of the loch. To examine changes in prey selection with time, up to a maximum of 20 fresh pellets were collected from the roost site approximately every two weeks between April 1990 and March 1991. Pellets were dissolved in biological washing powder and the undigested otoliths recovered and then air dried (Johnson *et al.*, 1990). Prey otoliths were then identified by comparing the recovered otoliths with a reference collection of known origin. No pellets were collected in June and July when only a few non-breeding cormorants feed on Loch Lomond, nor October 1990 when very high water levels dispersed the birds to other roosting sites.

To enable a comparison of predation rate with fish activity in the littoral zone, data are presented showing the annual cycle of fluctuation of littoral activity for the two commonest species in the diet of cormorants feeding on Loch Lomond, ruffe and roach. These data are derived from monthly collections of fish using a standard, multi-panel gill-net (40 m long, bar mesh size 19–50 mm) set in 1–6 m depth, the depth range most used by foraging cormorants (van Dobben, 1952) twice monthly, at five sites selected from throughout Loch Lomond from November 1988 to October 1989. All catches were corrected for fishing effort (see Adams & Tippett, 1991 for more complete details).

Results

Cormorants preyed upon a wide range of fish species (Fig. 1). Of the over 3000 fish otoliths identified during the year, 85% were ruffe. Roach was the second most common prey species making up 5.3% followed by powan *Coregonus lavaretus* (5.2%) and flounder *Platichthys flesus* (2%). Dace, *Leuciscus leuciscus*,

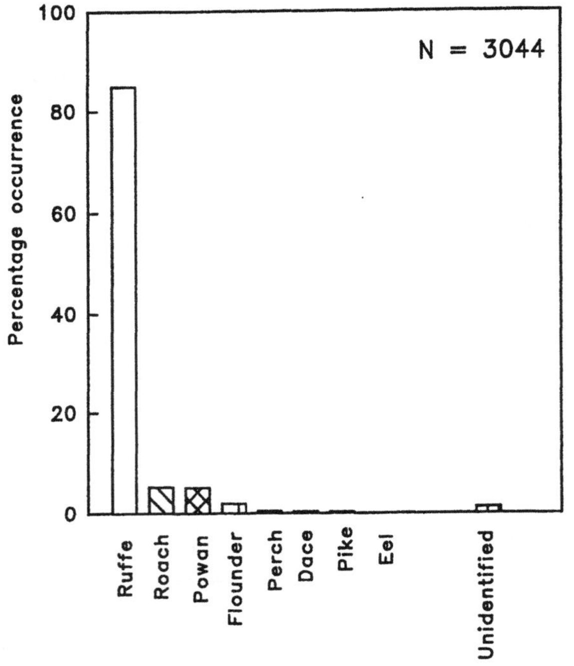

Fig. 1. The percentage occurance of otoliths of fish prey recovered from regurgitates of cormorants feeding in Loch Lomond 1988-89.

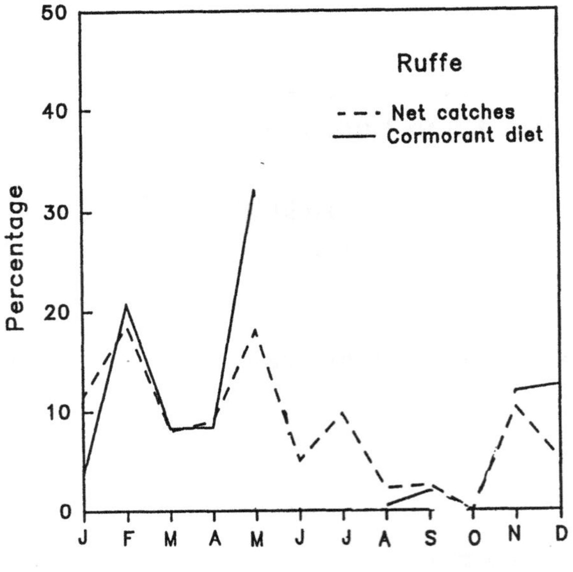

Fig. 2. The monthly capture rate of ruffe (expressed as a percentage of total captures) in gill nets and their monthly occurrence in the diet of cormorants (expressed as a percentage of annual intake).

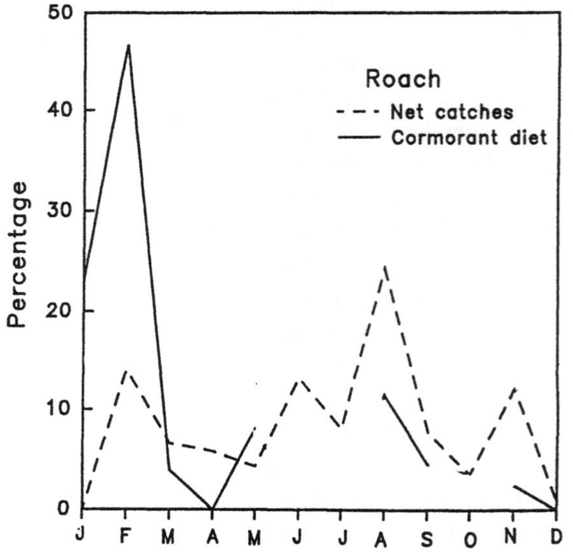

Fig. 3. The monthly capture rate of roach (expressed as a percentage of total captures) in gill nets and their monthly occurrence in the diet of cormorants (expressed as a percentage of annual intake).

pike *Esox lucius*, perch *Perca fluviatilis*, and eel *Anguilla anguilla* were also recorded but made up less than 1% each of the recovered otoliths. Although each fish preyed upon has 2 otolith sagittae, captive feeding studies (Adams, unpublished data) have shown that possibly up to around 40% of otoliths are never recovered from pellets. This suggests that the 3000 otoliths examined here may have emanated from as many as 2500 prey items.

It is well established that, in common with other large lakes, there are well defined annual cycles of fish activity in the littoral zone of Loch Lomond (Maitland *et al.*, 1981; Giles & Tippett, 1987; Adams & Tippett, 1990; Maitland & Adams, in prep). The majority of loch species spawn in shallow water and during their respective breeding periods, congregations of fish occur in the littoral zone. For the two species examined here spawning occurs in April/May, for ruffe and July/August for roach. In addition to spawning migrations, some species make non-breeding related migrations into shallow water often to feed (Giles & Tippett, 1987). Figures 2 and 3 show monthly mean net catches corrected for catch-per-unit-effort, expressed as a proportion of total catch of that species (all sites combined), for ruffe and roach respectively. The cyclicity of activity of these two species shows a high degree of concordance between collection sites (Kendal coefficient of concordance: ruffe $W=0.60$, d.f.$=7$, $p<0.01$; roach $W=0.56$, d.f$=7$, $p<0.01$), suggesting that these inshore migration patterns are stable.

The frequency of occurrence of prey species in the diet of cormorants feeding on Loch Lomond also varied seasonally. Figures 2 and 3 also show the mean monthly consumption rate of each prey species expressed as a proportion of total annual consumption of that prey (corrected for differing monthly sample sizes).

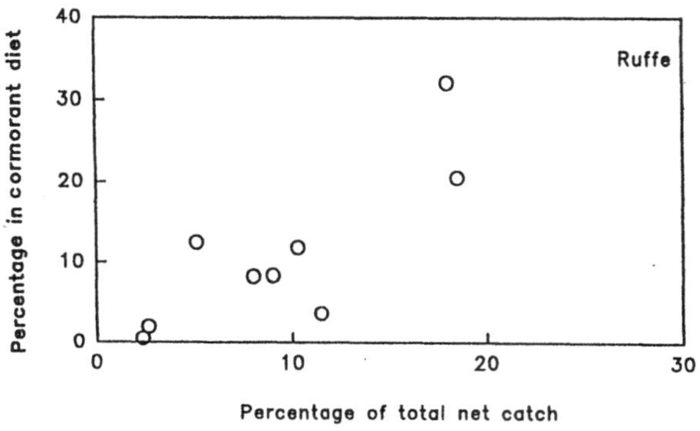

Fig. 4. The relationship between ruffe activity in the littoral zone (net catches) and their occurrence in the diet of cormorants (otolith recoveries from pellets) (Spearman Rank correlation coefficient, $r_S = 0.775$, $p < 0.025$).

The occurence of ruffe in gill net captures was highly correlated with their rate of predation by cormorants (Spearman Rank Correlation Coefficient, r_S=0.775 p<0.025) (Fig. 4). A similar although slightly less positive relationship was also found in roach (r_S=0.754, p<0.025).

Discussion

The cormorant population resident on Loch Lomond numbers around 40–50 birds (Keay unpublished data) and feeds apparently exclusively on the freshwater fish in the catchment (Fig. 1). The main species preyed on during this study was ruffe making up around 85% of all prey species identified from otolith regurgitates. As ruffe have only recently been introduced to Loch Lomond (Maitland & East, 1989) the cormorant colony feeding on Loch Lomond can only have recently switched their diet from native prey to feed on introduced ruffe. However, as there have been no studies on the diet of this colony of cormorants prior to the introduction of ruffe, we do not know what prey species comprised their diet at this time.

Using fish abundance in gill-net captures as an indicator of fish activity in the littoral zone, we describe the seasonal migrations into and out of the littoral zone. Data presented here show that these migrations are consistent between sites. From fish capture data collected over many years for the species examined here and others we know that such migrations are relatively stable from year to year in Loch Lomond (Maitland *et al.*, 1981; Giles & Tippett, 1987; Adams & Tippett, 1990; Maitland & Adams, in prep) and also that similar migration patterns occur in other lakes (see Bean & Winfield, 1992 for description of roach migration patterns in Lough Neagh).

Here we have shown that high activity in the littoral zone increases the risk of predation by cormorants, particularly for ruffe but also for roach.

Ruffe and roach both utilise the littoral zone to spawn (Johnsen, 1965; Slack *et al.*, 1957; Wheeler, 1969) and high abundance in the littoral during their respective breeding seasons, is certainly the result of spawning congregation. This suggests that spawning carries costs in addition to the direct energetic costs of reproduction, in the form of increased predation risk. Thus it would seem that during the spawning period these species face conflicting demands. The need to enter the littoral zone to spawn and the need to minimise predation risk. How these conflicting demands are resolved is unknown, however several potential mechanisms may be available to them to reduce predation risk, such as increased group vigilance and the predation dilution effects that accrue from schooling and synchronous spawning (Seghers, 1974; Taylor, 1984; Krebs & Davis, 1987).

References

Adams, C. E. & R. Tippett, 1990. The Status of the Fish Community of Loch Lomond. Report to the Nature Conservancy Council, Hope Terrace, Edinburgh, 31 pp.

Adams, C. E. & R. Tippett, 1991. Powan *Coregonus lavaretus* (L) ova predation by newly introduced ruffe, *Gymnocephalus cernuus* (L.), in Loch Lomond, Scotland. Aqua. Fish. Mgmnt. 22: 239–246.

Bean, C. W. & I. J. Winfield, 1992. Influences of the tapeworm *Ligula intestinalis* (L.) on the spatial distributions of juvenile roach *Rutilus rutilus* (L.) and gudgeon *Gobio gobio* (L.) in Lough Neagh, Northern Ireland. Neth. J. of Zool. 42: 416–429.

van Dobben, W. H., 1952. The food of the cormorant in the Netherlands. Ardea 40: 1–63.

Giles, N. & R. Tippett, 1987. Annual migration and diet of perch *Perca fluviatilis* L. in Loch Lomond, Scotland. Glasg. Nat. 21: 287–295.

Johnsen, P., 1965. Studies on the distribution and food of the ruffe (*Acerina cernua* L.) in Denmark, with notes on other aspects. Meddelelser fra Danmarks Fiskeri-og Havundersogelser 4: 137–156.

Johnstone, I. G., M. P. Harris, S. Wanless & J. A. Graves, 1990. The usefulness of pellets for assessing the diet of adult Shags *Phalacrocorax aristotelis*. Bird Study 37: 5–11.

Krebs, J. R. & N. B. Davis, 1987. An Introduction to Behavioural Ecology. Second edition, Blackwell, Oxford, 389 pp.

Maitland, P. S. & C. E. Adams, in prep. Ruffe in Loch Lomond: the first 12 years.

Maitland, P. S., B. D. Smith & S. M. Adair, 1981. The fish and fisheries. In The Ecology of Scotland's Largest Lochs. (Maitland, P. S. ed). Dr W. Junk Publishers, The Hague, 297 pp.

Metcalfe, N. B., F. A. Huntingford & J. E. Thorpe, 1987a. Predation risk impairs diet selection in juvenile salmon. Anim. Behav. 35: 931–933.

Metcalfe, N. B., F. A. Huntingford & J. E. Thorpe, 1987b. The influence of predation risk on feeding motivation and foraging strategy of juvenile Atlantic salmon. Anim. Behav. 35: 901–911.

Milinski, M., 1984. A predators cost of overcoming the confusion-effect of swarming prey. Anim. Behav. 32: 1157–1162.

Milinski, M. & R. Heller, 1978. Influence of a predator on the optimal foraging behaviour of sticklebacks (*Gasterosteus aculeatus* L.) Nature 275: 642–44.

Mills, D. H., 1965. The distribution and food of the cormorant in Scottish inland waters. Freshwater and Salmon Fisheries Research Publication No. 35, Department of Agriculture and Fisheries for Scotland, H.M.S.O., Edinburgh.

Seghers, B. H., 1974. Schooling behaviour in the guppy (*Poecilia reticulata*): an evolutionary response to predation. Evolution 28: 486–489.

Slack, H. D., F. W. K. Gervers & J. D. Hamilton, 1957. The biology of the powan. In Studies on Loch Lomond I (Slack, H. D. ed). Blackie, Glasgow, 133 pp.

Taylor, R. J., 1984. Predation. Chapman and Hall, London, 166 pp.

Wheeler, A., 1969. The fishes of the British Isles and N. W. Europe. Macmillan, London, 537 pp.

Hydrobiologia **290**: 139–151, 1994.
K. J. Murphy, M. C. M. Beveridge & R. Tippett (eds), The Ecology of Loch Lomond.

The ecology of the River Endrick: present status and changes since 1960

C. R. Doughty[1] & P. S. Maitland[2]
[1]*Clyde River Purification Board, Rivers House, Murray Road, East Kilbride, Glasgow G75 0LA, UK*
[2]*Fish Conservation Centre, Easter Cringate, Stirling FK7 9QX, UK*

Key words: River Endrick, water quality, fish, benthic invertebrates, conservation

Abstract

Long-term changes in water quality, sources of pollution, and the fish and benthic macroinvertebrate communities in the River Endrick catchment are described. The conservation status and management of the river are discussed and the need for research on nutrient loadings and the impact of recent fish introductions is emphasized.

Introduction

The River Endrick is the major inflow to Loch Lomond, Scotland's largest freshwater loch. It is also the richest inflow, both chemically and biologically, and is the main spawning and nursery area for the loch's salmonid fish. Most of the Loch Lomond catchment's human population and agriculture is concentrated in the Endrick sub-catchment. The pressures on water quality are thus greater here than in other parts of the catchment. The maintenance of the River Endrick's high water quality and the conservation of its natural resources are essential for the well-being of Loch Lomond itself.

Although there are a number of early references to the biology of the River Endrick, mostly concerned with its fish fauna (e.g. Ure, 1795; Brown, 1891; Hunter *et al.*, 1959), it was not until the early 1960s that a systematic study of the biology of the river was carried out. A general account of this work is given by Maitland (1966a). A further series of papers describes quantitative studies on invertebrates (Maitland, 1964a), the effects of water abstraction and sewage effluent discharges (Maitland, 1962a), feeding relationships of fish (Maitland, 1965a) and details of the biology of particular invertebrate species or groups (Maitland, 1962b, 1963, 1965b, 1965c, 1966b, 1966c, 1967a, 1967b; Hunter *et al.*, 1963).

The aims of this paper are to review the present ecological status of the River Endrick and to examine changes which have taken place over the last 30 years. Aspects of the management of the river's natural resources are discussed and research needs identified.

The physical environment

Geology & topography

These aspects have been described by Maitland (1966a), so only a brief account is given here.

The catchment of the River Endrick lies almost entirely to the south of the Highland Boundary Fault, within the Midland Valley of Scotland. Although the upper part of the catchment lies on basaltic lavas of the Clyde plateau, most of the catchment consists of Upper and Lower Old Red Sandstone rocks, together with some calciferous sandstone of Carboniferous age. Therefore, in contrast to the remainder of the Loch Lomond catchment, the rocks are predominantly base-rich.

The River Endrick rises as Mary Glyn's Burn at an altitude of 457 m in the Gargunnock Hills and flows in a mainly westerly direction for 49 km before entering Loch Lomond south of Balmaha (Fig. 1). The largest tributary, the Blane Water, joins the Endrick near Killearn, 16 km upstream of the loch. The river has a conventional profile, with the upper reaches above Fintry being characterized by a steep gradient, fast current and a rocky bed. In the middle reaches, between Fintry and Drymen, the river takes a more meandering course, with progressively longer pool stretches. Downstream of Drymen, the river enters its flood plain and has a convoluted course and sluggish flow between unstable sandy banks. There are two waterfalls of note. Some 5 km east of Fintry, the river plunges over the 30 m high Loup of Fintry. At the Pot of Gartness, about 1 km above the confluence with the Blane Water, there is a smaller fall about 2 m high. Both falls have an important role in controlling fish distribution in the Endrick.

Hydrology & water use

The Endrick catchment has an area of 264 km^2 (Maitland, 1981), about a third of the total area of the Loch Lomond catchment. Since 1963, the Clyde River Purification Board (CRPB) has operated a flow gauging station at Gaidrew, 2 km upstream of Drymen. The long term average flow is 7.1 $m^3\ s^{-1}$. Annual rainfall over the catchment is about 1400 mm (J. C. Curran, pers. comm.). There is evidence of a 25% increase in annual mean rainfall over the last 20 years. Over the same period, annual mean flow has increased by 36% (Curran & Poodle, 1992).

Water is abstracted for potable supply at two locations in the Endrick catchment. Since 1954, water has been abstracted from the upper reaches at a highly variable rate to supply Carron Valley Reservoir. The compensation flow below the reservoir is 0.05 $m^3\ s^{-1}$. The supply can be augmented by water taken from the Boquhan Burn in the adjacent River Forth catchment and discharged to the upper reaches of the Backside Burn in the headwaters of the River Endrick. In the south of the catchment, Burncrooks Reservoir supplies a compensation flow of 0.02 $m^3\ s^{-1}$ to the Carnock Burn, a tributary of the Blane Water.

Land use

About 50% of the catchment is rough ground, 35% is farmland, 12% forestry and just 2% urban (Maitland, 1981). The main changes over the last 30 years have been an increase in the area devoted to forestry and growth in the populations of the main villages. The implications of these are discussed below. Industrial activity remains at a low level.

Water quality

Present water chemistry

The CRPB currently monitors water chemistry at seven sites on the Endrick and six on the Blane Water. Mean 1990 values for selected determinands at three sites in the upper, middle and lower reaches of the Endrick are shown in Table 1. Owing to the predominance of base-rich rocks in the catchment, there is sufficient buffering capacity, even in the upper reaches, to ensure that the Endrick is free from the acidification problems which affect other parts of the Loch Lomond catchment (CRPB, 1984). Water quality is excellent, with high dissolved oxygen concentrations and low ammonia at all times.

Sources of pollution

There are currently 11 sewage treatment works (STWs), three of them privately operated, discharging to the River Endrick and its tributaries. The largest, at Strathblane, is designed to serve a population of 3800 and has a dry weather flow of 6.7 $l\ s^{-1}$. Although Fintry STW is volumetrically overloaded and Balfron STW is operating at design capacity, none of the STWs has a significant impact on the river.

There is very little industry in the catchment. The waste from a small distillery is treated by oxidation ditch before discharge to a small tributary of the Blane Water. The performance of the treatment plant has been very variable, but no adverse effects on the Blane Water have been recorded. The upper reaches of the Endrick receive a discharge of filter backwash water from a water treatment works, but again, without noticeable impact on the river.

Pollution by farm waste, particularly silage liquor and cattle slurry, is an intermittent problem in the catchment. Historically, the worst affected areas have been the Balfron area and a tributary, the Catter Burn. During the period 1978–1991, four incidents resulted in serious fish kills in the Catter Burn and the Machar Burn, Balfron, involving a total of over 4000 salmonid fish. It is probable that over the same period at least a similar number of fish have been lost through the deliberate introduction of poisons by poachers. In one such incident in August 1990, thousands of Atlantic salmon, sea trout and brown trout were killed over a 5 km stretch of river in the Balfron area.

Over the last 30 years, STWs have been upgraded to cope with the growing population. Balfron

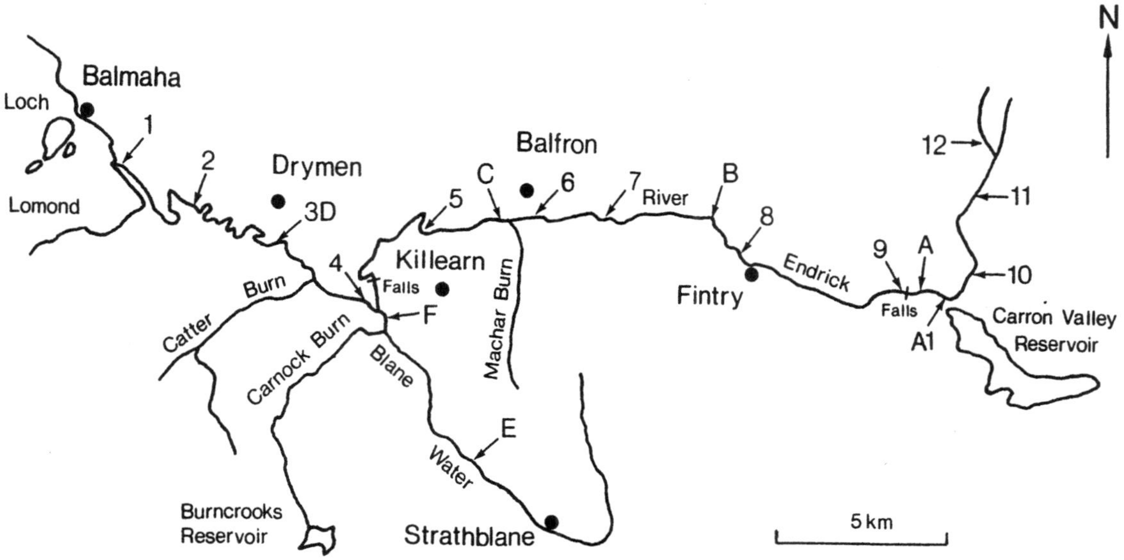

Fig. 1. Map of River Endrick catchment showing locations of CRPB chemical (A1, C, 2) and biological (A–F) monitoring sites and sites used to study long-term changes in invertebrate communities (1–12).

Table 1. Water chemistry at selected sites on upper, middle and lower reaches of the River Endrick (1990 annual means). Units mg l^{-1} except for conductivity ($\mu S\ cm^{-1}$).

Determinand	Site Todholes (A1)	Balfron (C)	Woodend (2)
Suspended solids	2	3	6
BOD	2.3	3.1	3.0
Dissolved oxygen	10.8	11.5	10.0
Total ammonia (N)	0.035	0.062	0.047
Nitrite (N)	0.009	0.016	0.018
Nitrate (N)	0.10	0.57	0.56
Soluble reactive P	0.005	0.057	0.048
pH	7.11	7.57	7.25
Alkalinity (as $CaCO_3$)	21	41	47
Total hardness (as $CaCO_3$)	28	59	65
Chloride	7	10	14
Conductivity	73	138	156

STW, which provided only primary treatment in 1962, was re-built to provide full biological treatment and was further modified in 1979 and 1987 to keep pace with housing developments. Strathblane STW, which served a population of 850 in 1962, now has a capacity of 3800 and with the aid of tertiary treatment, produces a high quality effluent. The works serving Killearn Hospital closed in the early 1970s. The total population served by the major STWs in the Endrick valley has increased from around 3600 in the early 1960s (Maitland, 1962a) to around 7800 in 1990 (CRPB, 1990).

Changes in water quality 1960–1990

In 1960, water quality in the Endrick was generally high. Although there was visible evidence of pollution downstream of STWs, the effects were localized and there was no discernible impact on fish populations (Maitland, 1962a). There has been an obvious improvement in the visual appearance of the Endrick and Blane in the intervening period. There is now no evidence of the turbidity and sewage fungus growths which were formerly present below Balfron STW and in the lower reaches of the Blane Water. According to the chemical classification scheme used for describing water quality in Scotland (Scottish Office Environment Department, 1992), the Endrick and its tributaries are Class 1 ('Rivers unpolluted and recovered from pollution').

In recent years there has been increasing concern, particularly among anglers, about possible eutrophication in the southern basin of Loch Lomond. Algal blooms in the loch and clogging of microstrainers at a water treatment works have been reported. Although it has been suspected that increasing inputs of phosphorus to the loch are to blame, this has been difficult to prove from existing chemical data. Water quality in the Endrick has been monitored by the CRPB since the 1960s, but the analytical methods initially used were too insensitive to determine the low levels of phosphorus present with sufficient accuracy. Hence in examining the data for possible trends, only the last 15 years have been considered. An additional problem has been the inadequate number of samples taken (only 6–12 annually), consequently the precision of estimates of annual mean concentrations and annual loadings are very poor. Annual mean concentrations of phosphorus in the lower reaches of the Endrick have varied from 0.024–0.059 mg l^{-1} as soluble reactive P over the period 1976–1990. Estimates of annual loadings (calculated from the product of mean flow-weighted concentrations and annual mean flow) over the same period range from 4.98–15.72 tonnes soluble reactive P. In addition to the poor precision of these estimates, soluble reactive phosphorus accounts for only a proportion of the potentially bioavailable phosphorus. Unfortunately, no data are available for other forms of phosphorus. There is no indication of any upward trend in phosphorus concentrations or loadings since 1976, but in view of the inadequacies of the available data, considerably greater sampling and analytical effort would be required to establish beyond doubt that no such trends exist. Since 1987, the CRPB has intensified monitoring of nutrient concentrations in Loch Lomond (Best & Traill, 1994), but owing to year to year variability, it is probable that monitoring will need to be continued for several more years to demonstrate the existence of any trend. Given the increase in catchment population over the thirty year period, it might be expected that the phosphorus loading to the loch would increase. However, such an assumption cannot be made without first establishing whether per capita phosphorus contributions and inputs from diffuse sources have changed.

The problems of attempting to identify trends in water quality from inadequate chemical data can often be overcome by using biological monitoring techniques. Biological communities can integrate the effects of episodic pollution events which may be missed by infrequent chemical monitoring and may also reflect the impact of pollutants which are not routinely determined chemically, or are present at concentrations below detection limits.

The CRPB has monitored the River Endrick and Blane Water using benthic macroinvertebrates communities since 1976. Figure 2 shows annual mean Average Score per Taxon (ASPT) (Armitage *et al.*, 1983) over the period 1976–1990. At Gartcarron and Fintry on the Endrick, there is no evidence of a significant trend. Low ASPT in 1980 at Fintry was thought to be due to physical disturbance of the site. At the other four sites, there appears to have been an increase in ASPT between 1976 and the mid-1980s, since when scores have levelled off or declined slightly. At Drymen, there was a significant linear increase ($P<0.02$) of 22% between 1976 and 1987. At Blane Bridge, the increase in ASPT was 16% over the same period.

It is difficult to account for these trends on the basis of the available chemical data. At Balfron, there is a significant negative correlation ($r = -0.590$, $P<0.05$) between ASPT and five-day biochemical oxygen demand (BOD) in the river. At Drymen, there is a highly significant negative correlation ($r = -0.672$, $P<0.01$) between ASPT and total ammonia. As both these sites are downstream of STWs, it is tempting to suppose that the macroinvertebrate communities were responding to changes in effluent quality and/or available dilution in the river. However, no such correlations were evident at Fintry, Dumgoyach and Blane Bridge, all of which are downstream of STWs, although in all three cases, the distance between the STW and the sampling site is greater than at Drymen and Balfron. At Drymen, the dilution available for the STW effluent in the river under dry weather flow conditions is over 300 times. It seems unlikely therefore that ammonia from the STW could be having a significant impact, even if STW performance was poor over a prolonged period.

An alternative explanation is that the improvement in ASPT is a reflection of the increased dilution of pollutants resulting from the observed increase in annual

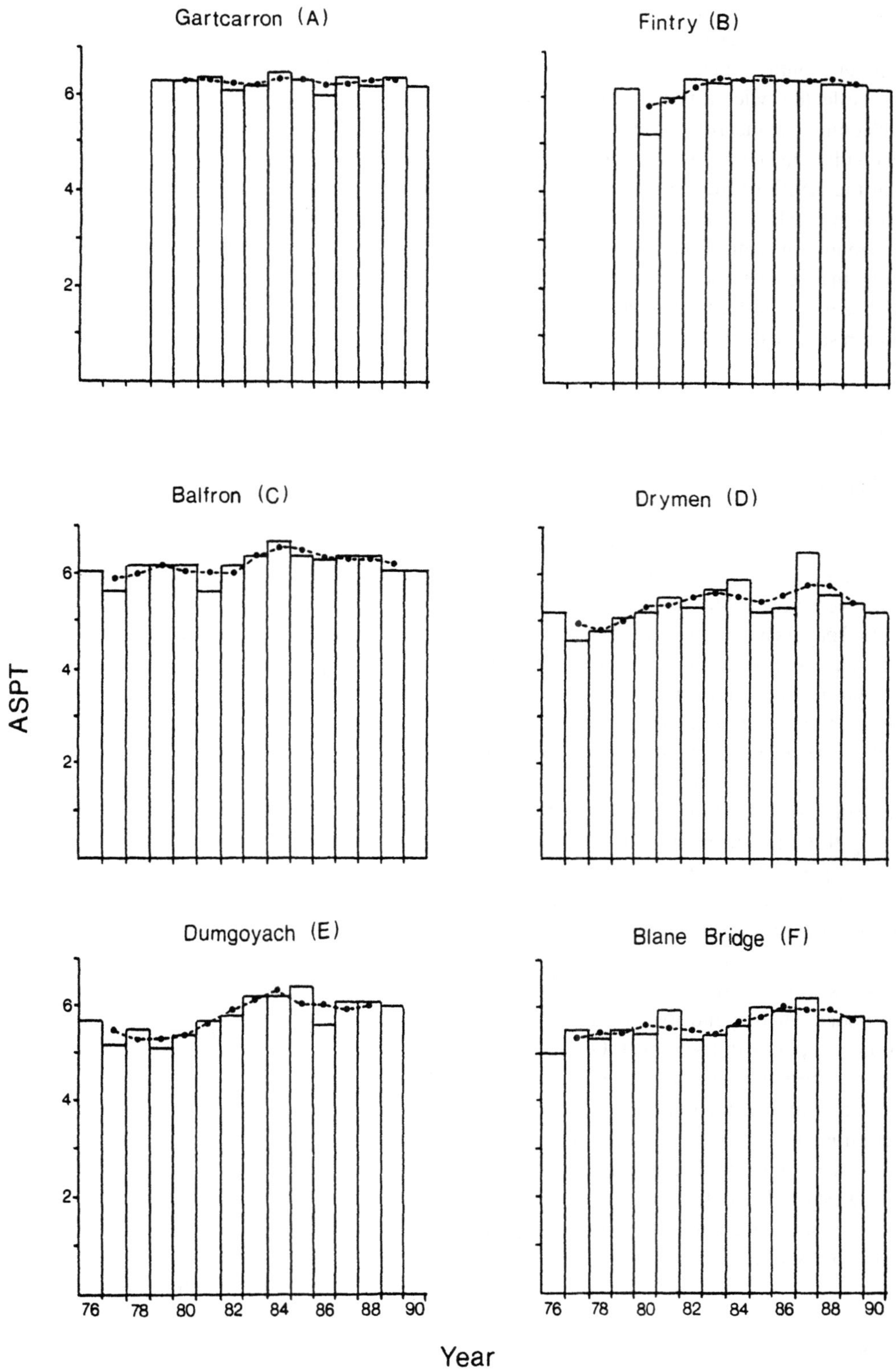

Fig. 2. Annual mean ASPT for sites on the River Endrick and Blane Water, with three-year rolling mean (- - -).

mean flow over the period (Curran & Poodle, 1992). This increase in flow was due to an increase in winter rainfall. There is no evidence of any increase in flow over the summer months. Macroinvertebrate samples are normally collected twice a year, in spring and in late summer/autumn. If macroinvertebrate communities were responding to an increase in flow, it would be expected that the ASPT of the spring samples would increase, while the ASPT of the summer/autumn samples would remain constant. There is no evidence of this happening. In fact at Drymen, the observed upward trend in ASPT seems to be due to an improvement in the summer/autumn samples rather than the spring ones.

Macrophytes

Much less is known about the flora of the River Endrick than the fauna. Observations made during 1959–61 were not carried out in a systematic and detailed manner, and only submerged higher plants common at the sampling sites were noted (Maitland, 1966a). A number of individual papers have been published on various aspects of the aquatic flora within the catchment, but none of these has been comprehensive.

The only detailed work known to have been carried out is the survey of the main river by Holmes (1983, 1989), as part of a comprehensive survey of the flora of British rivers. Six community types distributed among three major community groups were recognized, but species lists were not published. The lower reaches of the River Endrick were classified as being in Group B (sandstone and hard limestone). The middle reaches were in Group C (resistant geology) and the upper reaches were classified as Group D (upland and nutrient poor). Mosses and liverworts are by far the most important plants in rivers with Group C and D communities, where the substrates are characteristically much coarser and boulder-strewn sections are a feature. In contrast, Group B communities have more truly aquatic flowering plants. The species assemblage found in the lowest reaches of the Endrick was unusual in being more characteristic of mesotrophic – eutrophic rivers in north-east Scotland.

Table 2. Fish species in the River Endrick in 1960 and 1990.

Species	1960	1990
Sea lamprey		
Petromyzon marinus L.	+	?
River lamprey		
Lampetra fluviatilis (L.)	++	++
Brook lamprey		
Lampetra planeri (Bloch)	+++	+++
Atlantic salmon		
Salmo salar L.	+++	+++
Brown trout		
Salmo trutta L.	+++	+++
Powan		
Coregonus lavaretus (L.)	+	?
Pike		
Esox lucius L.	++	++
Crucian carp		
Carassius carassius (L.)	–	+
Gudgeon		
Gobio gobio (L.)	–	+++
Minnow		
Phoxinus phoxinus (L.)	+++	+++
Roach		
Rutilus rutilus (L.)	++	++
Chub		
Leuciscus cephalus (L.)	–	++
Dace		
Leuciscus leuciscus (L.)	–	++
Stone loach		
Noemacheilus barbatulus (L.)	+++	+++
Eel		
Anguilla anguilla (L.)	+++	+++
Three-spined stickleback		
Gasterosteus aculeatus L.	+++	+++
Perch		
Perca fluviatilis L.	++	++
Ruffe		
Gymnocephalus cernua (L.)	–	++
Flounder		
Platichthys flesus (L.)	+	+

Invertebrates

The macroinvertebrate fauna of the River Endrick was surveyed at 12 sites in October 1959, February 1960 and June 1961 (Fig. 1, Table 3), and the distributions and relative abundances of different species were compared (Maitland, 1966a). The same sites were re-surveyed in August 1989 and March/April 1990 with the aim of determining whether there had been any detectable changes over the thirty year period. Although the sampling technique used was similar in 1959–61 and 1989–90, sampling intensity was reduced in the later survey. Invertebrates were identified to species where possible, although only family-level data are presented here. A detailed analysis of the species-level data will be published elsewhere.

The total number of invertebrate families recorded in 1989–90 was 70, which compares closely with the 73 recorded in 1959–61. Of those present in 1959–61, six families (Dendrocoelidae, Planorbidae, Gerridae, Chrysomelidae, Sisyridae and Lepidostomatidae) were not found in 1989–90. Four families found in 1989–90 (Ephemeridae, Phryganeidae, Brachycentridae and Dolichopodidae) were not represented in 1959–61. The number of families recorded at each site is shown in Table 3. The richest fauna in 1959–61 and the joint richest in 1989–90 was found at Drymen, the site with the greatest habitat diversity (Maitland, 1966a). The rank order of sites in 1989–90 was similar to that in 1959–61 (Spearman rank correlation coefficient 0.846, $P<0.02$), the most obvious difference being that site 12 had the lowest number of families in 1959–61, but ranked seventh in 1989–90, a 33% increase in the number of families. Of the six sites which showed a change of 10% or more in the number of families present, an increase was observed only at site 12. The apparent loss of families at some sites in 1989–90 may be due in part to the reduced level of sampling, particularly for rare families. The acquisition of new families in 1989–90 may therefore be more significant.

The composition of the invertebrate communities at each site in 1959–61 and 1989–90 was compared using the Jaccard similarity coefficient (Hellawell, 1978) (Fig. 3). The greatest differences were at sites 1 and 12, which also had the largest changes in the number of families present, and site 5 in the middle reaches of the river. There is no obvious explanation for the changes at sites 1 and 12. Families from all the major groups were lost and gained, although at both sites, three families of Trichoptera were gained. A comparison of photographs of all sites taken from the same vantage points in 1959–61 and 1989–90 shows that significant physical changes have taken place at site 5. Although the south bank here has changed little, the north bank has collapsed for several hundred metres, doubling the width of the river at high water levels and leaving extensive beds of gravel (absent in 1959–61) exposed at low water. Again, changes in the invertebrate community at this site involved families from a wide range of groups, but of the 12 families lost, six were Trichoptera, and of the 12 families gained, five were Diptera.

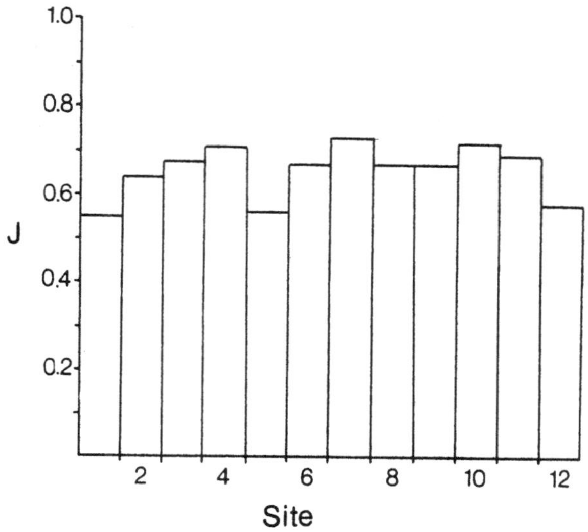

Fig. 3. Comparison of invertebrate community composition 1959–61 and 1989–90 using the Jaccard similarity coefficient (J).

The results suggest that the ranges of some families have extended over the 30 year period (e.g. Sphaeriidae, Perlidae, Limnephilidae, Sericostomatidae and several Diptera), while others appear to have contracted (e.g. Planariidae, Capniidae, Veliidae, Hydrophilidae, Scirtidae). To some extent, sampling differences may account for these apparent changes. In 1959–61, as a result of the greater sampling intensity, there would have been more chance of catching families with more specialized habitats, such as the surface-dwelling Veliidae and those Coleoptera families with species which tend to frequent river margins.

Sites were classified by two-way indicator species analysis (TWINSPAN) (Hill, 1979a) and the 1959–61 and 1989–90 classifications compared (Fig. 4). In both cases, the order in which the sites were arranged in the level 3 end-groups was highly correlated with distance from source, indicating that the composition of the River Endrick's invertebrate communities is, and was, determined primarily by factors which display a longitudinal gradient (e.g. slope, substrate type, current velocity, temperature and various chemical determinands). Although the two classifications were identical at level 1, differences were apparent at levels 2 and 3. In 1989–90, sites in the upper reaches of the river were more clearly separated from those in the middle reaches. These differences are accounted for by apparent changes in the distributions of indicator species over the thirty year period e.g. Hydrophilidae, an indicator at level 2 in 1959–61, which was then present at six

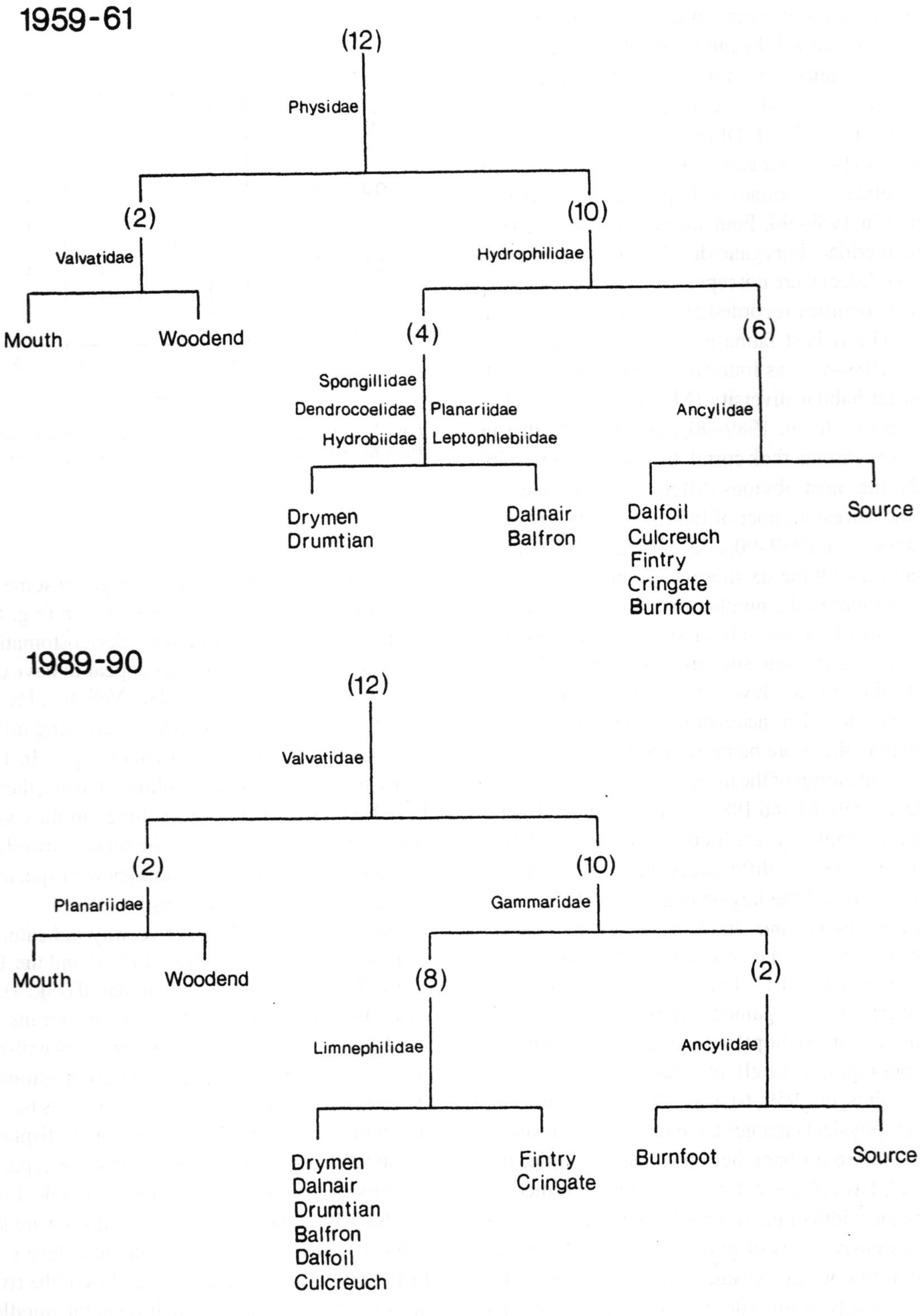

Fig. 4. TWINSPAN classification of sites showing indicator species (maximum of three) at each division, positive to the left, negative to the right. The number of sites in each group is shown in brackets.

Table 3. Numbers of invertebrate families in the River Endrick 1959-61 and 1989-90.

Site		No. of families		Families gained	Families lost	Net change	% change
		1959-61	1989-90				
1	Mouth	36	28	6	14	−8	−22
2	Woodend	40	34	5	11	−6	−15
3	Drymen	52	47	7	12	−5	−10
4	Dalnair	47	47	8	8	0	0
5	Drumtian	43	43	12	12	0	0
6	Balfron	42	43	9	8	+1	+2
7	Dalfoil	38	40	7	5	+2	+5
8	Culcreuch	40	42	9	7	+2	+5
9	Fintry	37	33	5	9	−4	−11
10	Cringate	34	28	2	8	−6	−18
11	Burnfoot	30	31	6	5	+1	+3
12	Source	27	36	13	4	+9	+33

sites, but was found at just two sites in 1989–90, only one of which had yielded the family in 1959–61.

Ordination of sites by detrended correspondence analysis using DECORANA (Hill, 1979b) gave similar results, with axis 1 being correlated with distance from source for both the 1959–61 and 1989–90 data.

Considering the River Endrick as a whole, classification, ordination and rank correlation of family richness all suggest that changes in the invertebrate community over the last 30 years have been relatively minor. Comparisons of the invertebrate communities at individual sites using the Jaccard coefficient have shown the degree of change to vary from site to site. The greatest changes have been in the extreme upper and lower reaches of the river, and at one site in the middle reaches which has undergone physical change. There is no real evidence that other changes known to have taken place in the catchment during the 30 year period have had any impact on invertebrate communities. Pollution from STWs had only very localized effects in 1959–61. Regular upgrading of STWs to keep pace with an expanding population has ensured that overall water quality has been maintained, although it is likely that an increase in nutrient concentrations has taken place. Any effects of increased afforestation or changes in the water abstraction regime would have been most noticeable at site 9 (sites 10–12 are upstream of all forestry and the Loch Carron abstraction point). The degree of change at this site was unexceptional. It is particularly difficult to account for the changes at site 12, which is upstream of all known impacts. Townsend *et al.* (1987) found significant changes in stream invertebrate communities over an eight year period, but noted that sites in the uppermost reaches tended to change least. The observed changes cannot all be attributed to sampling error: families such as Glossiphoniidae, Corixidae and Limnephilidae were common at site 12 in 1989–90. It seems unlikely that they could have been missed in 1959–61 given the greater sampling effort employed. There is no evidence linking changes in the extreme lower reaches of the river with any one factor, and it may be that these changes reflect the cumulative effects of the various changes known or suspected to have taken place in the catchment during the period e.g. in flows, nutrient status, land management and the fish community. It is well known that fish introductions or losses can affect the structure of invertebrate communities in lakes e.g. Brooks & Dodson (1965), Macan (1966), Bendell & McNicol (1987). However, it is unclear whether the addition or removal of fish can have a similar effect in running waters. In a review of this subject, Allan (1983) concluded that on balance, fish do not play a major role in the structuring of invertebrate communities in running waters. The work of Townsend *et al.* (1987) supports this conclusion. In a recent study, Bechara *et al.* (1993) attributed the observed lack of fish predatory impact in streams to the effectiveness of bed interstices as refugia for infauna.

Fish

There is considerable evidence that, prior to 1970, the fish fauna of the River Endrick was extremely stable, the same species having occurred there for several hundred years with no extinctions or additions. The accounts of Ure (1795), Brown (1891), Lumsden & Brown (1895), Scott & Brown (1901), Lamond (1931) and Hunter *et al.* (1959), though not agreeing in every detail, indicated that a total of 15 native species occurred in Loch Lomond and that most of these could be found in the River Endrick. Although other species had been introduced to the loch from time to time (Maitland, 1972), none had managed to establish successfully.

Work carried out in 1960 verified that 14 of the 15 species recorded from Loch Lomond were also present in the River Endrick (Table 2); the only species recorded from Loch Lomond but not from the River Endrick being the ten-spined stickleback (Maitland, 1966a). Most of the Endrick species were common in parts of the river at least, with the exception of sea lamprey and powan which appeared largely as vagrants from the loch. In addition to these species, occasional escaped rainbow trout could be found in the river (Maitland, 1966d) and carp were known from a small loch in the catchment (Maitland, 1964b).

As previously noted, there are two important waterfalls on the main river at the Pot of Gartness and the Loup of Fintry. The first of these can be surmounted at suitable water levels by adults of both Atlantic salmon and sea trout, which are common in the river above this point. However, it appears to be impassable to most other species. River lamprey, pike, gudgeon, roach, chub, dace, perch and ruffe, which are relatively common in the river below the falls, are absent above. The Loup of Fintry is a high (about 30 m), complex waterfall which appears to be impassable to all migratory fish except eels. These are probably able to reach the upper river through the Balmenoch Burn and Loch Walton system, which has gentler gradients.

Stocking has been a regular feature of the activities of game anglers, and both Atlantic salmon and trout (sometimes from native stock and sometimes from other systems) have been released at various times and at various sites over the last 100 years. Little is known of the impact of such stocking, but it does account for the presence of young Atlantic salmon above the Loup of Fintry from time to time.

Since 1960, changes have occurred to the fish community in the River Endrick. In 1970, gudgeon were introduced to Culcreuch Loch near Fintry (Burkel, 1971) and by 1980 these had established themselves widely in the Endrick system, being caught in traps at Woodend and Drymen Bridge in the main river and in the Blane Water near Blane Bridge. Over 100 adult fish were caught in a single lamprey trap here in 1984. However, the fact that gudgeon do not apparently occur above the Pot of Gartness casts some doubt on the origin of the Endrick stock and it is possible that, like the other species discussed below, it may have originated from independent stocking by anglers in the lower river or Loch Lomond.

In 1982, a few ruffe were found in Loch Lomond (Maitland *et al.*, 1983) and this species soon became abundant throughout the loch (Maitland & East, 1989). It was previously unknown in Scotland. The origin of the stock is unknown but it is assumed to have come from livebait released by pike anglers from England. By 1985, ruffe were also abundant in the lower reaches of the River Endrick and were taken there in traps at Woodend and Drymen Bridge.

Other species soon appeared in the river and in 1986, young chub and dace were electrofished in some numbers at Drymen Bridge (Maitland & East, 1989). Gudgeon, ruffe, chub and dace are all well established now in the lower reaches of the River Endrick and in some places are the commonest species present.

During this period also, minnows gained access to the river above the Loup of Fintry and are common there now for several kilometres above the waterfall.

The process of change is likely to continue with the discovery of a roach × bream hybrid in Loch Lomond in 1990 (Adams & Maitland, 1991) and crucian carp in ditches joining the lower River Endrick in 1991 (Adams & Mitchell, 1992). What effect the newcomers may be having on the native fish community is uncertain, for relatively little research is being carried out at present.

Conservation and management

Rivers are limited resources which are subject to often conflicting demands from a wide range of users. Rivers therefore need to be actively managed in such a way as to balance the needs of the various users and to ensure that both the quantity and quality of the resource are maintained. The River Endrick is used for potable water supply, livestock watering, land drainage, sewage effluent disposal and angling, and has a strong nature conservation interest and intrinsic aesthetic value. The main areas of potential conflict are conservation, angling and the use of the river as a conduit for sewage effluent and catchment run-off. In the case of the River Endrick, the river cannot be considered in isolation: the consequences of management decisions for Loch Lomond must also be taken into account.

The area around the mouth of the Endrick was declared a National Nature Reserve in 1962 and has since been extended to include the whole of the river below Woodend. Part of the reserve was designated a wetland of international importance under the Ramsar Convention. The reserve is particularly noted for its woodland and wetland plant communities and its breeding and migratory birds. The conservation value of the River Endrick was recognized in the Nature Conservation Review (Ratcliffe, 1977), the first comprehensive review of sites of conservation interest in Britain.

The main criteria used for identifying rivers of conservation interest in Britain are diversity of habitats and species, naturalness of catchment and river corridor and representativeness of rivers of a particular type (Newbold *et al.*, 1983; Boon, 1991). Other factors to be taken into account are rarity (of river type, species or communities), size or extent, fragility of habitats and communities, and geographical position. Evaluation of the conservation value of a river therefore requires a good knowledge of not only its animal and plant communities, but also its place in a national classification of river types. Classification of British rivers has in the past relied heavily on the use of plant communities. Of the six distinct plant community types recognized in the River Endrick, five score highly for representativeness and three are above average species richness when compared with communities of the same type on other rivers (Holmes, 1983). The atypical species assemblage found in the lower reaches of the river is of particular note. The conservation interest of the River Endrick as defined by its plant communities therefore appears to be high.

The assessment of the conservation value of rivers from their invertebrate communities has been more problematical because of the under-recording of many groups (Nature Conservancy Council, 1990). Even where the invertebrate fauna has been recorded in detail, as in the River Endrick, the lack of comparable data from elsewhere has made it difficult to place the river in a national context. A classification of British rivers based on macroinvertebrates has now been devised (Wright *et al.*, 1984) and computer software (RIVPACS) has been developed to enable new sites to be classified (Wright *et al.*, 1989). Application of RIVPACS to the data obtained from the 1959–61 survey suggests that the River Endrick has seven distinct community types. Those found in the upper and middle reaches are similar to the community types found elsewhere in upland Britain, but the invertebrate communities of the lower reaches are more typical of sluggish rivers in central and eastern England. Unfortunately it is not possible to determine with certainty whether the community types have changed since 1959–61, as the classification technique requires data from three seasons, rather than two as in 1989–90.

The degree of change noted in the composition of an apparently undisturbed invertebrate community in the headwaters of the River Endrick over the thirty year period indicates that even in the absence of anthropogenic impact, the long-term stability (persistence) of such communities may be low, with clear implications for their conservation.

The recent introduction of non-native species to the Loch Lomond catchment, probably by pike anglers, means that the fish community of the River Endrick, at least downstream of the falls at Gartness, can no longer be considered natural. Although it is hoped that byelaws will soon be introduced prohibiting the release of non-native species, the prevention of further introductions will be difficult to enforce in practice. There is a clear need for research on the effects of these introductions on the native fish community. However, the Endrick's fish community retains at least one feature of outstanding conservation interest in its population of freshwater-feeding river lampreys (Maitland, 1980), the only known population in Britain.

Some of the long-term physical changes along the river seem likely to have resulted from adjacent agricultural practices, and to a lesser extent, riparian management by anglers. Certainly some of the most obvious changes known to have occurred (e.g. at site 5) are where stock have been allowed unrestricted access to the river and its banks. In some years, and in the same places, ploughing has taken place to within a metre of the river bank. Serious erosion has resulted from both these practices. Riparian landowners and farmers need to be encouraged to take active and ecologically sound measures to prevent such erosion.

Organic pollution from sewage and agricultural wastes has never posed a serious threat to the River Endrick. The CRPB has taken pains to ensure that development in the catchment has not outstripped sewage treatment capacity, and although intermittent pollution incidents involving farm wastes have occurred, their effects have been short-lived. Of far greater concern is the risk of eutrophication of Loch Lomond arising from inputs of phosphorus from the Endrick catchment. There is an urgent need for research to identify the main sources of phosphorus in the catchment, to quantify loadings to the loch and if necessary, to recommend appropriate control measures. At present, evidence of eutrophication is largely anecdotal and insufficient to justify changes in land management or the introduction of phosphorus removal at STWs. However, the risk must be taken very seriously, since eutrophication of Loch Lomond would jeopardize its future as a public water supply and watersports centre and would threaten the existence of its unique fish community.

Summary

The River Endrick, the major inflow to Loch Lomond, drains a catchment of predominantly base-rich rocks and is therefore well buffered against the acid episodes which affect other rivers in the Loch Lomond catchment. Water quality is excellent, with no evidence of the localized pollution from STWs which was apparent thirty years ago. Long-term trends in water quality, including nutrient loadings to Loch Lomond, are difficult to detect from the inadequate chemical data, but there is biological evidence of an overall improvement since 1976 at some sites. Four new species of fish have become established since 1970. The ecological effects of these introductions are unknown. Taking the River Endrick as a whole, changes to the invertebrate fauna as measured by family richness, classification and ordination have been minor. However, marked changes in community composition were noted at certain sites. With the exception of one site which had undergone physical change, these shifts in composition are difficult to account for, although in the lower reaches, they may to some extent reflect the additive effects of changes known, or suspected to have taken place in the catchment over thirty years. The Endrick has a generally high conservation value, although this has been partly degraded by the introduction of non-native fish. There is an urgent need for research to quantify, and if necessary, to identify effective means of controlling nutrient inputs from the river to Loch Lomond. Research is also needed to remedy the current lack of information on primary producers in the river.

Acknowledgments

We wish to thank Dr A. J. N. Haig for his valuable comments on the final draft of the typescript, and Prof D. Hammerton, Director of the Clyde River Purification Board, for permission to publish this paper.

References

Adams, C. E. & P. S. Maitland, 1991. Evidence of further invasions of Loch Lomond by non-native fish species with the discovery of a roach × bream, *Rutilus rutilus* (L.) × *Abramis brama* (L.), hybrid. J. Fish. Biol. 38: 961–963.

Adams, C. E. & J. Mitchell, 1992. Introduction of another non-native fish species to Loch Lomond: crucian carp (*Carassius carassius* L.). Glasg. Nat. 22: 165–168.

Allan, J. D, 1983. Predator-prey relationships in streams. In Barnes, J. R. & G. W. Minshall (eds). Stream Ecology – Application and Testing of General Ecological Theory. Plenum, New York.

Armitage, P. D., D. Moss, J. F. Wright & M. T. Furse, 1983. The performance of a new biological water quality score system based on macroinvertebrates over a wide range of unpolluted running-water sites. Wat. Res. 17: 333–347.

Bechara, J. A., G. Moreau & L. Hare, 1993. The impact of brook trout (*Salvelinus fontinalis*) on an experimental stream benthic community: the role of spatial and size refugia. J. anim. Ecol. 62: 451–464.

Bendell, B. E. & D. K. McNicol, 1987. Fish predation, lake acidity and the composition of aquatic insect assemblages. Hydrobiologia 150: 193–202.

Best, G. A. & I. Traill, 1994. The physico-chemical limnology of Loch Lomond. Hydrobiologia 290: 29–37.

Boon, P. J., 1991. The role of Sites of Special Scientific Interest (SSSIs) in the conservation of British rivers. Freshwat. Forum 1: 95–108.

Brooks, J. L. & S. I. Dodson, 1965. Predation, body size and composition of plankton. Science 150: 28–35.

Brown, A., 1891. The fishes of Loch Lomond and its tributaries. Scott. Nat. 10: 114–124.

Burkel, D. L., 1971. Introduction of fish to new water. Glasg. Nat. 18: 574–575.

Clyde River Purification Board, 1984. Annual Report. CRPB, East Kilbride.

Clyde River Purification Board, 1990. Annual Report. CRPB, East Kilbride.

Curran, J. C. & T. Poodle, 1992. Hydrological characteristics of Loch Lomond – are they changing? In Dickinson, G., Pender, G. & G. E. Jones (eds.) Loch Lomond 1991. Universities of Glasgow & Strathclyde.

Hellawell, J. M., 1978. Biological Surveillance of Rivers. Water Research Centre, Stevenage.

Hill, M. O., 1979a. TWINSPAN – a FORTRAN program for arranging multivariate data in an ordered two-way table by classification of the individuals and attributes. Cornell University, Ithaca.

Hill, M. O., 1979b. DECORANA – a FORTRAN program for detrended correspondence analysis and reciprocal averaging. Cornell University, Ithaca.

Holmes, N. T. H., 1983. Typing British Rivers According to their Flora. Focus on Nature Conservation No. 4. Nature Conservancy Council, Peterborough.

Holmes, N. T. H., 1989. British rivers: a working classification. British Wildlife 1: 20–36.

Hunter, W. R., H. D. Slack & M. Hunter, 1959. The lower vertebrates of the Loch Lomond district. Glasg. Nat. 18: 84–90.

Hunter, W. R., P. S. Maitland & P. K. Yeoh, 1963. *Potamopyrgus jenkinsi* in the Loch Lomond area, and an authentic case of passive dispersal. Proc. malac. Soc. Lond. 36: 27–32.

Lamond, H., 1931. Loch Lomond. Glasgow.

Lumsden, J. & A. Brown, 1895. A Guide to the Natural History of Loch Lomond and Neighbourhood. Glasgow.

Macan, T. T., 1966. The influence of predation on the fauna of a moorland fish pond. Arch. Hydrobiol. 61: 432–452.

Maitland, P. S., 1962a. The fauna of the River Endrick in relation to local water use. Water Waste Treatment 9: 78–86.

Maitland, P. S., 1962b. *Bathynella natans* new to Scotland. Glasg. Nat. 18: 175–176.

Maitland, P. S., 1963. *Hemiclepsis marginata* and *Batracobdella paludosa* in Stirlingshire, with notes on the ecology and morphology of the latter species; Glasg. Nat. 18: 219–227.

Maitland, P. S., 1964a. Quantitative studies on the invertebrate fauna of sandy and stony substrates in the River Endrick, Scotland. Proc. r. Soc. Edinb. 68: 277–301.

Maitland, P. S., 1964b. A population of common carp (*Cyprinus carpio*) in the Loch Lomond district. Glasg. Nat. 18: 349–350.

Maitland, P. S., 1965a. The feeding relationships of salmon, trout, minnows, stone loach and three-spined sticklebacks in the River Endrick, Scotland. J. anim. Ecol. 34: 109–133.

Maitland, P. S., 1965b. The distribution, life cycle, and predators of *Ephemerella ignita* (Poda) in the River Endrick, Scotland. Oikos 16: 48–57.

Maitland, P. S., 1965c. Notes on the biology of *Ancylus fluviatilis* in the River Endrick, Scotland. Proc. malac. Soc. Lond. 36: 339–347.

Maitland, P. S., 1966a. Studies on Loch Lomond 2. The Fauna of the River Endrick. Blackie, Glasgow.

Maitland, P. S., 1966b. Notes on the biology of *Gammarus pulex* in the River Endrick. Hydrobiologia 28: 142–152.

Maitland, P. S., 1966c. The distribution, life cycle and predators of *Amphinemura sulcicollis* (Stephens) (Plecoptera) in the River Endrick, Scotland. Entomologist 99: 72–81.

Maitland, P. S., 1966d. Rainbow trout, *Salmo irideus* Gibbons, in the Loch Lomond district. Glasg. Nat. 18: 421–423.

Maitland, P. S., 1967a. The ecology of four species of Elminthidae in a Scottish river. Arch. Hydrobiol. 63: 104–122.

Maitland, P. S., 1967b. The ecology of the Simuliidae in a Scottish river. J. anim. Ecol. 36: 179–206.

Maitland, P. S., 1972. Loch Lomond: man's effects on the salmonid community. J. Fish. Res. Bd. Can. 29: 849–860.

Maitland, P. S., 1980. Scarring of whitefish (*Coregonus lavaretus*) by European river lamprey (*Lampetra fluviatilis*) in Loch Lomond, Scotland. Can. J. Fish. aquat. Sci. 37: 1981–1988.

Maitland, P. S. (ed.), 1981. The Ecology of Scotland's Largest Lochs. Dr W. Junk Publishers, The Hague.

Maitland, P. S. & K. East, 1989. An increase in numbers of ruffe, *Gymnocephalus cernua* (L.), in a Scottish loch from 1982 to 1987. Aquacult. Fish. Mgmt. 20: 227–228.

Maitland, P. S., K. East & K. H. Morris, 1983. Ruffe *Gymnocephalus cernua* (L.), new to Scotland, in Loch Lomond. Scott. Nat. 1983: 7–9.

Nature Conservancy Council, 1990. A Brief Assessment of the Nature Conservation Importance of Scottish Freshwaters. Nature Conservancy Council, Edinburgh.

Newbold, C., J. Purseglove & N. Holmes, 1983. Nature Conservation and River Engineering. Nature Conservancy Council, Peterborough.

Ratcliffe, D. A. (ed.), 1977. A Nature Conservation Review. Cambridge University Press, Cambridge.

Scott, T. & A. Brown, 1901. The marine and freshwater fishes. In Elliot, G. F. S., M. Laurie & J. B. Murdoch (eds), Fauna, Flora and Geology of the Clyde Area. Glasgow.

Scottish Office Environment Department, 1992. Water Quality Survey of Scotland 1990. HMSO, Edinburgh.

Townsend, C. R., A. G. Hildrew & K. Schofield, 1987. Persistence of stream invertebrate communities in relation to environmental variability. J. anim. Ecol. 56: 597–613.

Ure, D., 1795. Parish of Killearn. Statistical Account of Scotland. 16: 100–129.

Wright, J. F., D. Moss, P. D. Armitage & M. T. Furse, 1984. A preliminary classification of running-water sites in Great Britain based on macroinvertebrate species and the prediction of community type using environmental data. Freshwat. Biol. 14: 221–256.

Wright, J. F., P. D. Armitage & M. T. Furse, 1989. Prediction of invertebrate communities using stream measurements. Regul. Rivers: Res. & Mgmt. 4: 147–155.

Hydrobiologia **290**: 153–166, 1994.
K. J. Murphy, M. C. M. Beveridge & R. Tippett (eds), The Ecology of Loch Lomond.

The ecology of pelagic communities in Lochan Dubh

Roger Tippett
Department of Zoology, The University, Glasgow, G12 8QQ, UK

Key words: phytoplankton, zooplankton, seston, sediment, energy flow

Introduction

Lochan Dubh is a small monomictic, nutrient-poor dystrophic lake which exhibits summer stagnation and lies near the east shore of Loch Lomond at the Ross Peninsula. It lies within the SSSI which covers much of the oak woodland on the east shore of Loch Lomond and at its nearest lies 50 m from the University Field Station at Rowardennan, established by Glasgow University in 1965.

Because of its proximity to the Field Station as well as its intrinsic interest, Lochan Dubh has been a subject of limnological study over the past 40 years. This includes work on the fish community (Shafi, 1969) and studies on specific species of zooplankton including *Holopedium gibberum* (Hamilton, 1958) and vertical migration of *Chaoborus flavicans* (Goldspink & Scott, 1971). Stewart, Walker & Dickson (1984) described the pollen profile of sediment cores as part of a description of the vegetation history of the area.

This paper is concerned with the pelagic environment and communities involved. Klarer (1978) studied the seasonal changes and interactions of the phytoplankton and zooplankton standing crop and productivity and Islam (1987) extended the work to follow the flow of energy to the sediment through the detrital pathway. McDonnell (1988) contributed a detailed study of the surface biota (neuston). This paper will inevitably be a review of the work done, and no attempt is made to present detailed methods or results. For any further details required, the reader is referred to the original theses upon which this review is based, copies of which are available from Glasgow University Library.

Description of study area

Lochan Dubh is situated at national grid ref NS 377 964, Lat. 56 ° 7′N, Long. 4 ° 35′ W. The bedrock of the catchment is mainly schistose grits of the Dalradian series and forms part of the Tay nappe (MacDonald, 1994). Land use is primarily softwood plantation, dominated by Norway Spruce (*Picea abies*). There is a single natural inflow, halfway along the east shore, and the lochan also receives treated domestic sewage effluent from the University Field Station. The outflow is to Loch Lomond through a reed bed to the northwest. Table 1 presents morphometric data and Fig. 1 is a map of the lochan.

Pelagic environment

Temperature and oxygen

Figure 2 presents depth profiles of temperature and oxygen for the period September 1975 to October 1976 (Klarer, 1978). The water column is isothermal from mid October to mid April and cools to below 4 °C during February. Temporary ice cover can occur (e.g. in 1985, see Islam, 1987). From mid April to September thermal stratification develops with the metalimnion starting at about 3 m depth, deepening to 5–6 m by the end of August. Maximum hypolimnetic temperature at the deepest part of the lake rises to about 9 °C.

Dissolved oxygen stays at above 80% saturation throughout the water column, and remains at this level in the epilimnion throughout the summer. However in the hypolimnion, oxygen depletion begins to develop almost immediately thermal stratification has started, so the by mid July oxygen is undetectable in the water

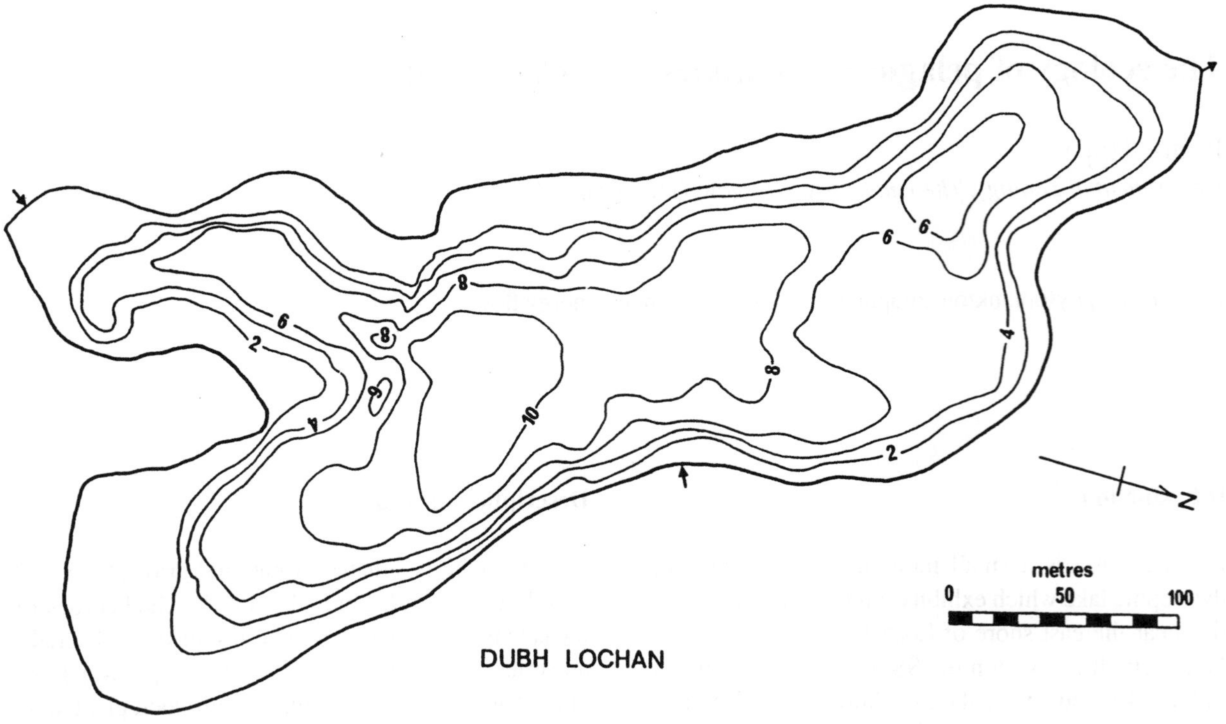

Fig. 1. Bathymetric map of Lochan Dubh showing inflows and outflow.

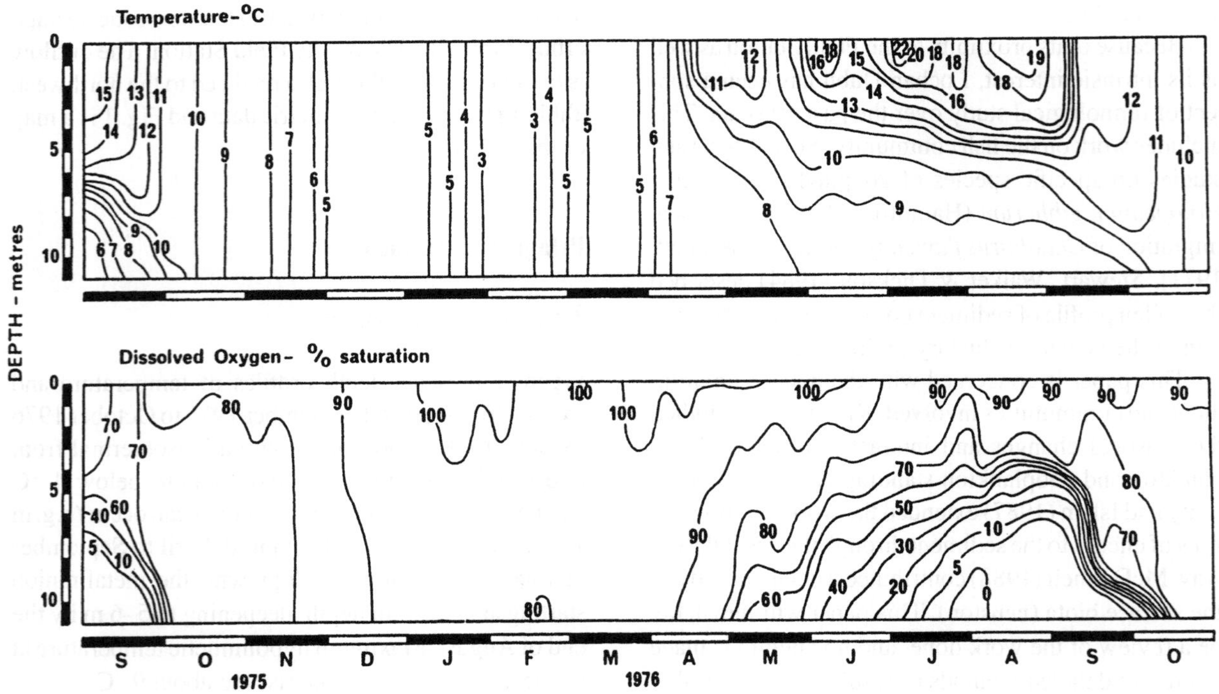

Fig. 2. Seasonal changes in temperature and oxygen (percentage saturation) over the period September 1975 to October 1976.

above the deepest sediment, and remains at less than 5% saturation until the end of September, when the water column again becomes isothermal.

This complete oxygen depletion affects the bottom 3 m of the water column and so overlays approximately one fifth of the benthic sediment area. Conditions such

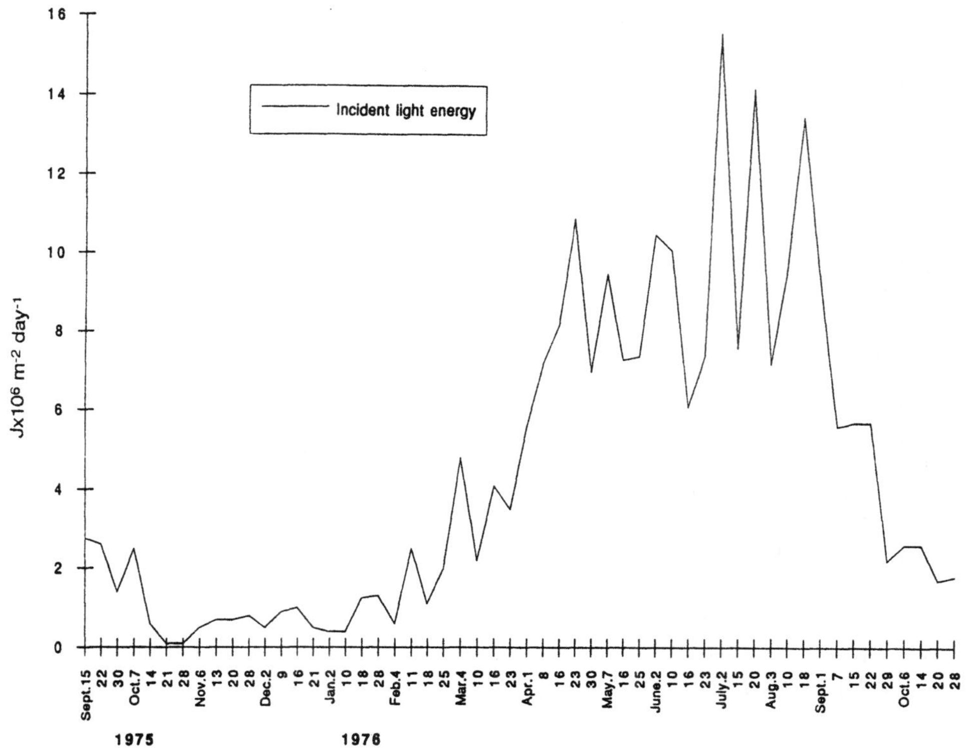

Fig. 3. Seasonal changes in measured incident light energy reaching the surface of the loch covering the period September 1975 to October 1976.

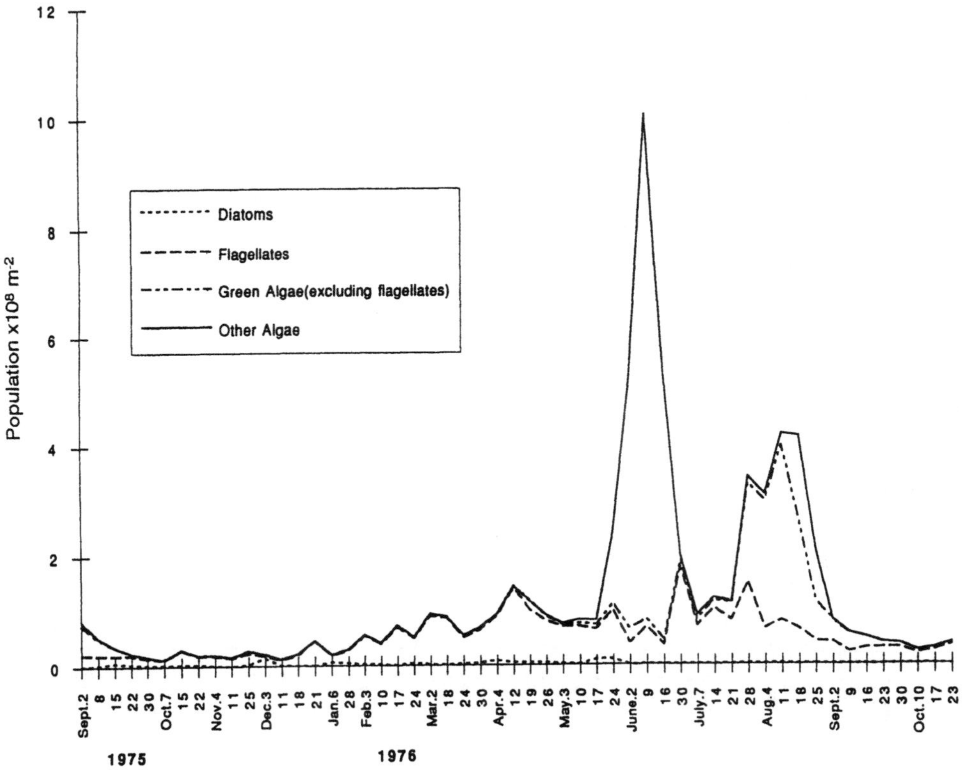

Fig. 4. Seasonal changes in phytoplankton abundance for the major groups covering the period September 1975 to October 1976.

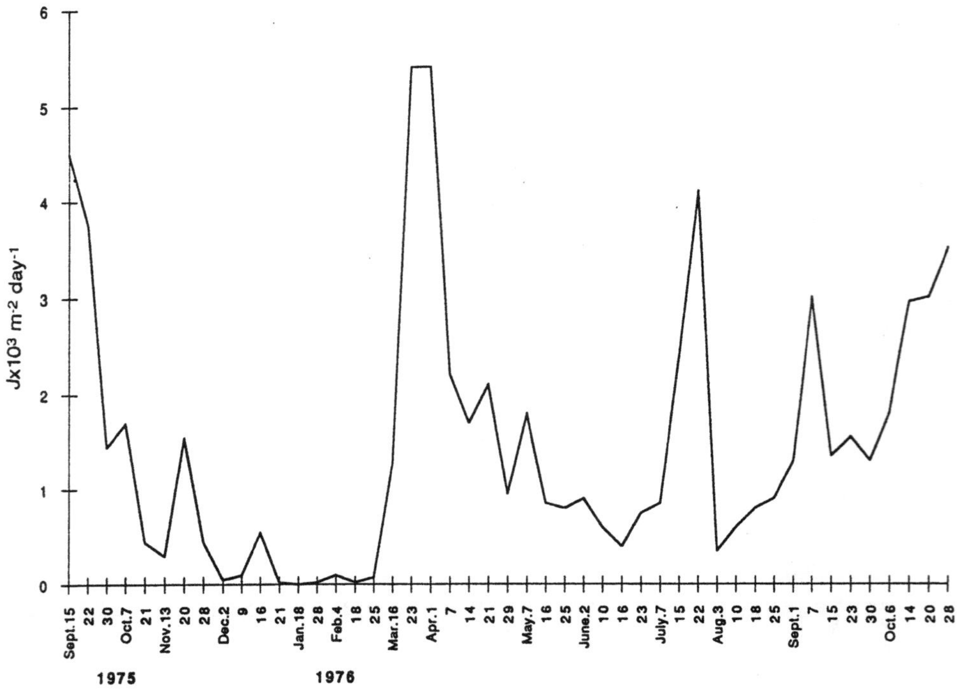

Fig. 5. Seasonal changes in phytoplankton primary production covering the period September 1975 to October 1976.

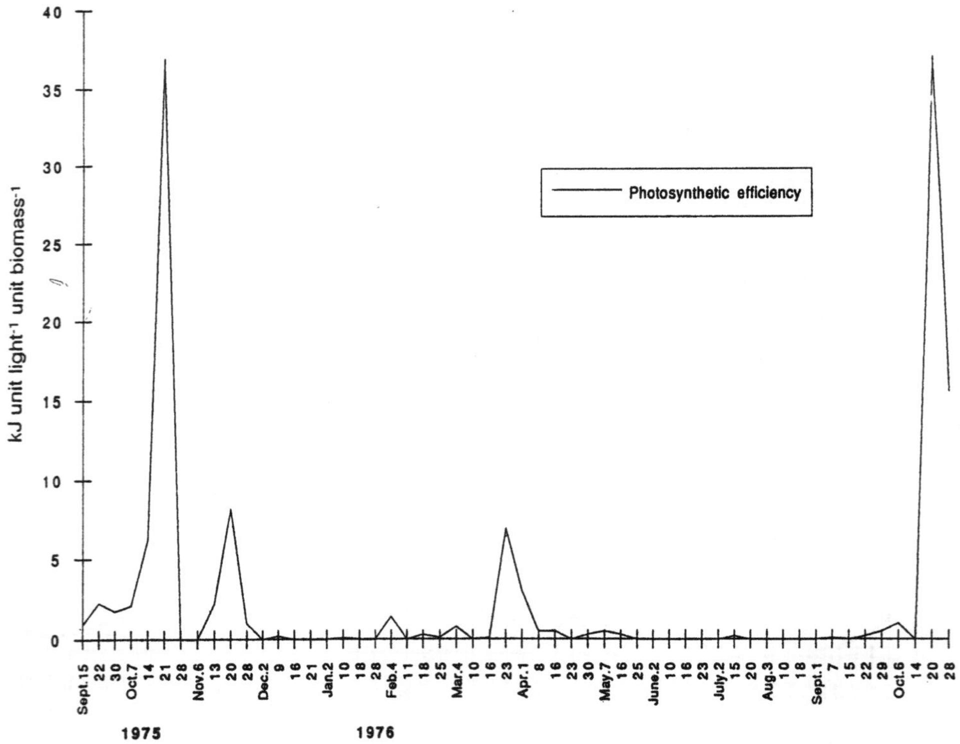

Fig. 6. Seasonal changes in phytoplankton efficiency in use of available energy covering the period September 1975 to October 1976.

Table 1. Morphometry of Lochan Dubh (Calculations according to Hutchinson, 1957)

Surface area	7.06×10^4 metres^2
Total volume	3.38×10^5 metres^3
Maximum depth	11.1 metres
Mean depth	4.8 metres
Length	550 metres
Mean breadth	128 metres
Shoreline	1550 metres
Shoreline development	1.645
Volume development	1.301
Catchment area	1.1 Kilometres^2

Table 2. Summary of nutrient data in January and August 1976

	End January		End August	
	1m	10m	1m	10m
pH	5.2	5.2	5.8	6.2
Alkalinity (m-mol l^{-1})	55	55	45	65
Total diss. carbon (mg l^{-1})	6	6	2	6
PO_4-P ($\mu g\ l^{-1}$)	1	1	2	15
Si (mg l^{-1})	1.4	1.4	0.25	1.25
NO_3-N ($\mu g\ l^{-1}$)	80	80	40	100
NO_2-N ($\mu g\ l^{-1}$)	1	1	1	8
NH_4-N ($\mu g\ l^{-1}$)	20	20	20	300
Cl ($\mu g\ l^{-1}$)	500	500	300	400
SO_4 ($\mu g\ l^{-1}$)	300	300	400	300

Cations

	1m		9m	
	Mean	Range	Mean	Range
Ca (mg l^{-1})	1.68	1,39-1.9	1.66	1.44-1.90
Mg (mg l^{-1})	0,98	0.88-1.1	1.02	0.98-1.14
Fe ($\mu g\ l^{-1}$)	279	100-630	625	150-2200
Mn ($\mu g\ l^{-1}$)	241	192-322	450	208-712
Cu ($\mu g\ l^{-1}$)	14	5-58	30	47-100
Zn ($\mu g\ l^{-1}$)	117	34-382	77	47-124

Table 3. Depth (metres) at which light is 1% of surface illumination

White	5.5
Red	7.0
Green	6.0
Blue	2.0

as these will cause reduction of the otherwise oxidised microlayer on the sediment surface (Hutchinson, 1957) and lead to the possibility of nutrient release from the sediment into the water column.

Nutrients

Table 2 summarises mineral nutrient data for the top and bottom of the water column at the end of January in the middle of the period of isothermal conditions, and at the end of August, towards the end of stratification and summer stagnation. During winter chemical conditions are uniform throughout the water column. pH at 5.2 is low, and this is reflected in the low alkalinity (0.025 m-mols l^{-1}) although total dissolved carbon in surface waters is at its highest level in the year.

Of minerals likely to limit phytoplankton growth (Phosphorus, nitrogen and silicon), phosphate phosphorus was low, nitrate and ammonia quite high and silicate was also high. Nitrite was virtually undetectable. At the end of August however, the effects of thermal stratification and hypolimnetic oxygen depletion are easy to observe. pH In the epilimnion is slightly above the winter level, with alkalinity also increased. Significant surface depletion of nitrate and silicate is observed, with little change in phosphate and ammonia. Beneath the metalimnion however, there is a significant increase in pH, alkalinity, phosphate, nitrite and ammonia, associated with the reduction of the sediment surface. This release is also apparent in the significant release of iron and manganese and to a lesser extent of copper during stratification as demonstrated by the wide range in values at the 9 m depth. Other major ions, both anions and cations, do not exhibit these seasonal differences in the deep waters.

After the breakdown of stratification in Autumn there is no evidence of there being an increase in orthophosphate concentration in the water column indicating that the elevated hypolimnetic levels are quickly removed from solution in the well oxygenated environment either by biological uptake or by physical removal (precipitation). Similarly there is no evidence of sudden enrichment of surface waters for nitrate or ammonia.

Light

Phytoplankton production is, in addition to nutrients, limited by light availability, and this has two components, that which reaches the surface of the water and is dependent on time of year and meteorological conditions (Fig. 3), and the absorption of the light through the water column. Table 3 presents the depth at which illumination is 1% of surface for various sections of the visible spectrum. From this it can be seen that red wavelengths penetrate furthest, and blue light is the most swiftly attenuated. Absorption is by three components, the water itself, particulate and dissolved matter. This spectrum indicates poor penetration of light resulting from suspended particulate matter, and the rapid attenuation of blue light indicating water which is heavily pigmented with dissolved organic matter.

Phytoplankton standing crop

There have been two studies of the phytoplankton (Klarer, 1978; Islam, 1987) approximately 10 years apart. They used comparable methods for study (sedimentation with Lugol's iodine and enumeration using an inverted microscope). Between the two studies, a total of 192 taxa were recorded. Table 4 presents a list of the 26 species which made significant contributions to the community at one time of the year or another. No significant changes in plankton species composition were detected in the 10 years between the two studies.

In each of the 5 years of sampling by Klarer & Islam over the period 1974 to 1985 clear patterns of occurrence were observed. For clarity, the results from one representative year (1976) only are presented in Fig. 4. This year was chosen because of the availability of concurrent productivity results. In all years of sampling the community reached low numbers by the end of September or beginning of October and stayed at this low level throughout the winter. A slow rather irregular increase commenced in all years towards the beginning of February and continued to a spring peak during April. A second peak occurred in each year in June and a third peak in August in some years. However the relative importance of these peaks differed from year to year, so that in 1975 the April peak was greatest, June peak next in importance and the August peak smallest. In 1976 however, the April peak was smallest and the June peak most important. The peaks of numbers were normally of short duration, lasting 4–5 weeks at the most.

There is also a consistent seasonal succession with respect to the species composition of the phytoplankton. Although a small number of species were frequently present, diatoms never made an important contribution to the community. *Eunotia pectinalis* var. *minor* and *E. lunaris* were the most consistent representatives. The spring peak was always composed mainly of flagellates including *Cryptomonas* spp. and two species of small (3–7μ m long) green flagellates which remain unidentified.

Although flagellate populations remained high throughout the summer until they declined at the end of August, the later peaks were composed of *Dinobryon divergens* (June peak in all years) and *Sphaerocystis schroeteri* in August, followed by *Oocystis lacustris* in August 1976.

Standing crop measurements using chlorophyll were not attempted by Klarer (1978) or Islam (1987)

Table 4. The phytoplankton species which make significant contributions to biomass.

Chlorophyta
Botryococcus braunii Kutz
Chlamydomonas globosa Snow
Chlorella ellipsoidea Gerneck
Closterium toxon West
Oocystis lacustris Chod.
Oocystis rhomboidea Fott
Oocystis solitaria Wittr.
Sphaerocystis schroeteri Chod.
Flagellate 1
Flagellate 2
Euglenophyta
Trachelomonas volvocina Ehr
Chrysophyta
Dinobryon divergens Imhof
Eunotia lunaris (Ehr.) Grun.
Eunotia pectinalis var. *minor* (Kutz.) Rabh.
Frustulia rhomboides (Ehr.) de Toni var. *crassinervia* (Breb. ex W. Sm.) Ross
Mallomonas acaroides Perty
Navicula cryptocephala Kutz.
Navicula cuspidata Kutz
Navicula exigua (Gregory) O. Muller
Navicula subtilissima Cl.
Peronia heribaudi Brun and Per.
Tabellaria flocculosa (Roth.) Kutz
Cryptophyta
Cryptomonas marssoni Skuja
Cryptomonas ovata Ehr.
Cyanobacteria
Aphanocapsa delicatissima West and West
Coelosphaerium naegelianum Unger.

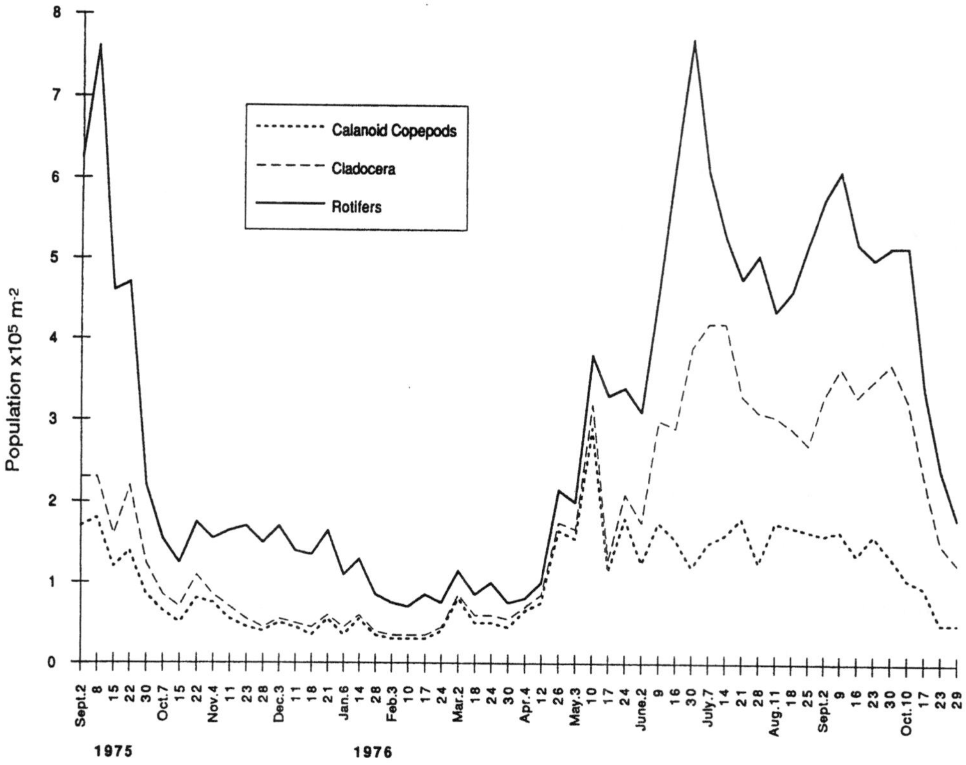

Fig. 7. Seasonal changes in zooplankton abundance for the major groups covering the period September 1975 to October 1976.

because the low populations were at the limit of detection with the equipment available, but McDonnell (1988) obtained values for surface and water at 1 m depth for the period July 1980 to September 1982. Values ranged from 0.42 to 3.9 μ g l^{-1} chlorophyll *a* in 1980, 0.11 to 18.5 μ g l^{-1} in 1981 and 0.33 to 8.1 μ g l^{-1} in 1982.

Primary production

Primary production was measured using the ^{14}C technique with *in situ* incubation at 0, 1, 3, 5 and 10 m depths and 4 hour incubation periods. Results are available from Klarer (1978) and Islam (1987) with a less complete set available from McDonnell (1988). Results for 1976 are presented in Fig. 5. After virtually undetectable production in January and February a great increase occurred during the month of March to reach the largest peak of the year in early April, at the time when the phytoplankton was dominated by flagellates. After an irregular decline in production to June a second peak occurred at the end of July, about a month later than the increase in *Dinobryon* standing crop. The third and fourth peaks in mid September and at the end of October were at times of declining standing crop. Incident light energy was measured over the period (Fig. 5) and photosynthetic efficiency expressed as energy (kJ) unit light^{-1} unit biomass^{-1} (Fig. 6). Efficiency was very low over most of the summer, being high only at the beginning of the period of growth in March and particularly at the end of the season in October after the autumn overturn.

During the winter production levels were at the limits of sensitivity of the method at less than 0.04 mg C m^{-2} d^{-1} at the spring peak with the mid July peak reaching 0.99 mg C m^{-2} d^{-1}, about four times greater than the average summer production rate.

Klarer (1978) recorded a total production for the year 1976, when expressed on an energy basis of

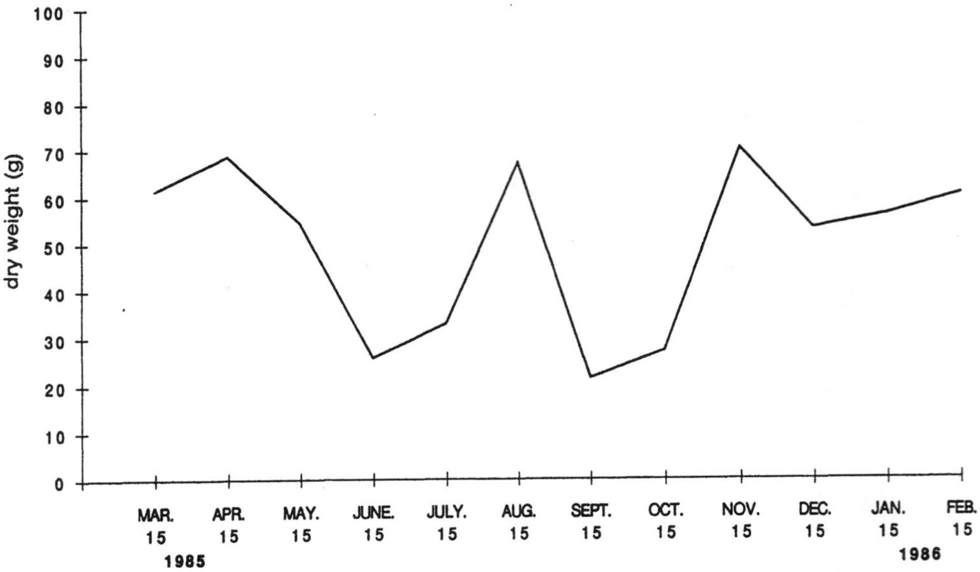

Fig. 8 a. Seasonal changes in seston composition over the period March 1985 to February 1986. a. Dry weight per metre^2 per month.

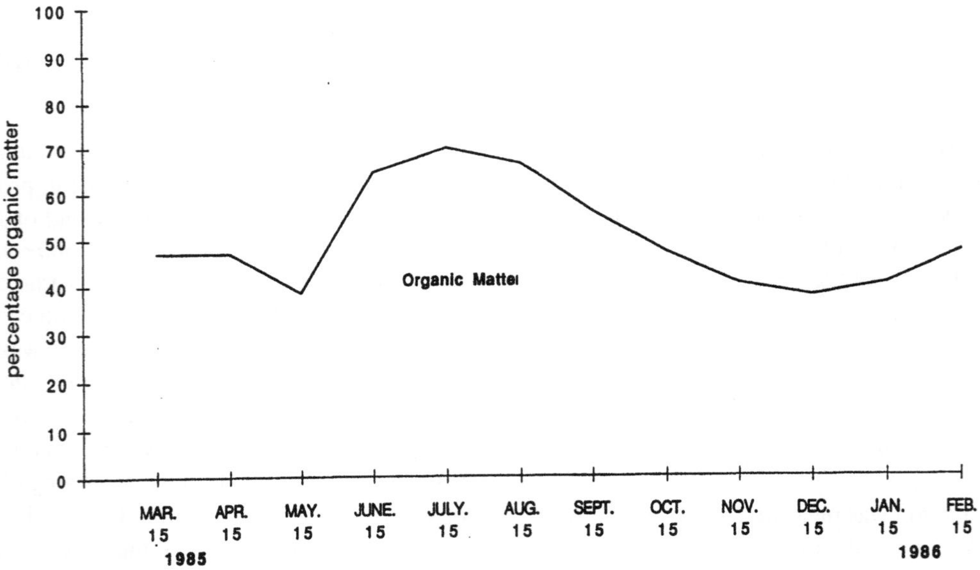

Fig. 8 b. Percentage organic matter.

36.07×10^5 KJ m^{-2} y^{-1}. Islam (1987) recorded annual values more than twice this at 92.323×10^5 KJ m^{-2} y^{-1}. It is possible that there are problems of accuracy inherent in sampling waters of this type, where so much of the production results from growth of small, delicate unicells such as nano-flagellates which may

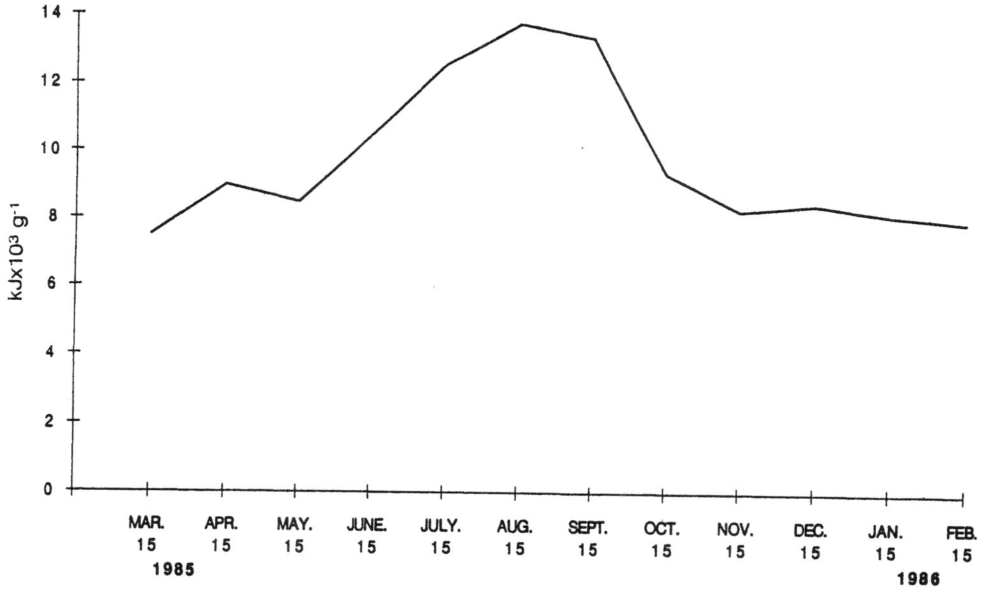

Fig. 8 c. Energy content per gram dry weight.

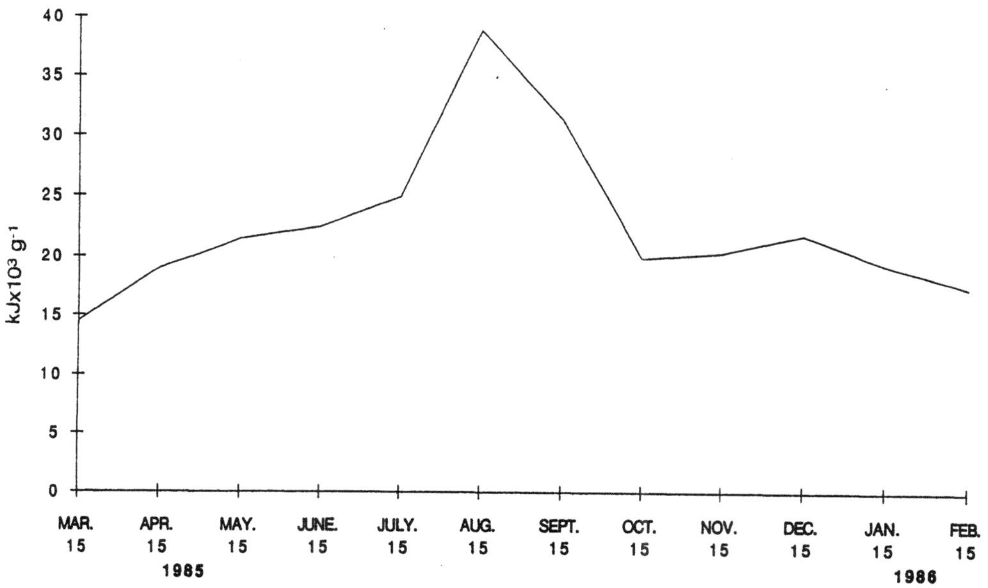

Fig. 8 d. Energy content per gram ash-free dry weight.

easily be ruptured and destroyed during the filtration process.

Zooplankton

The zooplankton community was studied concurrently with the phytoplankton by Klarer (1978), Islam (1987)

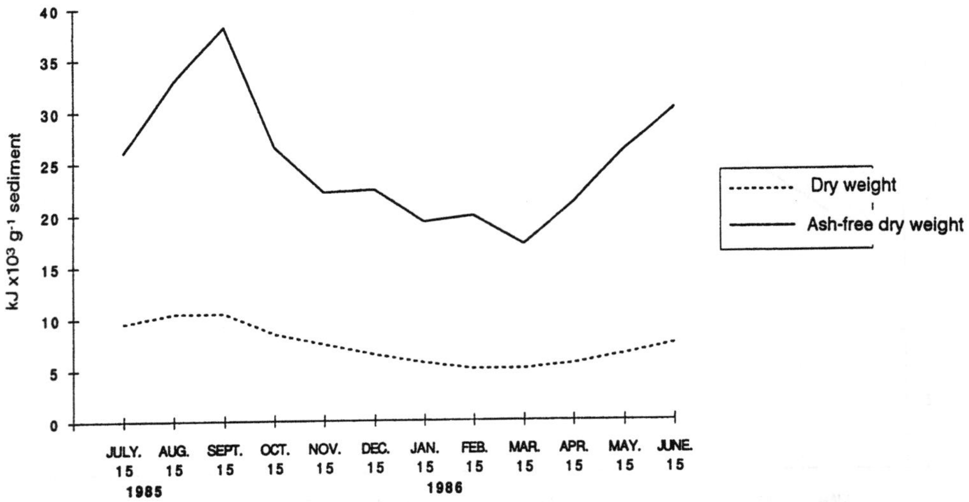

Fig. 9. Seasonal changes in energy content of sediment over the period March 1985 to February 1986 on a dry weight and ash-free dry weight basis.

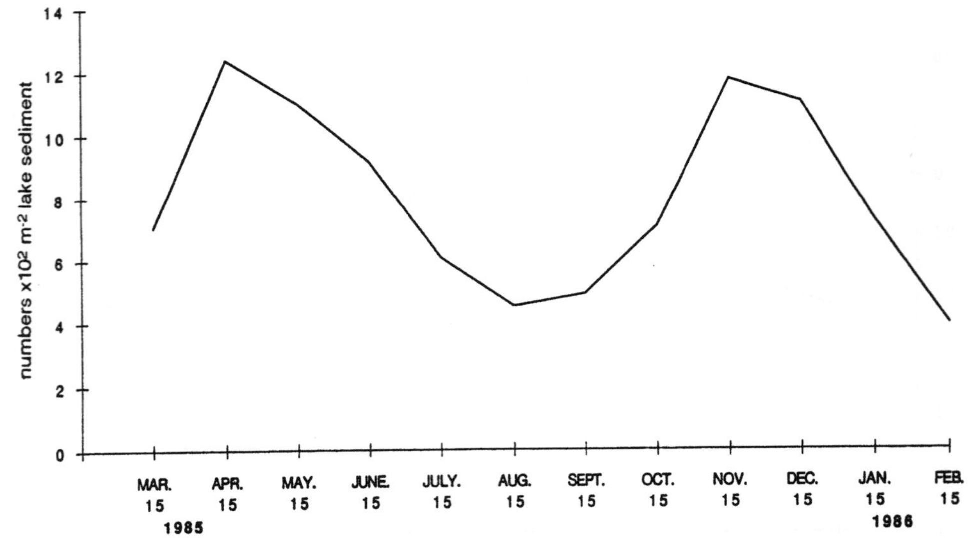

Fig. 10. Seasonal changes in benthos population over the period March 1985 to February 1986.

and as part of the investigation on neuston by McDonnell (1988). 61 species of zooplankton were recorded in the pelagic environment by Klarer (1978) and Islam (1987). McDonnell (1988) added a further 15 species, but all of these were restricted to the littoral sites. Of these species, 1 copepod, 4 cladocera and 9 rotifera made significant contributions to the community and are recorded in Table 5. One species, *Chaoborus flavicans*, deserves special mention since it forms a significant contribution to the zooplankton for much of the year during hours of darkness (Goldspink & Scott,

Table 5. The zooplankton species which contribute significantly to biomass.

Aschelminthes
Rotifera
Asplanchna priodonta Gosse
Ascomorpha ecaudis Perty
Conochilus unicornis Rousselet
Kellicottia longispina Kellicott
Keratella quadrata Muller
Keratella quadrata var. *curvicornis* Ehr.
Keratella quadrata var. *testudo* Ehr.
Polyarthra vulgaris Carlin
Trichocerca similis Wierzejski
Arthropoda
Crustacea
Copepoda
Eudiaptomus gracilis Sars.
Cladocera
Bosmina corregoni Baird
Ceriodaphnia reticulata Jurine
Diaphanosoma brachyurum Lieven
Holopedium gibberum Zaddach

Table 6. Energy measurements on zooplankton

Respiration rates (μg O^2 individual^{-1} hr^{-1} at 10 °C)	
Asplanchna	0.0038
Cladocera	0.009
Holopedium	0.032
Eudiaptomus	0.019
Eudiaptomus (nauplius)	0.0025
Caloric values (Joules individuals^{-1})	
Rotifera	
Kellicottia	2.09×10^{-3}
Keratella	1.38×10^{-3}
Polyarthra	2.55×10^{-3}
Ascomorpha	1.75×10^{-3}
Asplanchna	4.18×10^{-3}
Cladocera	
Ceriodaphnia	2.717×10^{-2}
Holopedium	2.926×10^{-1}
Bosmina	2.048×10^{-2}
Copepoda	
Eudiaptomus eggs	4.43×10^{-3}
Eudiaptomus nauplius	3.34×10^{-3}
Eudiaptomus adult	1.13×10^{-1}
Assimilation efficiency (%)	
Rotifers	74.6
Holopedium	57.9
Ceriodaphnia	44.6
Eudiaptomus nauplius	74.7
Eudiaptomus adult	47.4

1971) but was never recorded during any daytime sampling.

The seasonal pattern of growth for the zooplankton community was less variable than for the phytoplankton from year to year. Figure 7 presents the results for the year 1976. During the last three months of 1975 there was a steady decline in numbers for all major groups with Cladocera representing a very small proportion of the community. The community was at its lowest in February and from that time there was a slow increase in Copepod numbers which accelerated during the latter half of April to reach a peak in mid May, approximately one month after the spring peak of phytoplankton. At that time this was mainly composed of small flagellates such as *Cryptomonas* sp. and *Eudiaptomus gracilis* was the only species making a significant contribution to the zooplankton. From May through to mid September the Copepod population stayed more or less constant before declining to low levels in October.

From May onwards first Rotifers and then Cladocera increased in numbers to reach peak populations in late June and early July respectively. Both showed a second peak at the end of August before declining sharply in numbers from late September onwards. The Rotifer community was much more diverse than the others, with six species making major contributions to the community. *Kellicottia longispina* was by far the most important, with a large population being maintained over a long period. *Keratella cochlearis* was the dominant rotifer throughout the winter, reaching a low population in April. From then onwards the increase in rotifers to the June peak comprised *Conochilus*, *Keratella* and mainly *Kellicottia*. In the succession of species *Asplanchna* followed in early August, *Polyarthra* and *Ascomorpha* in late August and *Kellicottia* again in early September. *Polyarthra* formed a second peak at the end of September before the seasonal decline in the community.

Ceriodaphnia dominated the cladoceran contribution to the community, being twice as abundant as the next most important species (*Diaphanosoma*). At the end of May *Bosmina* was the first species to increase in numbers, followed by *Ceriodaphnia* and *Diaphanosoma* at the end of June, with *Holopedium* early in July and a second peak of *Diaphanosoma* in late July. *Ceriodaphnia* and *Holopedium* formed second peaks in September.

Zooplankton energetics

Klarer (1978) measured a number of aspects of the energy relations of individual species within the zooplankton community in order to attempt a budget for the flow of energy through the plankton community. Respiration rates of representatives of the three main components of the zooplankton were measured using a micro-Winkler technique (Golterman & Clymo, 1971) at temperatures ranging from 3 to 20 °C. The caloric content of representatives of the three groups comprising the zooplankton was measured using a Scott-Marlow micro-bomb calorimeter and the assimilation efficiency of the major zooplankton species was determined using the method outlined by Sorokin (1968). This involved feeding the isolated single species with ^{14}C labelled *Chlorella* and following the loss of radioactivity from the animals with time. A summary of the results from these measurements is presented in Table 6. These data together with biomass production calculated from the regular seasonal counts of zooplankton standing crop. This enabled a table of the energy budget for the main groups of the zooplankton the be presented (Table 7).

From this we can see that on a biomass basis, the single species *Eudiaptomus gracilis* represents more than half (57%) of the total zooplankton production, with cladocera accounting for a further 39% and rotifers, although numerically well represented, only 3.7% of production because of their small size. However on an efficiency basis the crustaceans are very similar with just less than 7% of the energy ingested ending up as biomass production more than 50% passing through the gut unassimilated. For rotifers on the other hand, nearly 22% of the energy ingested goes to biomass production and only just over one quarter is unassimilated. Of the energy assimilated, for rotifers 29% goes to biomass production and the rest to respiration, but for the crustaceans, both groups respire about 85.5% of the energy assimilated, with only 14.5% going towards production.

Seston

Islam (1987) studied the flow of energy through the decomposition pathway to the sediment and hence benthic production. Kolkwitz (1912), quoted in Hutchinson (1967) defined seston as all particulate matter present in the free water, whether of autochthonous or allochthonous origin. In this paper all particulate matter caught in a trap will be considered as seston.

In this study, a trap consisting of straight-sided tubes with a height to diameter ratio of 6.8 (Hargrave & Burn, 1979) was suspended about 1 m above the sediment in the deepest part of Lochan Dubh (subjected to stagnation in August and September), and emptied every month throughout the period March 1985 to February 1986. The material caught in the trap was inspected visually, dry weighed, and organic, mineral, nitrogen, phosphorus and energy content analysed. In this review I propose to discuss only the weight, organic, mineral and energy content of the seston (Figs 8a to d).

From this it can be seen that for much of the year supply of seston was normally raining down at a rate of about 60 g m^{-2} sediment surface, except for the months of June and July at the beginning of stratification, and September and October at the beginning of the autumn overturn, when supply was reduced to about 20 g m^{-2} sediment surface. The organic content of the seston was about 50% for much of the year, being a little lower (40%) for the winter months November to January, and somewhat higher at 70% from June to August. The energy content per gram dry weight showed an increase in July to September when it was nearly double the values for the rest of the year. However this value is influenced both by changing quality of organic matter and changing proportions of mineral matter, so if results are expressed on an ashfree dry weight basis (Fig. 8d) it can be seen that there is a slow, steady increase in energy content to July, with elevated values in August and September, followed by more of less constant quality in October to December, and slow decline in January and February.

The low energy quality of the seston in winter can be accounted for by its being mainly of low energy allochthonous origin, and the increase in quality through to July represents the increase in autochthonous supply from the phytoplankton. The sudden increase in energy content in July and August coincides with the period of anoxia in the hypolimnion and the presumably less efficient anaerobic decomposition processes.

Total	Ingested	Assimilation	Fate of assimilated energy
Total energy ingested by zooplankton **6.65×10^4**	Ingested by rotifers **8.06×10^2** (1.2% of total)	Unassimilated **$2,05 \times 10^2$** (25.4%)	Respired (52.8% of ingested) **4.26×10^2** (71% of assimilated)
		Assimilated **6.01×10^2** (74.6%)	Biomass production (21.7% of ingested) **1.75×10^2** (3.7% of total) (29% of assimilated)
	Ingested by cladocera **2.66×10^4** (40% of total)	Unassimilated **1.41×10^4** (53.25%)	Respired (39.83% of ingested) **1.06×10^4** (85.2% of assimilated)
		Assimilated **1.25×10^4** (46.75%)	Biomass production (6.91% of ingested) **1.83×10^3** (39% of total) (14.8% of assimilated)
	Ingested by copepods **3.91×10^4** (58.8% of total	Unassimilated **2.06×10^4** (52.53%)	Respired (40.58% of ingested) **1.59×10^4** (85.5% of assimilated)
		Assimilated **1.85×10^4** (47.47%)	Biomass production (6.8% of ingested) **2.7×10^3** (57.3% Of total) (14.5% of assimilated)

Table 7. An energy budget for the main groups of zooplankton.

Sediment

Over the same period (March 1985 to February 1986) Islam (1987) also sampled the energy quality of the sediment at the deepest part of Lochan Dubh. He used a Jenkin corer to obtain the surface (flocculent) sediment in as undisturbed a condition as possible. The sediment down to a depth of 2 cm was between 25 and 35% organic matter throughout the year. Figure 9 presents the energy content of the sediment on a dry weight and ash free dry weight basis. The sediment has highest energy value in July and August on a dry weight basis, falling consistently to low values in February and March, before increasing again. On an ash-free basis a similar pattern is maintained, although there is just a single peak in September and the percentage difference is greater, indicating the better quality organic matter in late summer and Autumn compared with late winter and early spring (160% compared with 100% increase).

Benthos

As potential consumers of pelagic production Islam (1987) studied the community of macro-benthic fauna living in the sediment. Using an Eckman grab he sampled the sediment in the deepest part of Lochan Dubh every month over the period March 1985 to February 1986, and also a shallower water site (5 m depth) for comparison. Although *Chaoborus flavicans* was present at both sampling sites, because it is presumed to feed in the plankton by night, and because he was interested in the energy relationships within the sediment Islam (I think unwisely) excluded this species from his results.

At the deep site only one other species, *Chironomus plumosus* was found. The population was found to exhibit two peaks in April and November, with low population in February and March, and again in July to September (Fig. 10). Emergence of this species is in early Spring and the summer drop in population can be accounted for by migration of the larvae to shallower waters to avoid the developing anoxic environment in the hypolimnion. At the shallower site, *C. plumosus* was recorded only during the months June to September. At the shallow site, other *Chironomus* species, *Procladius* sp. and *Pisidium* sp. were the main components of the community, with *Sialis lutaris* being present in low numbers throughout the year and *Sigara* sp. also being recorded.

This review has, through lack of space, been able to present only a very general overview of some of the hitherto unpublished information available on the complex interactions of plant and animal communities within and outside the ecosystem known as Lochan Dubh.

Dedication

This paper is dedicated to the memory of A. K. M. Sirajul Islam who provided much of the information reported here, and who died suddenly, in 1990.

References

Goldspink, C. R. & D. B. C. Scott, 1971. Vertical migration of *Chaoborus flavicans* in a Scottish loch. Freshwat. Biol. 1: 411–421.

Golterman, H. L. & R. S. Clymo, 1971. Methods for Chemical Analysis of Freshwaters. I.B.P. Handbook 8. Blackwell, Oxford, 188 pages.

Hamilton, J. D., 1958. On the biology of *Holopedium gibberum* Zaddach (Crustacea: Cladocera). Verh. int. Ver. Limnol. 13: 786–788.

Hargrave, B. T. & N. M. Barnes, 1979. Assessment of sediment trap collection efficiency. Limnol. Oceanogr. 24: 1124–1136.

Hutchinson, G. E., 1957. A Treatise on Limnology. Volume 1: Geography, Physics and Chemistry. John Wiley & Sons Inc., New York. 1015 pp.

Hutchinson, G. E. 1967. A Treatise on Limnology. Volume 2: Introduction to lake biology and the limnoplankton. John Wiley & Sons Inc., New York, 1115 pp.

Islam, A. K. M. S., 1987. The Role of Plankton in the Energy Transfer to the Sediment in Dubh Lochan. Ph.D. Thesis, University of Glasgow, 273 pp.

Klarer, D. M., 1978. Some Studies on the Plankton of the Dubh Lochan. Ph.D. Thesis, University of Glasgow, 217 pp.

MacDonald, J. G., 1994. Geology of the Loch Lomond Catchment. Hydrobiologia 290: 13–20.

McDonnell, E. A. A Seasonal Study of the Surface Biota of the Dubh Lochan. Ph.D. Thesis, University of Glasgow, 143 pp.

Shafi, M., 1969. Comparative Studies of Populations of Perch (*Perca fluviatilis* L.) and Pike (*Esox lucius* L.) in Two Scottish Lochs. Ph.D. Thesis, University of Glasgow, 143 pp.

Sorokin, J. I., 1968. The use of ^{14}C in the study of nutrition of aquatic animals. Mitt. int. Ver. Limnol. 16: 41 pp.

Stewart, D. A., A. Walker & J. H. Dickson, 1984. Pollen diagrams from Dubh Lochan, near Loch Lomond. New Phytol. 98: 531–549.

Hydrobiologia **290**: 167–170, 1994.
K. J. Murphy, M. C. M. Beveridge & R. Tippett (eds), The Ecology of Loch Lomond.

Report on the discussion

Angela K. Turner
School of Biological Sciences, University of Sussex, Brighton BN1 9QG, UK

The conference ended with a Discussion on the conflicts of interest in the Loch Lomond Catchment. The Discussion was chaired by Dr Gordon Dickinson, of the Department of Geography and Topographic Science, Glasgow University. The panel represented a range of interests: Mr Alexander McCreadie of the Central Scotland Water Development Board; Mr William Dalrymple of the Loch Lomond Park Authority; Mr John Burlison of Scottish Natural Heritage; and Mr Jack Bisset, of the Loch Lomond Users Association. Several areas of concern, actual or potential, had been identified by speakers at the conference.

1. Abstraction of water and the Leven barrage. Although concern has been expressed, Curran & Poodle concluded that so far these have had little effect on water levels.
2. Water pollution, e.g. from sewage works and farms. Water quality in the catchment is generally excellent, although there are intermittent problems with farm waste, and there is no evidence of enrichment affecting phytoplankton (Best; Tippett).
3. Changes in land use, especially an increase in recreational pressure. Tourist pressure is very high locally, especially at weekends and the large number of boats on the Loch is now a serious problem (Dickinson; Adams & Tippett).
4. Road construction and improvement. This has damaged invertebrate communities locally and in the short-term (Frame *et al.*).
5. Bank erosion. Heavy tourist use contributes to damage caused by floods (Dickinson & Pender).
6. Pressure on sensitive plant communities and rare species. Drainage, grazing and agricultural runoff, for example, affect some valley bogs (Murphy *et al.*).
7. Pressure on sensitive animal species such as the endemic race of the river lamprey (Maitland & Morris).
8. Invasions by alien organisms, introducing problems of competition, parasites, hybridisation, predation on native species and changes in ecology. Ruffe are a particular problem as they are now major predators of the rare powan (Adams).

The Discussion focussed on two issues encompassing these areas of concern. (1) Has the Loch already been profoundly affected by human activities and is it still vulnerable? (2) What would be the value of a more integrated approach to studying its problems?

The Chairman introduced the Discussion by saying that the conference had been very constructive and that he hoped the Discussion would capitalise on the extra knowledge gained by the interaction of the specialists who had come together for the meeting.

The first theme of the Discussion was the vulnerability of the Loch. The Chairman posed a number of questions: has Loch Lomond already been profoundly affected; how vulnerable is it; what evidence is there for change within the Loch, such as the increase in numbers of ruffe; is the quality of the Loch's water likely to change; and what biological and physical pressures on the Loch can be recognised?

Tom Huxley of Scottish Natural Heritage started the questions session by asking what effect would a threat such as the Craig Royston pump storage scheme proposed a decade ago have on the mixing of surface water half way up the Loch. Sandy McCreadie replied that the effect of mixing was not known but that it would probably be minimal as the water quality is good at the top of the Loch. When it was first proposed, the potential ecological effects of the Craig Royston pump storage scheme were investigated jointly by the Institute of Terrestrial Ecology and the Hydro Board. The five reports that came out of the investigation remain confidential but the scientific results have been published and have provided much useful information, as Peter Maitland, who was involved in producing the reports, pointed out; much of the information was indeed heard at the conference. William Dalrymple added that the scheme has only been shelved and that at the time it was considered an important development. Gordon Dickinson said it was shelved for economic reasons and he could imagine it coming back again. The threat of a pump storage scheme being implemented therefore remains real.

Tom Huxley felt that the Loch was already on a knife's edge and wondered whether it could cope if its biological and physical parameters were pushed by additional impacts of human activities such as a new Craig Royston scheme; its buffering capacity might be insufficient to continue to counteract the effects of such activities. Continuing this theme, Gordon Dickinson considered the redevelopment of the east side road as a major threat, adding greatly to the recreational pressure, present traffic levels being just about in balance with the carrying capacity of the recreational sites. He wondered whether, although present conditions might be stable, small pushes could be important and eventually catastrophic, or whether there is a capacity to absorb continued change. Kevin Murphy said that people should not be complacent: although Loch Lomond is in good condition at present, it may be threatened for example by eutrophication in the future, and is currently threatened by invasions of fish. Small additional factors could therefore be important.

Boating emerged as a particular problem, with Loch Lomond almost at capacity for this activity. Members of the Loch Lomond Users Association view the situation with a great deal of concern. Boating pressure may increase, as users of Lake Windermere may move to Loch Lomond if a planned 10 mph speed limit is put on boats on Windermere, as Jack Bisset warned. A pilot scheme has been set up by the Loch Lomond Users Association and Park Authority to prepare a policy for recreational activity on the Loch which might include registration of boat users. Jack Bisset said there was a strong environmental and ecological support group among the users of the Loch. David Hill, of Dumbarton District Council, asked if there had been any studies of the damage caused by boats on lakes. On Windermere, the main problem is the number of people using the lake, as Kathleen Atkinson, of the Institute of Freshwater Ecology, reported, rather than boats per se. On Loch Lomond the potential effects of boating activity include disturbance and damage to wildlife, a deterioration in water quality, erosion of banks and a decline in recreational quality and safety (Adams & Tippett). The Chairman said that research, particularly in Canada, showed that the recreational carrying capacity of a water body depends on the expectations and requirements of the users. He referred in particular to Joy Tivy's study in the middle 1970s of the recreational use of Scottish lochs.

Sandy McCreadie said that the Clyde River Purification Board was not complacent. Although there was no chemical deterioration in the catchment, the Board had been worried about, for example, potential increases in algae and a proposed fish farm, now cancelled. The Board is very conscious, however, of the problem of deterioration in water quality and in the future water from the catchment would have to be treated more intensively.

The Chairman then turned the discussion to the question of future studies. He said that a number of detailed investigations have suggested the need for more integrated work. An example is the concern over phytoplanktonic blooms. The study of phytoplankton in the Loch has provided a wealth of information, whereas we still know little about the effect of zooplankton grazing on blooms. He asked whether more can be gained by bringing together specialist knowledge.

Tony Bailey-Watts, of the Institute of Freshwater Ecology, pointed out that longterm intensive work is very useful and necessary, as seen elsewhere. He also said that we must not be complacent that Loch Lomond, despite being such a large body of water, is not being affected by our activities. The Chairman asked whether the size of Loch Lomond made it more resilient than other water bodies. Brian Moss, of the University of Liverpool, said that early symptoms of problems will show up on the bottom of the loch, not in the plankton and that a number of cases have shown the problems of eutrophication to be difficult to solve once it has occurred. It is best to tackle the situation before the problem has arisen. The water of the Endrick has substantial levels of nitrate and phosphate, and Brian Moss thought that phosphate stripping should be implemented on the sewage treatment plants in the Endrick catchment, as the input of phosphate into the water is only likely to increase.

Fish farms are a potential problem but so far there has been only a single application to start one. Roger Tippett suggested that Loch Lomond is vulnerable to applications for fish farms and wondered whether the Park Authority would consider giving permission for fish farms in future. William Dalrymple replied that the Loch Lomond Park Authority has no policy on such farms. Sandy McCreadie explained that the one application made so far was rejected by Dumbarton District Council because the shore-based part of the structure was unacceptable rather than from a consideration of its potential impact on the Loch. The effect fish farms would have on water quality in Loch Lomond is not clear, but the legislation has now been amended so new planning applications would have to go to the River Boards, which would consider the impact of the proposal on water quality. William Dalrymple and David Hill both pointed out that planners and decision makers, who are not experts on scientific matters, need more access to the findings of scientific studies on such factors as water quality, so that decisions about planning applications can be made on a sounder footing. Ian Smith returned the discussion to the question of the resilience of the Loch and the lack of knowledge of how waters mix within it; planners need to know, for example, how much River Endrick water travels north in the Loch. Peter Maitland said that most conservation issues fall into the 'nail in the coffin syndrome': single applications for a development may affect only a tiny part of the loch; planners need to consider what the catchment as a whole can sustain.

John Burlison felt that conflicting statements were being made about pressures on Loch Lomond. Thus it was pointed out that planners may not have been given the information they need in a manner they can digest and also that the situation in the Loch is 'okay'. He suggested that the questions to be asked have not yet been identified so research cannot be targeted onto particular areas of water quality, conservation, and so on. Malcom Beveridge said we tend to have a 'Not in My Back Yard' attitude and we should consider a wider agenda, *i.e.* the importance of, and impacts on, Loch Lomond in relation to issues elsewhere. The Chairman also pointed out that more comparative work and a specific agenda for action are needed. Kathleen Atkinson suggested it was important to consider the future development of the local population and its effects on Loch Lomond. There was a consensus that future work must include all of the Loch Lomond catchment area. Peter Maitland, for example, pointed out that the present Working Group on Loch Lomond and the Trossachs does not cover any of the Endrick valley and Jim Hamilton also emphasized the importance of considering the catchment as a whole.

As the Discussion drew to a close, the Chairman asked the panel to explain the priorities of their various organisations.

Jack Bisset pointed out that, looking at Loch Lomond and its catchment area in a wider context, it has great scientific value but it is also the widest stretch of water in Britain for recreation and is the second most important tourist resource in Scotland. The Users' Association represents a variety of interests, not just boat enthusiasts but also, for example, bird watchers. Thus there has to be a balance: Loch Lomond cannot be kept for the benefit of just one group. The Secretary of State's Working Party is looking at not only the Trossachs and Loch Lomond but also other water areas to see if the load can be spread between them. Jack Bisset also reiterated the need for the Working Party to include all of the catchment area in its considerations.

Sandy McCreadie said the main concern of the Central Scotland Water Development Board was maintaining water quality for the up to one million people who drink Loch Lomond water. He considered the biggest threat to be increasing eutrophication and suggested N/P limits may need to be put on sewage discharges.

William Dalrymple said that the Loch Lomond Park Authority has three objectives: conservation; the promotion of sustainable enjoyment of the Loch Lomond area; and its socioeconomic well-being. Conservation has top priority, however, in situations where these objectives conflict. The Authority has been in existence for only three years but its data base, for example on visitor numbers, is old arising from surveys undertaken by the Countryside Commission in 1973. In addition, the Loch Lomond policy document stems from a management group set up in 1970 and the Loch Lomond Plan was drawn up in the late 1970s and early 1980s. The data base and policy document thus need updating. Surveys in 1991 in the Loch Lomond area showed that recreational pressure is now very high, with 1.9 million visits in 15 sites over a seven-month period. Boat numbers are increasing; although the problem overall is not as severe, certain areas do experience densities as high as on Lake Windermere (also see Adams & Tippett). Bye-laws are needed to control visitor pressure, especially the use of boats (also see Pennington). The Park Authority is keen to consult with as many people as possible about controlling the use of the Loch and a bye-law advisory group with a wide membership is being formed for this purpose. The Park Authority can then be sure that any further controls that are introduced do serve a need and can be enforced. William Dalrymple said, however, that he was please to see from the conference that 'we are not over the brink as far as Loch Lomond is concerned'.

John Burlison said that the role of Scottish Natural Heritage is to conserve and enhance our natural heritage, and to increase knowledge, and promote the use, of that heritage. A natural resource is inevitably going to be used by many interests. A balance must be sought, however, so that the use is sustainable even if that means banning or limiting some activities. This will work only if there is agreement between the interested parties. Hence all user bodies must be involved in planning the future of a site, but as yet there is a lack of knowledge of what the right balance is.

The Chairman acknowledged a note of congruence among the panel and added that closer cooperation is needed; scientific and policy management groups must come together and define the questions to be asked.

Ian Smith said that it is the scientists' job to understand the processes going on in Loch Lomond but perhaps they do not explain it well to others. Even if the processes are not fully understood, the knowledge we do have still needs to be communicated so that planners and decision makers can cope with the accumulation of small effects that are already happening. Eventually we need a picture of all the processes controlling Loch Lomond.

There was a consensus by the end of the Discussion that planners, decision makers, scientists and user bodies often do not know what is going on amongst other groups and do not approach each other for information, especially if the problem is not a high profile one such as eutrophication. This situation should be rectified. Tom Huxley left delegates with the message not to 'underestimate how ignorant people are, even people who you think would know'. The Chairman ended the discussion optimistically by saying that Loch Lomond is a 'remarkable scientific as well as human asset on our doorstep and ... the range of expertise and interest represented here bodes well for the future'.

References

Adam, C. E., 1994. The fish community of Loch Lomond: its history and rapidly changing status. Hydrobiologia 290: 91–102.

Adams, C. E. & R. Tippett. Recreation craft activity on Loch Lomond.

Best, G. A. & I. Traill, 1994. The physico-chemical limnology of Loch Lomond. Hydrobiologia 290: 29–37.

Curran, J. C. & T. Poodle, 1994. Aspects of the hydrology and hydrography of Loch Lomond. Hydrobiologia 290: 21–28.

Dickinson, G., 1994. Vegetation and land use in the Loch Lomond catchment. Hydrobiologia 290: 53–61

Dickinson, G. & G. Pender, 1994. Natural and human factors in recent bank erosion.

Frame, J. H., C. R. Doughty, M. J. Turner & I. C. Simpson, 1994. The effects of road construction on freshwater communities in the Loch Lomond catchment.

Maitland, P. S., K. H. Morris & K. East, 1994. The ecology of lampreys (Petromyzonidae) in the Loch Lomond area. Hydrobiologia 290: 105–120.

Murphy, K. J., K. D. Hudson & J. Mitchell, 1994. Freshwater and wetland plant communities of Loch Lomond. Hydrobiologia 290: 63–74.

Pennington, N. C., 1994. Bye-laws for Loch Lomond.

Tippett, R., 1994. The phytoplankton community of Loch Lomond.

MIX
Papier aus verantwortungsvollen Quellen
Paper from responsible sources
FSC® C105338

If you have any concerns about our products,
you can contact us on
ProductSafety@springernature.com

In case Publisher is established outside the EU,
the EU authorized representative is:
Springer Nature Customer Service Center GmbH
Europaplatz 3, 69115 Heidelberg, Germany

Printed by Libri Plureos GmbH
in Hamburg, Germany